THE GENETICS OF ALGAE

BOTANICAL MONOGRAPHS

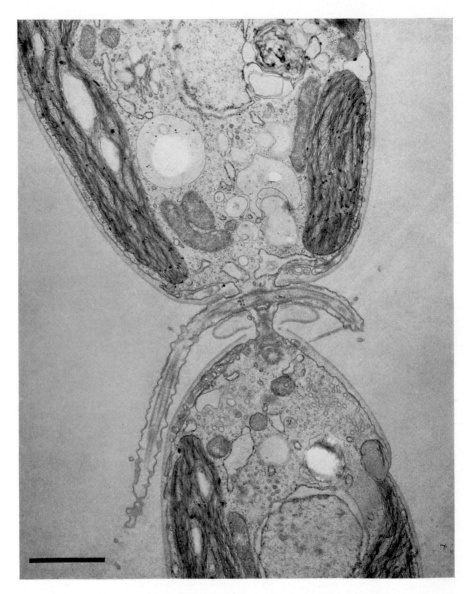

Gametes of *Chlamydomonas moewusii* Gerloff *in copula*: longitudinal section through proto-plasmic bridge joining the papillae of the two cells. Electron-micrograph by Dr R. E. Triemer. Scale bar = 1μm.

BOTANICAL MONOGRAPHS · VOLUME 12

THE GENETICS OF ALGAE

EDITED BY

RALPH A. LEWIN

PhD, ScD

Scripps Institution of Oceanography,
University of California, San Diego

UNIVERSITY OF CALIFORNIA PRESS

BERKELEY AND LOS ANGELES

UNIVERSITY OF CALIFORNIA PRESS
Berkeley and Los Angeles, California

ISBN: 0-520-03149-0
Library of Congress Catalog Card Number: 75-27297

© 1976 Blackwell Scientific Publications

Printed in Great Britain

CONTENTS

Contributors   vii

Acknowledgements   ix

Introduction   1

1   Genetics of Blue-Green Algae   7
    S. F. DELANEY, M. HERDMAN and N. G. CARR

2   Formal Genetics of *Chlamydomonas reinhardtii*   29
    G. A. HUDOCK and H. ROSEN

3   Genetic Determinants of Flagellum Phenotype in *Chlamydomonas*
    reinhardtii   49
    SIR JOHN RANDALL and D. STARLING

4   Genetics of Cell Wall Synthesis in *Chlamydomonas reinhardtii* 63
    D. R. DAVIES and K. ROBERTS

5   Plastid Inheritance in *Chlamydomonas reinhardtii*   69
    G. M. W. ADAMS, KAREN P. VANWINKLE-SWIFT,
    N. W. GILLHAM and J. E. BOYNTON

6   Genetics of Chloroplast Ribosome Biogenesis in *Chlamydomonas*
    reinhardtii   119
    ELIZABETH W. HARRIS, J. E. BOYNTON and N. W. GILLHAM

7   Genetics of *Chlamydomonas moewusii* and *Chlamydomonas*
    eugametos   145
    C. SHIELDS GOWANS

8   Genetic Aspects of Sexuality of Volvocales   174
    L. WIESE

9   Genetics of Zygnematales   198
    P. BIEBEL

10 Genetics of Charophyta   210
    V. W. PROCTOR

11 Genetics of Multicellular Marine Algae   219
    A. FJELD and A. LØVLIE

12 Approaches to the Genetics of *Acetabularia*   236
    BEVERLEY R. GREEN

13 Inheritance and Synthesis of Chloroplasts and Mitochondria of
    Euglena gracilis   257
    G. W. SCHMIDT and H. LYMAN

    Appendix A: Publications by A. Pascher on the Genetics of Algae 300
    Translations and Commentaries by R. E. REICHLE

    Appendix B: Publications by Franz Moewus on the Genetics of
    Algae   310
    C. SHIELDS GOWANS

    Author Index   333

    Species Index   344

    Subject Index   348

CONTRIBUTORS

G. M. W. ADAMS, Department of Zoology, Duke University, Durham, North Carolina 27706, U.S.A.

P. BIEBEL, Department of Biology, Dickinson College, Carlisle, Pennsylvania 17103, U.S.A.

J. E. BOYNTON, Department of Botany, Duke University, Durham, North Carolina 27706, U.S.A.

N. G. CARR, Department of Biochemistry, University of Liverpool, P.O. Box 147, Liverpool L69 3BX, England.

D. R. DAVIES, John Innes Institute, Colney Lane, Norwich NR4 7UH, England.

S. F. DELANEY, Department of Biochemistry, University of Liverpool, P.O. Box 147, Liverpool L69 3BX, England

A. FJELD, Zoologisk Laboratorium, Universitetet I Oslo, Post Box 1050, Blindern, Oslo-3, Norway

N. W. GILLHAM, Department of Zoology, Duke University, Durham, North Carolina 27706, U.S.A.

C. S. GOWANS, Division of Biological Sciences, College of Arts and Science, University of Missouri-Columbia, 105 Tucker Hall, Columbia, Missouri 65201, U.S.A.

BEVERLEY R. GREEN, Botany Department, University of British Columbia, Vancouver V6T 1W5, B.C. Canada

ELIZABETH W. HARRIS, Department of Botany, Duke University, Durham, North Carolina 27706, U.S.A.

M. HERDMAN, Service de Physiologie Microbienne, Institut Pasteur, 25 Rue du Dr. Roux, 75 Paris, France.

G. A. HUDOCK, Department of Zoology, Indiana University, Jordan Hall 224, Bloomington, Indiana 47401, U.S.A.

A. LØVLIE, Zoologisk Laboratorium, Universitetet I Oslo, Post Box 1050, Blindern, Oslo-3, Norway

H. LYMAN, Department of Cellular and Comparative Biology, Division of Biological Sciences, State University of New York at Stony Brook, Stony Brook, New York 11790, U.S.A.

V. W. PROCTOR, Department of Biology, Texas Tech University, P.O. Box 4149, Lubbock, Texas 79409, U.S.A.

SIR JOHN RANDALL, Department of Zoology, University of Edinburgh, King's Buildings, West Mains Road, Edinburgh EH9 3JT, Scotland

R. E. REICHLE, Department of Biology, California State University, Sacramento, California 95819, U.S.A.

K. ROBERTS, John Innes Institute, Colney Lane, Norwich NR4 7UH, England.

H. ROSEN, Department of Botany, California State College—Los Angeles, 5151 State College Drive, Los Angeles, California 90032, U.S.A.

G. W. SCHMIDT, Department of Cellular and Comparative Biology, Division of Biological Sciences, State University of New York at Stony Brook, New York 11790, U.S.A.

P. C. SILVA, Department of Botany, University of California, Berkeley 94720, U.S.A.

D. STARLING, Department of Zoology, University of Edinburgh, King's Buildings, West Mains Road, Edinburgh EH9 3JT, Scotland.

KAREN P. VAN WINKLE-SWIFT, Department of Zoology, Duke University, Durham, North Carolina 27706, U.S.A.

L. WIESE, Department of Biological Science, Florida State University, Tallahassee, Florida 32306, U.S.A.

ACKNOWLEDGEMENTS

A compilation like this, I feel, should be a cooperative affair, with a certain amount of give and take between authors and editors—and sometimes friends and relations, too. I thank my collaborators, the original contributors to this volume, not only for the time and effort they devoted to the work, but also for generously considering, and generally accepting, so many suggestions from colleagues as well as from the editor. I am indebted to Eileen Ottoson and Gayle Kidder for compiling the three indexes. Dr Paul C. Silva kindly checked the specific names of all the algae mentioned in this book, and the author for each in the taxonomic index. Others who aided the editorial work in various ways, and whom I would also like to thank, include Drs A. R. O. Chapman, R. W. Hoshaw, S. H. Howell, R. C. Starr, J. G. Stewart and, of course, Lanna Cheng.

Ralph A. Lewin

This volume is affectionately dedicated to Ann, Bob, Connie, Frances, Frederica, Lanna, Lou, Nan, Tye and others—to our teachers, spouses, colleagues and students, past and present (the categories are not necessarily mutually exclusive), who have provided guidance, solace, enlightenment and frustration (not always in that order).

INTRODUCTION

An introduction to a book of this kind should start with definitions. What are algae, and what is genetics?

The algae are a heterogeneous assemblage of photosynthetic plants—essentially all of those that are not classifiable as land plants (bryophytes and tracheophytes). Algae range in length from 1 μm to a few tens of metres, and in complexity from an outwardly simple spherical cell to a highly differentiated plant like that of *Sargassum*. In cell organization some (the blue-greens) are prokaryotic, but most are eukaryotic. Sexually, they range from an apparently strict celibacy (as in *Chlorella*) to a complex life-cycle with two mating types or sexes and two or even three dissimilar generations (Figs. 1–4). They may have one circular chromosome, demonstrated for certain blue-green algae, or a few hundred little rod-like ones, as in some dinoflagellates and desmids (Godward 1966). Indeed, apart from a general inclination to occupy wet places, almost the only feature that the algae have in common is their ability to fix carbon from CO_2 and to evolve gaseous oxygen when suitably illuminated. The algae are thus more of a guild than a clan, united more by common aptitude than by common ancestry.

The word 'genetics' is equally hard to delimit. We have chosen to interpret it generously here, to embrace not only studies of the inheritance of differential characteristics from one generation to another (counting sex as one kind of differentiation, although it is not readily subject to mutation) but also investigations relating to the mechanisms underlying such differences. Thus, in keeping with the popularly accepted meaning of the word, we have included studies on the nature and replication of various algal nucleic acids, and the translation of their information into gene products. We accordingly consider studies of the DNA of even non-sexual *Chlorella*, *Ochromonas* and *Euglena* species (Hellman & Kessler 1974; Gibbs & Poole 1973; Lyman & Schmidt Chapter 13) as a sort of 'pro-genetics'.

Some of us have tried to indicate new and potentially fruitful research directions for algal geneticists, who tend to suffer from a regrettably blinkered view of biology and too slavishly follow trails blazed by bacteriologists. Many of these directions were far from obvious 75 years ago, when Mendelism was rediscovered, or even 25 years ago, when we resumed serious work on *Chlamydomonas* (Lewin 1949). Nobody then could have foreseen the tangle of ways in which our knowledge of the genetics of this genus has more recently evolved. Twenty-five years ago there was barely enough information to justify even a slim review chapter on the genetics of all algae: here, now, we have a 400-page book: 25 years hence . . . I shudder to think of the prospect!

Surely we shall soon have a lot more information on the genetics of dinoflagellates, especially now that some have been domesticated for laboratory work and have been shown to exhibit heterothallic sexual reproduction and genetic recombination (von Stosch 1973; Beam & Himes 1974; Tuttle & Loeblich 1974). Certainly there will soon be interesting stories unfolding in the realm of diatom genetics, too, and perhaps even among the ostensibly chaste Chlorococcales if preliminary reports of sexuality in *Scenedesmus* and *Chodatella* (Trainor & Burg 1965; Ramaley 1968) can be confirmed. Doubtless we shall soon have more clock-like chromosome maps to illustrate the cryptic genetic behaviour of cyanophytes, like that already produced for *Anacystis* (see Chapter 1). Hopefully, someone will soon take up the challenge of *Pedinomonas*, and will investigate the behaviour of mitochondrial genes in a type of cell which has clearly only a single mitochondrion (Manton 1959). (Or is this perhaps not so unusual after all? Do many other cell types have only one mitochondrion, albeit a highly reticulate organelle?)

If red algal spermatia, which somehow manage to reach and adhere to the correct conspecific trichogynes, have some kind of specific surface agglutinins—as one may suspect—then the nature and role of such substances must eventually be tackled and elucidated by a combined biochemical and genetic approach. In due course, someone will successfully open up the biochemical genetics of accessory photosynthetic pigments in red and brown algae, perhaps taking advantage, for genetic analyses, of the large meiotic tetrads produced by some of these seaweeds (Lewin 1956). By cell fusion and organelle transplantation, someone will successfully 'hybridize' unrelated algal cells, as one has already integrated tissue-cultured cells of mice and men, and will thereby enable us to graft various kinds of chloroplasts into heterologous cytoplasts. Some evolution-minded geneticists will look into the fascinating problems of algal evolution along sea-shores, where, in distribution ranges thousands of kilometres long yet only a few metres wide, gene flow is effectively restricted to a 1-dimensional system. Someone eventually will offer a satisfactory explanation for the paradoxical differences in mutation spectra among algae and other organisms (Li *et al.* 1967). Someone will manage to elucidate genetic bases for the morphological differentiation of normal haploid filaments and haploid dwarf males in nannandrous *Oedogonium* spp., or diploid carposporophytes and diploid tetrasporophytes in the Florideae, and for the unexpected biochemical differences that diploids and haploids may exhibit in cell-wall composition, etc. (Chen *et al.* 1973; McCandless *et al.* 1973; Pickmere *et al.* 1973). Some biochemical geneticists will look into nucleo-cytoplasmic relations in the developing cystocarps of red algae, where diploid nuclei unaccountably migrate in and out of haploid parental cells in the developing cystocarps. Sooner or later we shall need answers for such questions.

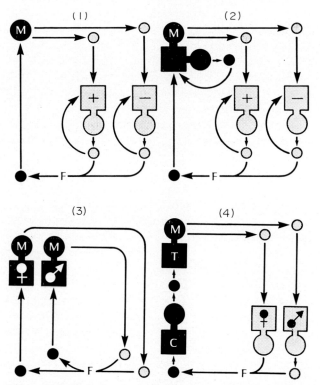

Fig. 1. Life cycle typical of most freshwater green algae (e.g. *Chlamydomonas, Spirogyra*). All haploid (shown here in grey) except for the zygote (shown here in black), in which meiosis (M) occurs. Haploid plants may be heterothallic (+/−). They may recycle asexually. Gamete fusion indicated at F.

Fig. 2. Life cycle typical of most brown algae (e.g. *Ectocarpus, Laminaria*) and of many marine green algae (e.g. *Ulva*), as well as of land plants (Bryophyta and Tracheophyta). Alternating diploid (black) and haploid (grey) generations, which may be vegetatively similar or dissimilar. Haploid plants may be heterothallic (+/−). They may recycle asexually. Gamete fusion indicated at F; meiosis in meiosporangium at M.

Fig. 3. Life cycle typical of some brown algae (e.g. *Fucus*) and some marine green algae (e.g. *Caulerpa*). (Almost all animals have this kind of life cycle.) All diploid (black) except for the gametes (grey). Meiosis (M) occurs in gametangia: male and female gametes may be borne on separate plants (as shown), or plants may be monoecious. Gamete fusion indicated at F.

Fig. 4. Life cycle typical of most red algae (e.g. *Polysiphonia*). Separate free-living haploid (grey) and diploid (black) generations, which may be vegetatively similar or dissimilar, with an interposed diplophase (black). Fertilization (F) on the female plant gives rise to a more or less endophytic diploid phase, the carposporophyte (C), which produces carpospores; these grow into tetrasporophytes (T) bearing sporangia where meiosis (M) occurs.

These are some of the many possible topics that are *not* represented among the chapters of this compendium. Then what *do* we know now about algal genetics? Inevitably, perhaps, most of the data deal with *Chlamydomonas*, which as early as 1916 presented itself as a particularly favourable organism for genetic studies. We have therefore included English translations of two old but classical papers on the subject by Pascher, which should be of at least historical interest to geneticists in general (Appendix A). Pascher, being perhaps more a taxonomist than an experimentalist, didn't follow up these leads. Moewus did so, some 15 years later, presumably under Kniep's direction; but as he seems to have strayed from the path of strict veracity we have relegated a resumé of his publications to a separate part of this volume (Appendix B). It, too, may be of general interest, at least from the viewpoint of scientific ethics.

In this volume we have tried to present up-to-date reviews of the genetics of each type of alga for which information is available. The chapters were completed and submitted at various times between June 1974 and March 1975. The fact that they differ considerably in their levels of scientific sophistication is only partly a result of personal diversity among the various authors. Mostly it is a consequence of innate differences in the rates of reproduction and other sexual proclivities of the organisms in culture, matters which are still largely beyond experimental control.

We now have lots of good, if sometimes controversial, data on the inheritance of *Chlamydomonas* genes, in at least two species and at least three systems: nuclear, plastidic, and mitochondrial. Among the so-called nuclear genes, which behave in classical Mendelian fashion, those concerned with flagellar formation and action, and with cell wall synthesis, have chapters to themselves in this compendium. Perhaps, in the near future, we may have comparable bodies of information on the genetics of tactic movements, mating behaviour, cell division cycles, and internal clocks in *Chlamydomonas*, on which some information is already being gathered (Howell & Naliboff 1973; Howell 1974; Bruce 1972, 1974). We have reviewed here some useful if preliminary studies on the genetics of other green algae, notably in the Volvocales, Ulvales, Conjugales and Charales.

We have included a chapter on *Acetabularia*, for which there has long been a body of data on the inheritance and transmission of morphogenetic factors in grafts, and for which we have now a wealth of information on nuclear-cytoplasmic relations, thanks to the ease with which this paradoxical kind of unicell can be enucleated. Technological advances may soon enable us to extend such studies into the field of formal genetics. We have even added a chapter on *Euglena gracilis*; though they seem not to reproduce sexually, the cells (of at least one strain) have the ability to survive without their chloroplasts, and much is now known of the inter-organellar and intra-organellar biochemistry both of the original green strain and of derived white variants.

On the whole, the prospects for algal genetics seem bright. We hope the present book will not only indicate how far we have come in this fascinating field (or jungle!), but will also serve to illuminate some of the many interesting paths ahead.

Ralph A. Lewin

NOTE

I have not attempted to bring conformity to the diverse abbreviations used by various algal geneticists to designate mutant strains—or the mutations responsible for the changed genotypes—and have generally accepted the current laboratory jargon in which the words 'mutant' and 'mutation' tend to be used interchangeably. Strictly, a mutant is a type of cell, culture or organism bearing a mutation or mutated gene; but it is often a convenient ellipsis to state briefly that a specific mutant occurred, or reverted, without explaining in each case that it was the responsible gene which behaved in that way.

REFERENCES

BEAM C.A. & HIMES M. (1974) Evidence for sexual fusion and recombination in the dinoflagellate *Crypthecodinium (Gyrodinium) cohnii. Nature* **250**, 435–6.

BRUCE V.G. (1972) Mutants of the biological clock in *Chlamydomonas reinhardi. Genetics* **70**, 537–48.

BRUCE V.G. (1974) Recombinations between clock mutants of *Chlamydomonas reinhardi. Genetics* **77**, 221–30.

CHEN L.C.-M., MCLACHLAN J., NEISH, A.C. & SHACKLOCK P.F. (1973) The ratio of kappa-to lambda-carrageenan in nuclear phases of the rhodophycean algae, *Chondrus crispus* and *Gigartina stellata. J. Mar. biol. Ass. (U.K.)* **53**, 11–16.

GIBBS S.P. & POOLE R.J. (1973) Autoradiographic evidence for many segregating DNA molecules in the chloroplast of *Ochromonas danica. J. Cell Biol.* **59**, 318–28.

GODWARD M.B.E. (Ed.) (1966) The Chromosomes of the Algae. St. Martin's Press, N.Y. 212 pp.

HELLMAN V. & KESSLER E. (1974) Physiologische und biochemische Beiträge zur Taxonomie der Gattung *Chlorella.* VIII. *Die Basenzusammensetzung der DNS. Arch. Microbiol.* **95**, 311–18.

HOWELL S.H. (1974) An analysis of cell cycle controls in temperature sensitive mutants of *Chlamydomonas reinhardi.* In *Cell Cycle Controls,* eds. Padilla J.M., Cameron I.L. & Zimmerman A., pp. 235–49.

HOWELL S.H. & NALIBOFF J.A. (1973) Conditional mutants in *Chlamydomonas reinhardtii* blocked in the vegetative cell cycle. I. An analysis of cell cycle block points. *J. Cell Biol.* **57**, 760–72.

LEWIN R.A. (1949) Genetics of *Chlamydomonas*—paving the way. *Biological Bulletin* **97**, 243–4.

LEWIN R.A. (1956) Genetics and marine algae. In *Perspectives in Marine Biology,* ed. Buzzati-Traverso A.A. University of California Press, pp. 547–57.

LI S.L., RÉDEI G.P. & GOWANS C.S. (1967) A phylogenetic comparison of mutation spectra. *Molec. Gen. Genetics* **100**, 77–83.

B

6   INTRODUCTION

McCandless E.L., Craigie J.S. & Walter J.A. (1973) Carrageenans in the gametophytic and sporophytic stages of *Chondrus crispus*. *Planta* **112**, 201–12.

Manton I. (1959) Electron microscopical observations on a very small flagellate: the problem of *Chromulina pusilla* Butcher. *J. Mar. biol. Assn.* (*U.K.*) **38**, 319–33.

Pickmere S.E., Parsons M.J. & Bailey R.W. (1973) Composition of *Gigartina* carrageenan in relation to sporophyte and gametophyte stages of the life cycle. *Phytochem.* **12**, 2441–4.

Ramaley A.W. (1968) Sexuality in *Chodatella*. *Science* **161**, 809–10.

von Stosch H.A. (1973) Observations on vegetative reproduction and sexual life cycles of two freshwater dinoflagellates, *Gymnodinium pseudopalustre* Schiller and *Woloszynskia apiculata* sp. nov. *Br. phycol. J.* **8**, 105–34.

Trainor F.R. & Burg C.A. (1965) *Scenedesmus obliquus* sexuality. *Science* **148**, 1094–5.

Tuttle R.C. & Loeblich A.R. (1974) Genetic recombination in the dinoflagellate *Crypthecodinium cohnii*. *Science* **185**, 1061–62.

CHAPTER 1

GENETICS OF BLUE-GREEN ALGAE

S. F. DELANEY, M. HERDMAN* AND N. G. CARR

Dept. of Biochemistry, University of Liverpool, U.K.,
and Institut Pasteur,* Paris

1 Introduction 7

2 Mutants 9
  2.1 Resistance mutants 9
  2.2 Morphological and develop-
       mental mutants 10
  2.3 Pigment mutants 11
  2.4 Temperature-sensitive
       mutants 12
  2.5 Auxotrophs 12

3 Repair mechanisms 13

4 Cyanophages 15

5 Transformation 16

6 Temporal genetic mapping 19

7 References 23

1 INTRODUCTION

The inclusion in this volume of a section on the blue-green algae is, in part, a reflection of a historical view of these organisms as algae. The possession of the range of pigments characteristically associated with algae gave emphasis to their essentially plant-like mode of photosynthesis and led most botanists (with the notable exception of Cohn 1853) to set them completely apart from the photosynthetic bacteria. During the last few decades there has become established a quite different view of the phylogenetic relationships among photosynthetic micro-organisms. The blue-green algae and the bacteria are seen as a coherent group, the prokaryotes, which characteristically lack membrane-bound sub-cellular organelles such as mitochondria and chloroplasts and which do not have their genetic material located on more than one chromosome in a membrane-bound nucleus (Stanier & van Niel 1962). Succeeding years have seen the close relationship between these two groups confirmed and extended, especially on the basis of their nucleic acid biochemistry, fine structure and cell-wall chemistry (Echlin & Morris 1965; Whitton *et al.* 1971; Drews 1973). The distinction between the blue-green algae and eukaryotic algae now appears to be complete, in that no putative intermediate forms have withstood close examination.

The marked structural differences between prokaryotes and eukaryotes gave reason to expect that blue-green algae and eukaryotic unicellular algae would respond in different ways to selective pressure and to the expression of mutational change. There is no equivalent in blue-green algae of the organelle DNA of eukaryotes and hence no possibility of non-nuclear inheritance on this basis (Sager 1965). However, Kumar, Singh and Prakash (1967) 'cured' a streptomycin-resistant strain of *Anacystis nidulans*, suggesting that this character was carried on a plasmid. And in this same species, *A. nidulans*, the presence of a plasmid has been indicated indirectly: a small, circular DNA molecule, separate from the main part of the genome, has been observed by physical means, although no genetic function has yet been ascribed to it (Asato & Ginoza 1973).

In electron micrographs the nuclear regions of blue-green algal cells are seen to be very similar to those of bacteria, which they also resemble in their reactions to Feulgen staining and DNase treatment, and in the location of incorporated ^{3}H-thymidine (see Leach & Herdman 1973). Their DNA consists of fibrils 2·5 nm in diameter (Ris & Singh 1961) which may aggregate into larger units (Leak 1967). In common with other prokaryotic micro-organisms, and in contrast to most eukaryotes, *Anabaena cylindrica* has been shown to lack histones (Makino & Tsuzuki 1971). In two species of blue-green algae, Allen (1972) observed the presence of a mesosome-like structure with a presumptive role in genome segregation like that demonstrated in bacteria.

Genetic studies with blue-green algae have been considerably delayed because of three interrelated technical reasons (see Van Baalen 1973). First, the development of axenic strains of blue-green algae has been a relatively slow process, due, at least in part, to difficulties in devising suitable solid media. Secondly, efficiency of plating was generally poor—i.e. the quantitative growth of single cells on agar plates was not routinely successful—until the work of Van Baalen (1965a) and Allen (1968). Lastly, the difficulty in isolating mutants, especially auxotrophs, of the blue-green algae under study arrested genetic studies of this group for many years.

The first report of gene transfer among cyanophytes was by Kumar (1962), who mixed two antibiotic-resistant strains of *Anacystis nidulans* and reported the isolation of a doubly-resistant recombinant. Pikálek (1967) criticized this work on the grounds that, under the conditions employed, one of the antibiotics (penicillin) was unstable. Evidence for recombination in the filamentous blue-green alga *Cylindrospermum majus* was sought by Singh and Sinha (1965), but pure recombinant clones were not isolated. These workers grew a spore-forming, penicillin-resistant strain with a non-sporulating strain resistant to streptomycin, and isolated from the mixed culture a doubly-resistant strain, apparently a recombinant, that produced spores at low frequency. However, since these early studies were performed entirely in liquid media, in two-factor crosses it is not possible to distinguish clearly between recombination and back-mutation. Singh (1967), in further experiments involving crosses of different non-sporulating strains of *Anabaena doliolum*, isolated sporulating colonies which he attributed to the result of the transfer of genetic material. Gene transfer in *Anacystis nidulans* was demonstrated by Bazin (1968), who isolated stable clones resistant to both polymyxin B and streptomycin from a mixed culture of singly-resistant strains. The stringent conditions and more careful controls of Bazin's work have led to its general acceptance as the first unequivocal demonstration of genetic recombination in a blue-green alga. Gene transfer was also reported in *Nostoc muscorum* by Stewart and Singh (1975), who incubated a streptomycin-resistant, non-nitrogen-fixing mutant (nif^- st-R) with the wild type (nif^+ st-S) and could then isolate clones which were both streptomycin-resistant and able to fix nitrogen (nif^+ st-R). They occurred at frequencies 100 times the spontaneous mutation rate. Filament anastomosis during the developmental cycle of *N. muscorum*, as observed by Lazaroff and Vishniac (1962), is consistent with the idea of gene transfer by cellular contact.

2   MUTANTS

Mutations have been induced in only about ten species of blue-green algae, of which *Anacystis nidulans* has been the most popular. Many mutagens have been used, including various radiations, base analogues, alkylating agents, nitrosoamines and intercalating agents. Shestakov and Jevner (1968) commented on the efficiency of the nitrosoamines, nitrosomethyl urea (NMU) and N-methyl-N'-nitro-N-nitrosoguanidine (MNNG), in inducing mutations in blue-green algae, whereas the alkylating agents they tested, diethyl sulphate (DES), methyl methanesulphonate (MMS) and ethyl methanesulphonate (EMS), were less effective. Agents such as ultraviolet light (UV), X-rays, hydroxylamine, acridine orange and mitomycin C were considered poor mutagens (Shestakov 1972). A survey of mutants of blue-green algae, isolated as pure clones, is given below.

2.1   *Resistance mutants*

This class of mutants has been most frequently sought because of the ease with which they can be isolated by positive selection. Clonally pure mutants resistant to antibiotics, toxic chemicals, radiations and cyanophages have been isolated both after spontaneous mutation and after treatment with various mutagens. Antibiotic-resistant mutants include strains resistant to streptomycin (Singh *et al.* 1966), penicillin (Kumar 1962; Singh & Sinha 1965), polymyxin B (Bazin 1968), chloramphenicol (Shestakov 1972), cryneomycin (Shestakov 1972), kanamycin (Asato & Folsome 1970), erythromycin (Shestakov & Khyen 1970) and sulphanilamide (Kumar 1965). Mutants have also been isolated which are resistant to nitrofurazone (Singh *et al.*

1969), 5-fluorouracil (Bazin 1971), propionate (Smith & Lucas 1971), tyrosine (Herdman & Carr 1972), amino-acid analogues (Phares & Chapman 1975) and chlorate (Singh *et al.* 1972), all of which are toxic to the wild type.

Mutants altered in their response to UV light have been induced in a number of organisms. Most are UV-resistant (*uvr*), although some are UV-sensitive (*uvs*). In *A. nidulans*, *uvr* mutants have been induced using EMS (Kumar 1968a), MNNG (Kumar 1968b), MMS (Kumar 1968c), DES (Gupta & Kumar 1970), UV (Kumar 1963; Singh *et al.* 1966), mitomycin C (Kumar 1968d) and neomycin (Kumar 1968d). Asato (1972) and Mitronova, Shestakov and Zhevner (1973) induced *uvs* mutants of *A. nidulans* with MNNG, and Zhevner and Shestakov (1972) isolated *uvs* mutants of *Synechocystis aquatilis* using MNNG or NMU. Both *uvr* and *uvs* mutants of *Anabaena doliolum* were induced by a combined treatment with ultraviolet light and caffeine (Srivastava 1970).

With the discovery of a cyanophage capable of lysing *Plectonema boryanum*, spontaneous phage-resistant mutants of this organism have been isolated (Singh *et al.* 1972); others have been induced with MNNG, UV and 2-aminopurine (Singh *et al.* 1972). Likewise strains of *A. nidulans* resistant to cyanophage AS-1 have been found to occur spontaneously or as a result of EMS treatment (Delaney & Carr 1975).

2.2 *Morphological and developmental mutants*

The most extensively studied morphological mutants have been the filamentous strains of *Agmenellum quadruplicatum* induced with MNNG. Two distinct classes have been described (Ingram 1970; Ingram & Van Baalen 1970); those with cross-walls (septate) and those without cross-walls (non-septate). Ingram and Thurston (1970) and Brown and Van Baalen (1970) described these classes further; the septate mutants have regular cross-walls out the outer wall fails to disarticulate, leaving the daughter organisms connected, whereas the non-septate mutants are generally 'coenocytic' and divide only sporadically. Phenocopies of some of these mutants have been produced by growth of the wild type in the presence of chloramphenicol and penicillin G (Ingram, Thurston & Van Baalen 1972). Other filamentous mutants included one which failed to divide during rapid growth but divided into unit lengths as the culture density increased (Ingram & Fisher 1973a) and a temperature-sensitive filamentous mutant (Ingram & Fisher 1973b). In both of these mutants the wild-type morphology could be restored by the addition of an extract of spent medium from a wild-type culture (Ingram & Fisher 1973a, b), suggesting that a chemical factor was involved in the regulation of cell division.

In *Anacystis nidulans*, likewise, two classes of filamentous mutant have been isolated following MNNG mutagenesis. These were first characterized by Kunisawa and Cohen-Bazire (1970), and those independently isolated by

Asato and Folsome (1969), Herdman and Carr (1972), and Zhevner, Glazer and Shestakov (1973) fell into one or other of these classes. In class I mutants the cell length varies from one to ten times the normal length, unequal fission generally producing one long and one short daughter organism. Class II mutants are characterized by coenocytic, bent filaments of variable width, up to 100 times the normal length. In addition, UV (Asato & Folsome 1969), MMS (Shestakov & Khyen 1970) and DES (Gupta & Kumar 1970) have been used successfully to induce filamentous mutants of this species. Minute-colony mutants have been induced by MNNG or UV (Asato & Folsome 1969).

Various developmental mutants have been isolated among filamentous blue-green algae. Two distinct non-sporulating mutants of *Anabaena doliolum* were isolated following UV irradiation (Singh 1967). Similar mutants in *A. doliolum* (Singh 1967) and *Cylindrospermum majus* (Singh & Sinha 1965) arose spontaneously. Non-heterocystous mutants of a species of *Cylindrospermum* were reported to occur both spontaneously and after MNNG treatment (Singh *et al.* 1972). Such mutants are interesting in that they exhibit a concurrent loss in nitrogen-fixing ability, in agreement with the widely held view that the heterocyst is the site of nitrogen fixation. All but two of these non-heterocystous mutants have also lost their ability to form spores.

Anabaena cylindrica mutants with abnormal patterns in the spacing of heterocysts among the vegetative cells of the filaments, obtained after treatment with MNNG (Wilcox *et al.* 1975), may prove useful in testing models of developmental patterns in this alga. Other mutants which have been described include a UV-induced *Nostoc linckia* mutant with true branching (Singh & Tiwari 1969), a mutant of *A. doliolum* which undergoes spontaneous lysis (Sinha & Kumar 1973), an MNNG-induced mutant of *Plectonema boryanum* with short trichomes (Padan & Shilo 1969) and a spontaneous mutant of *Anabaena ambigua* with an altered filament morphology (Kale & Talpasayi 1969).

2.3. *Pigment mutants*

Mutants with altered pigment ratios have been described in *Anacystis nidulans* and *Agmenellum quadruplicatum*. In *A. nidulans* such mutations have been reported after treatment with MNNG (Van Baalen 1965b; Shestakov & Jevner 1968; Asato & Folsome 1969) or NMU (Shestakov & Jevner 1968). Shestakov and Jevner (1968) isolated five types of mutant altered in phyco-cyanin and carotenoid synthesis, as well as NMU-induced yellow-green pigment mutants. The yellow mutants described by Asato and Folsome (1969) have reduced levels of phycocyanin, chlorophyll and carotenoids. Blue mutants, similar to those first isolated by Asato and Folsome (1969), have been shown to have a reduced ratio of chlorophyll *a* to phycocyanin (Delaney & Evans, unpublished data). In *A. quadruplicatum* MNNG-induced yellow mutants have been reported to be blocked in the nitrate-reduction pathway (Stevens & Van Baalen 1969, 1970), the characteristics of the yellow phenotype presumably resulting from the breakdown of phycocyanin during nitrogen starvation (Allen & Smith 1969).

2.4. *Temperature-sensitive mutants*

Van Baalen (1965b) described a mutant of *A. nidulans* which grew at 30° (the permissive temperature) but not at 39° (the non-permissive temperature), while Kaney and Dolack (1972) isolated a large number of mutants of this species which grew at 35° but not at 45°. In none of these mutants was any growth-factor requirement demonstrated at the non-permissive temperature, suggesting that all the mutations were in vital functions which could not be repaired by external supplementation. A thiamine-requiring mutant of *A. nidulans* has now been isolated which is also temperature-sensitive. Auxotrophy and temperature-sensitivity appear to be separate mutations (Delaney 1973) since when cultured at 34°C the cells require thiamine for growth, but at 38°C no growth occurs even in the presence of thiamine. However, revertants to thiamine independence are no longer temperature-sensitive. After MNNG mutagenesis a temperature-sensitive filamentous mutant of *Agmenellum quadruplicatum* was isolated in which the usual relationships of permissive and non-permissive temperatures were inverted, the mutation being expressed at the lower temperature (35°) but not at the higher one (44°) (Ingram & Fisher 1973b).

2.5. *Auxotrophs*

Despite the progress made in the isolation of mutants of blue-green algae, yields of auxotrophs have been generally poor. Early work led to the isolation of mutants whose nutritional requirements could not be precisely defined, but more recently a number of mutants requiring an amino acid or other essential growth factor have been isolated. The first reports of auxotrophs of blue-green algae were those of Singh and Singh (1964a, b) in *Anabaena* species. UV mutagenesis of *Anabaena cyadeae* led to the isolation of a non-nitrogen-fixing mutant (Singh & Singh 1964a). In both *Anabaena cycadeae* and *Anabaena doliolum* mutants requiring glucose for growth have been isolated following UV irradiation (Singh & Singh 1964a, b). After UV irradiation of *Phormidium mucicola* Srivastava (1969) isolated a mutant with a glucose requirement which could also be satisfied by casein hydrolysate, acetate or a mixture of vitamins. In view of its non-specific requirements, it is difficult to see where its mutational block could be. Possibly these supplements acted as chelating or osmotic agents, indirectly supporting growth in this way. Several mutants defective in the reduction of nitrate have been characterized. Van Baalen (1965b) isolated a mutant of *A. nidulans*, after

MNNG mutagenesis, which was unable to reduce nitrate, and the induction of similar mutants of *A. nidulans* by MNNG and NMU was described by Shestakov and Jevner (1968). In *Agmenellum quadruplicatum*, Stevens and Van Baalen (1970) identified two types of MNNG-induced yellow mutants deficient in nitrate reduction: some were defective in nitrate reductase, others in nitrite reductase. A mutant of *A. nidulans* which lacks 6-phosphogluconate dehydrogenase and is unable to survive incubation in the dark has been isolated by Doolittle and Singer (1974).

Until recently the isolation of amino-acid-requiring auxotrophs in blue-green algae was achieved only by MNNG mutagenesis followed by enrichment with penicillin. Ingram, Pierson, Kane, Van Baalen and Jensen (1972) isolated a tryptophan-requiring auxotroph of *A. quadruplicatum* which they showed to be defective in the tryptophan-synthetase A protein. This mutant proved to be unstable, however, and was subsequently lost (Ingram, personal communication). A phenylalanine-requiring mutant of *Synechoccocus cedrorum* defective in prephrenate hydratase was isolated by Kaney (1973). In our laboratory several auxotrophs of *A. nidulans*, respectively requiring ammonia, acetate, biotin, methionine, phenylalanine or cysteine, were isolated following mutagenesis with MNNG (Herdman & Carr 1972).

The development of a chemically defined complete medium (Herdman *et al.* 1973) led to the isolation of an even more extensive range of auxotrophs following MNNG mutagenesis. By altering the incubation time in minimal medium prior to penicillin enrichment, or by increasing the number of cycles of enrichment, additional mutants requiring thiamine, adenine, uracil or p-aminobenzoic acid were recovered (Delaney *et al.* 1974). Some EMS mutants proved to require ammonia, thiamine or biotin. A double mutant was found, defective in intermediary carbon metabolism, which requires a mixture of a C_2 compound (acetate or pyruvate) and a C_4 compound (fumarate, malate or succinate) for growth. The requirement for a C_4 compound presumably results from decreased levels of malate dehydrogenase in the mutant (Delaney & Carr, unpublished).

3. REPAIR MECHANISMS

Several mechanisms for the repair of genetic damage operate in bacteria; photoreactivation, excision repair and post-replication repair have been described (see Hayes 1968 and Witkin 1969a, b). Recently a fourth process has been indicated, involving the resumption of interrupted replication (Bridges 1972). At least some of these mechanisms evidently occur also in blue-green algae. Photoreactivation, which in bacteria cleaves pyrimidine dimers produced in the DNA by UV light, has been reported in *Agmenellum quadruplicatum* (Van Baalen 1968; Van Baalen & O'Donnell 1972), *Fischerella muscicola* (Singh & Singh 1972) and *Anacystis nidulans* (Asato & Folsome

1969), and a DNA-photoreactivating enzyme has been purified from *Plecto-nema boryanum* (Werbin & Rupert 1968) and *A. nidulans* (Saito & Werbin 1970; Minato & Werbin 1972). The suggestion by Van Baalen (1968) that a photoreactivation system was responsible for repair of damage to the photo-synthetic apparatus, rather than to the genome, was supported by observa-tions of Wu, Lewin and Werbin (1967), who showed that in *Plectonema boryanum* UV-induced damage to the cells could be repaired by white, blue, red or 'black' light, whereas UV-induced damage to the cyanophage LPP-1 could be repaired by exposing the infected host to only white or blue light. Apparently in this organism there are two photoreactivation processes, only one of which can reverse damage to the virus, while the other is somehow involved in the repair of photosynthetic damage. However, in an *in vitro* system the virus DNA was photoreactivated by extracts of *Plectonema boryanum* even in 'black' light. Photoreactivation of damage produced by γ-radiation was reported in *A. nidulans* (Asato 1971). UV-sensitive mutants apparently deficient in photoreactivation have been isolated from *A. nidulans* (Asato 1972) and *Synechocystis aquatilis* (Zhevner & Shestakov 1972).

There is now good evidence from several different lines of research to suggest that there are also dark-repair mechanisms in blue-green algae like those of bacteria. Caffeine and acriflavine are believed to prevent excision of UV-induced damaged segments of the DNA (Setlow 1964; Clarke 1967), causing an increased sensitivity to the mutagen. Under conditions preventing photoreactivation, sensitivity to UV light was found to increase as a result of caffeine treatment of *A. doliolum* (Srivastava 1970; Srivastava *et al.* 1971) and *A. nidulans* (Asato 1972) and acriflavine treatment of *A. nidulans* (Singh *et al.* 1969) and *Fischerella muscicola* (Singh & Singh 1972). A UV-sensitive mutant of *A. nidulans* was shown to be insensitive to this effect of acriflavine, indicating that the mutant lacked such an excision-repair mechanism (Singh 1968). Stevens and Van Baalen (1969) showed that MNNG-induced damage of *Agmenellum quadruplicatum* could be repaired by subsequent incubation of treated cells in light too dim to support photosynthetic growth, suggesting the operation of a dark-repair system similar to that of bacteria. Zhevner and Shestakov (1972) isolated several UV-sensitive mutants of *Synechocystis aquatilis* which showed increased sensitivity to alkylating agents and X-rays, like certain *hcr* mutants of *E. coli* deficient in the excision-repair mechanism. Another mutant sensitive to the killing effect of UV light, though with reduced mutability, resembles *exr* mutants of *E. coli* lacking a second dark-repair process, post-replication repair (Witkin 1969a).

The existence of similar repair processes in *Anacystis nidulans* has been suggested by studies on the repair of EMS-induced damage in the presence of caffeine (Delaney & Carr 1975), which increases the mutagenic effect of EMS without altering its lethal effect. Since caffeine inhibits the excision of mutational damage, and since photoreactivation has no effect on the repair of EMS-induced damage, any repair of EMS-induced damage must be attributable to a second dark-repair process. This postulated process is evidently as efficient as excision repair in correcting lethal damage, although it increases mutagenesis (like the post-replication repair process described by Witkin (1969a) in *E. coli*).

4 CYANOPHAGES

Several viruses which lyse blue-green algae have been characterized during the last ten years, and there is now a considerable literature on their classification, structure, physiology and ecology (see Safferman 1973; Padan & Shilo 1973). Most information has been derived from studies of the LPP cyanophages. These normally cause lysis of the filamentous blue-green alga *Plectonema boryanum*; recently some lysogenic strains have also been isolated (Cannon *et al.* 1971; Padan *et al.* 1972). In all essential respects cyanophages are similar to bacteriophages, suggesting their possible use in genetic analysis of blue-green algae by means of a transduction process. However there has not as yet been any unequivocal demonstration of virus-mediated gene transfer, perhaps because of the lack of suitable genetic markers in susceptible strains. The isolation of a lytic cyanophage (see Fig. 1.1.), AS-1, infecting *Anacystis nidulans* (Safferman *et al.* 1972) has renewed hope of exploiting

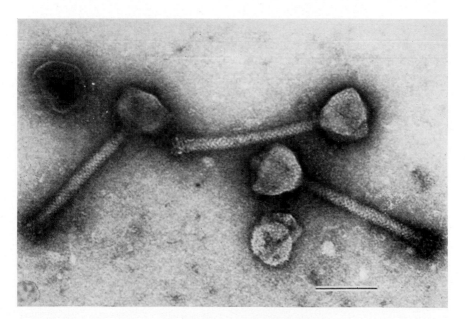

Fig. 1.1. Cyanophage AS-1 negatively stained with 4% aqueous uranyl acetate (length of bar, 0·1 μm). (Photograph kindly provided by Dr. R. S. Safferman.)

such a transduction process, although no genetic activity has yet been reported.

The isolation of temperature-sensitive mutants of the cyanophage LPP2-SPI has permitted the construction of a linkage map of this virus. Twenty-three mutants have been isolated following MNNG mutagenesis (Rimon & Oppenheim 1974). No physiological functions have been assigned to them, but complementation tests have shown that they belong to 14 different groups, probably representing different cistrons. Recombination between mutants from each group has been studied in two-factor crosses, and a genetic map has been constructed (Fig. 1.2). Map distances are additive, although where markers are closely linked (less than 2·5 recombination units apart) unexpectedly high recombination frequencies have been observed, indicating interference between adjacent cistrons.

Fig. 1.2 Linkage map of cyanophage LPP2-SPI (Rimon & Oppenheim 1974). No physiological function has yet been assigned to the cistrons, which are referred to arbitrarily by letters.

5 TRANSFORMATION

In early studies of recombination, only three mutant phenotypes (penicillin, streptomycin and polymyxin resistance) were available in *A. nidulans* and only two (penicillin and streptomycin resistance) in *Cylindrospermum majus*: thus multifactorial crosses could not be made. The isolation of a wide range of auxotrophic mutants (Herdman & Carr 1972) and the optimization of conditions for rapid and quantitative growth on agar surfaces (Herdman *et al.* 1973) facilitated further investigation of recombination in *A. nidulans*. A mechanism of gene transfer requiring cellular contact has not yet been demonstrated. There is no need to invoke such a direct 'sexual' process, since in experiments of Herdman and Carr (1971) recombination occurred at frequencies of up to $1·2 \times 10^{-3}$ when cell-free filtrates of one mutant strain were incubated with cultures of a second strain, suggesting that a process of transformation was operating. The transforming principle was sensitive to nucleases but not proteases, and operated after deproteinization of the medium. After concentration by ultrafiltration, nucleic acids isolated from the culture medium were purified by the procedure of Craig, Leach and Carr (1969). Quantities of the order of ng per ml of culture were obtained which retained considerable transforming activity. Preincubation of filtrates with DNase reduced transforming activity by only 60%. Surprisingly, RNase had a similar effect. Preincubation of a transforming preparation with both enzymes together did not further reduce its action. It was suggested that the transforming principle secreted into the culture medium was a DNA-RNA complex in which both nucleic acids were essential for efficient transformation.

The accumulation of nucleic acids in the medium is not unusual in bacterial cultures. DNA has been demonstrated in the culture media of *Brucella* (Mauzy *et al.* 1955), *Pseudomonas* (Catlin 1956), *Staphylococcus* (Catlin & Cunningham 1958), *Neisseria meningitidis* (Catlin 1960), *Bacillus subtilis* (Takahashi 1962) and *Pneumococcus* (Ottolenghi & Hotchkiss 1960). In the last three of these cases such nucleic acids were found to be active in transformation.

Transformation in *Pneumococcus*, mediated by RNA fractions chemically extracted from donor cells, is sensitive to both RNase and DNase (Evans 1964), and in this respect resembles transformation by extracellular nucleic acids of *A. nidulans*. However, in *Pneumococcus* the transformants initially behave as unstable heterozygotes in which ultimately the transferred characters are either lost or stably inherited, whereas in *A. nidulans* all of the transformants appear to be stable.

Transformation with DNA chemically extracted from donor cells has also been demonstrated in *A. nidulans* (Shestakov & Khyen 1970; Mitronova *et al.* 1973; Orkwiszewski & Kaney 1974). This system resembles the classical bacterial system of transformation, and differs from that mediated by extracellular nucleic acids, in that the transforming agent is completely resistant to RNase and completely sensitive to DNase (Herdman 1973a). The highest rates of transformation were observed when the recipient was in the late-exponential phase of growth and was exposed to donor DNA (20–40 μg/ml) for 4h at pH 8·0 or higher (Orkwiszewski & Kaney 1974).

In *A. nidulans*, linkage of markers has been demonstrated in transformations mediated by both extracellular nucleic acids and chemically extracted DNA (Herdman 1973b). In transformation with extracted DNA only the markers *bio-1* and *str-2* exhibited a high cotransfer index; in contrast, in transformation with extracellular nucleic acids more extensive linkage was observed, and a high proportion of the transformants inherited all of the donor markers employed in the cross. For example, the markers *nit-1* and *met-2*, apparently unlinked in transformation experiments with chemically extracted DNA, were cotransformed to 87% of the progeny by extracellular nucleic acids, suggesting that these are of a higher molecular weight than chemically extracted DNA (possibly because they embody RNA, too). In transformation of *Bacillus subtilis* with extracellular DNA, there is similarly increased linkage for several markers which appear unlinked in transformation by chemically extracted DNA (Ephrati-Elizur 1968).

In *A. nidulans* genetic experiments indicated the map order for five markers shown in Fig. 1.3 (Herdman 1973a), although variations in linkage values in different experiments did not permit the establishment of accurate map distances. Some of the variations could be attributed to lethal mutations, which arose at high frequency, causing the loss of certain classes of recombinants. Such mutations arose only in recombinants: the mutation was evidently associated with, and inseparable from, recombination, since variation in DNA concentration, treatment with DNase, and reduction of the molecular weight of extracellular nucleic acids from donor cells, all affected transformation and mutation simultaneously and to the same extent. Finally, it was found that such mutation occurred at the site of recombination: thus in *A. nidulans* the process of recombination appears itself to be mutagenic (Herdman 1973b).

Since lethal mutation during recombination prevents the recovery of some of the potential recombinants, and therefore interferes with the calculation of linkage values, a coherent genetic analysis using conventional procedures may not be possible in this species. However, it is possible to use

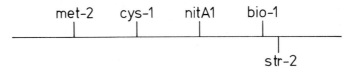

Fig. 1.3.   Recombination map of *Anacystis nidulans* (Herdman 1973a).

this mutagenic phenomenon itself to order markers, since the relative position of a new mutation may be calculated from the recombination of outside markers. For example, a cross of a strain carrying the markers *nit-1* and *str-1* with prototrophic DNA produced both double (*nit-1+ str-1+*) and single (*nit-1+ str-1*) recombinants. All strains of the latter group, produced by recombination between *nit-1* and *str-1*, were found also to be *bio* mutants, indicating that a *bio* gene lies between *nit-1* and *str-1*. The genetic map (Fig. 1.3) shows that this is indeed the case. Other markers can be mapped in the same way (Herdman 1973a).

Mutagenesis associated with recombination has been observed during meiotic recombination in *Saccharomyces cerevisiae* (Magni 1963), in transduction in *Salmonella typhimurium* (Demerec 1962), during transformation in *Bacillus subtilis* (Yoshikawa 1966) and during integration of the phage Mu-1 in *Escherichia coli* (Taylor 1963). In all of these cases recombination is closely associated with the mutating locus, but mutation occurs only at much lower frequencies than in *A. nidulans*.

The nature of the mutagenic process during recombination in *A. nidulans* remains to be elucidated. The high frequency of the phenomenon suggests that the mutation probably results from a fundamental error in the recombination process. The mutants revert to wild type at rates similar to those of MNNG-induced mutants (Herdman 1973b) and are therefore presumably due to single-base changes. Six recombination-induced mutants and five MNNG-induced mutants were caused to revert by MNNG, EMS or DES,

but not by 5-bromouracil, 2-aminopurine, proflavin, rifampicin or hydroxylamine, which suggested that all the mutants were transversions (purine ⇌
pyrimidine)—although the possibility that two of them were frameshifts
could not be excluded (Herdman & Carr, unpublished). Thus in *A. nidulans*
one or more steps of the recombination process seem to be error-prone and
liable to produce single-base changes.

6 TEMPORAL GENETIC MAPPING

Several attempts have been made to construct a genetic map of *A. nidulans*
by studying the sequential replication of the genome rather than by recombination analysis after gene transfer. This method is based on the change
in mutation frequency of a gene as it replicates and may most conveniently
be studied during synchronous growth. In this species synchrony of cell
division and DNA synthesis can be induced by a period of light deprivation
(Herdman *et al.* 1970) (Fig. 1.4). In MNNG or EMS mutagenesis of a culture
during synchronous DNA synthesis the maximum mutation frequency of a
gene corresponds to the time at which it is replicated (Fig. 1.5). By using
MNNG for sequential mutagenesis of synchronized cultures, Asato and
Folsome (1970) were able to construct a temporal genetic map for six markers
(Fig. 1.6); in a similar way, Herdman (1971) using UV produced a map for
three markers. EMS mutagenesis of synchronized cultures of *A. nidulans* has
enabled a temporal genetic map for 19 markers to be constructed (Delaney
& Carr 1975) as shown in Fig. 1.7. Some markers seem to be located at

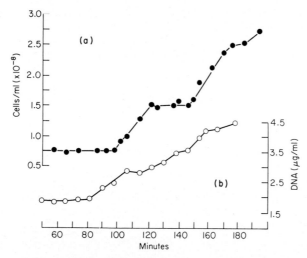

Fig. 1.4. (*a*) Growth of *Anacystis nidulans*, (● – ●) (as determined by increase
in cell numbers), and (*b*) DNA synthesis (○ – ○), after synchronization by light
deprivation (Herdman, Faulkner & Carr 1970).

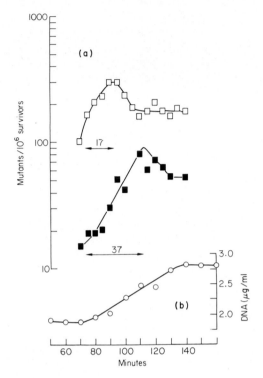

Fig. 1.5. (*a*) Numbers of mutants resistant to propionate (*prp*) (■ – ■) and to cyanophage AS-1 (*aso*) (□ – □) induced by EMS treatment at different stages in replication cycle (Delaney & Carr 1975). (Temporal map positions of *aso* and *prp* markers indicated by peaks in mutation frequency 17 and 37 min, respectively, after initiation of synchronous DNA replication.) (*b*) DNA content, showing synchronous DNA replication in the period 75 to 135 minutes.

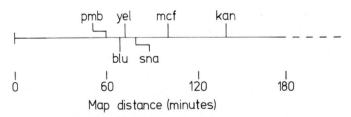

Fig. 1.6. Temporal genetic map of *Anacystis nidulans* obtained by MNNG mutagenesis (Asato & Folsome 1970). The time scale represents 180 min of the DNA replication cycle of 420 min.

more than one map position, indicating that mutations at more than one locus may result in the same phenotype. Genes for yellow and blue pigmentation and for resistance to polymyxin B show multiple map positions, and have therefore to be distinguished, for example as *yelA*, *yelB* and *yelC*. In addition, multiple map positions are observed for the *bio-2* marker when mapped by reversion. Revertants are of two types; large dark colonies resembling those of the wild type, and small yellow colonies. The latter appear to be suppressed mutants and they have accordingly been mapped at a locus designated *supA*. Suppressor mutations have not been observed in the mapping of any other auxotrophs.

The temporal genetic map of *A. nidulans* may be compared with the conventional genetic map with respect to four markers, and on this basis two models of replication can be proposed (Fig. 1.8). Either replication starts from a single initiation point and is bidirectional, or replication is unidirectional and starts from several initiation points. At present it is not possible to distinguish between these two hypotheses, but since bidirectional replication is now well established in other organisms such as *Escherichia coli* (Masters & Broda 1971), *Bacillus subtilis* (Gyurasits & Wake 1973) and simian virus 40 (Danna & Nathans 1972) as well as in mammalian cells (Huberman & Riggs 1968) the former model seems the more plausible.

Assuming bidirectional replication of the *A. nidulans* genome, the genetic organization in this organism is comparable with that of the Enterobacteriaceae. The temporal genetic map of *A. nidulans* shows remarkable similarities with a conventional genetic map of *E. coli* K12 (Taylor & Trotter 1972). The positions of ten markers on the *A. nidulans* map correlate well with analogous markers on the *E. coli* map, as shown in Fig. 1.9. The markers *cys-1*, *str*, *met-4*, *bio-1*, *bio-2* and *supA* exhibit very similar map positions to the corresponding markers in *E. coli*; the *fil* in *A. nidulans* occupies a similar position to the *lon* (long or filamentous form) in *E. coli*; while *nitA1*, *yelA* and *yelB* map in regions corresponding to *chl* markers in *E. coli* (which determine the responses to chlorate and specify the synthesis of nitrate reductase). This suggests that in *A. nidulans yelB* and *yelC* may be involved in the synthesis of nitrate reductase (see page 12).

Since mutations at more than one site can result in the same phenotype (for example, methionine requirement), and since these genes may be scattered throughout the genome, the map positions of some of these markers may correspond purely by chance. However, the correspondences between *A. nidulans* and *E. coli* maps are striking, and provide evidence for a degree of similarity in the organization of the genetic material between these two conventionally unrelated organisms, and perhaps within the prokaryota as a whole.

C

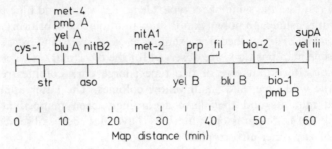

Fig. 1.7. Temporal genetic map of *Anacystis nidulans* determined by EMS mutagenesis (Delaney & Carr 1975).

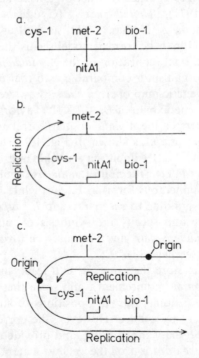

Fig. 1.8. The comparison of the temporal genetic map of *Anacystis nidulans* with the recombination map (Herdman 1973a) is consistent with two possible modes of replication. Only the gene orders are shown; map distances are not relative. *a*, temporal genetic map. *b*, recombination map showing bidirectional replication from a single origin near *cys-1*. *c*, recombination map showing unidirectional replication from two (or more) origins.

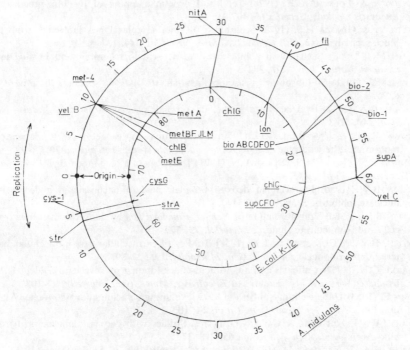

Fig. 1.9. Comparison of the temporal genetic map of *Anacystis nidulans* with the map of *Escherichia coli* K-12.

ACKNOWLEDGEMENT

The work at Liverpool on blue-green algal genetics was initiated in collaboration with Dr B. M. Faulkner, to whom the authors are indebted for much advice. The Science Research Council has provided financial support.

7 REFERENCES

ALLEN M.M. (1968) Simple conditions for growth of unicellular blue-green algae on plates. *J. Phycol.* **4**, 1–4.

ALLEN M.M. (1972) Mesosomes in blue-green algae. *Arch. Mikrobiol.* **84**, 199–206.

ALLEN M.M. & SMITH A.J. (1969) Nitrogen chlorosis in blue-green algae. *Arch. Mikrobiol.* **69**, 114–20.

ASATO Y. (1971) Photorecovery of gamma irradiated cultures of blue-green alga, *Anacystis nidulans*. *Radiation Bot.* **11**, 313–6.

ASATO Y. (1972) Isolation and characterisation of ultraviolet-light sensitive mutants of the blue-green alga *Anacystis nidulans*. *J. Bacteriol.* **110**, 1058–64.

ASATO Y. & FOLSOME C.E. (1969) Mutagenesis of *Anacystis nidulans* by N-methyl-N'-nitro-N-nitrosoguanidine and UV irradiation. *Mutat. Res.* **8**, 531–6.

ASATO Y. & FOLSOME C.E. (1970) Temporal genetic mapping of the blue-green alga, *Anacystis nidulans*. *Genetics* **65**, 407–19.

ASATO Y. & GINOZA H.S. (1973) Separation of small circular DNA molecules from the blue-green alga *Anacystis nidulans*. *Nature, New Biol.* **244**, 132–3.

BAZIN M.J. (1968) Sexuality in a blue-green alga: genetic recombination in *Anacystis nidulans*. *Nature, Lond.* **218**, 282–3.

BAZIN M.J. (1971) 5-Fluorouracil resistance in *Anacystis nidulans* Drouet. *Br. Phycol. J.* **6**, 25–8.

BRIDGES B.A. (1972) Evidence for a further dark repair process in bacteria. *Nature, New Biol.* **240**, 52–3.

BROWN R.M. & VAN BAALEN C. (1970) Comparative ultrastructure of a filamentous mutant and the wild type of *Agmenellum quadruplicatum*. *Protoplasma* **70**, 87–99.

CANNON R.E., SHANE M.S. & BUSH V.N. (1971) Lysogeny of a blue-green alga, *Plectonema boryanum*. *Virology* **45**, 149–53.

CATLIN B.W. (1956) Extracellular deoxyribonucleic acid of bacteria and a deoxyribonuclease inhibitor. *Science* **124**, 441–2.

CATLIN B.W. (1960) Transformation of *Neisseria meningitidis* by deoxyribonucleates from cells and from culture slime. *J. Bacteriol.* **79**, 579–90.

CATLIN, B.W. & CUNNINGHAM L.S. (1958) Studies of extracellular and intracellular bacterial deoxyribonucleic acids. *J. Gen. Mikrobiol.* **19**, 522–39.

CLARKE C.H. (1967) Caffeine- and amino acid-effects upon *trp*⁺ revertant yield in U.V.-irradiated *hcr*⁺ and *hcr*⁻ mutants of *E. coli* B/r. *Molec. Gen. Genet.* **99**, 97–108.

COHN F. (1853) Untersuchungen über die Entwickelungsgeschichte mikroscopischer Algen und Pilze. *Nov. Act. Acad. Leop. Carol.* **24**, 103–256.

CRAIG I.W., LEACH C.K. & CARR N.G. (1969) Studies with deoxyribonucleic acid from blue-green algae. *Arch. Mikrobiol.* **65**, 218–27.

DANNA K.J. & NATHANS D. (1972) Bidirectional replication of Simian Virus 40 DNA. *Proc. Natl. Acad. Sci. U.S.A.* **69**, 3097–100.

DELANEY S.F. (1973) Studies on the organisation of the genome in the blue-green alga *Anacystis nidulans*. Ph.D. Thesis, University of Liverpool, England.

DELANEY S.F. & CARR N.G. (1975) Temporal genetic mapping in the blue-green alga *Anacystis nidulans* using ethyl methanesulphonate. *J. Gen. Microbiol.* **88**, 259–68.

DELANEY S.F., HERDMAN M., FOULDS I.J. & CARR N.G. (1974) Mutational studies with *Anacystis nidulans*. *Proceedings of the Society for General Microbiology* II, 13–14.

DEMEREC M. (1962) "Selfers"—attributed to unequal crossovers in *Salmonella*. *Proc. Natl. Acad. Sci. U.S.A.* **48**, 1696–1704.

DOOLITTLE W.F. & SINGER R.A. (1974) Mutational analysis of dark endogenous metabolism in the blue-green bacterium *Anacystis nidulans*. *J. Bacteriol.* **119**, 677–83.

DREWS G. (1973) Fine structure and chemical composition of the cell envelopes. In *The Biology of Blue-Green Algae*, eds. Carr N.G. and Whitton B.A. pp. 99–116. Blackwell Scientific Publications, Oxford.

ECHLIN P. & MORRIS I. (1965) The relationship between blue-green algae and bacteria. *Biol. Rev.* **40**, 143–87.

EPHRATI-ELIZUR E. (1968) Spontaneous transformation in *Bacillus subtilis*. *Genet. Res.* **11**, 83–96.

EVANS A.H. (1964) Introduction of specific drug-resistant properties by purified RNA-containing fractions from *Pneumococcus*. *Proc. Natl. Acad. Sci. U.S.A.* **52**, 1442–9.

GUPTA R.S. & KUMAR H.D. (1970) Action of mutagenic chemicals on *Anacystis nidulans*. V. Diethyl sulphate. *Arch. Mikrobiol.* **70**, 313–29.

GYURASITS E.B. & WAKE R.G. (1973) Bidirectional chromosome replication in *Bacillus subtilis*. *J. Molec. Biol.* **73**, 55–63.

HAYES W. (1968) *The Genetics of Bacteria and their Viruses*, pp. 331–6. 2nd edn. Blackwell Scientific Publications, Oxford.

HERDMAN M. (1971) Genetic studies of the blue-green alga *Anacystis nidulans*. Ph.D. Thesis, University of Liverpool, England.

HERDMAN M. (1973a) Transformation in the blue-green alga *Anacystis nidulans* and the associated phenomena of mutation. In *Bacterial Transformation*, ed. Archer L.J. pp. 369–86. Academic Press, London.

HERDMAN M. (1973b) Mutations arising during transformation in the blue-green alga *Anacystis nidulans*. *Molec. Gen. Genet.* **120**, 369–78.

HERDMAN M. & CARR N.G. (1971) Recombination in *Anacystis nidulans* mediated by an extracellular DNA/RNA complex. *J. Gen. Microbiol.* **68**, xiv–xv.

HERDMAN M. & CARR N.G. (1972) The isolation and characterisation of mutant strains of the blue-green alga *Anacystis nidulans*. *J. Gen. Microbiol.* **70**, 213–20.

HERDMAN M., DELANEY S.F. & CARR N.G. (1973) A new medium for the isolation and growth of auxotrophic mutants of the blue-green alga *Anacystis nidulans*. *J. Gen. Microbiol.* **79**, 233–7.

HERDMAN M., FAULKNER B.M. & CARR N.G. (1970) Synchronous growth and genome replication in the blue-green alga *Anacystis nidulans*. *Arch. Mikrobiol.* **73**, 238–49.

HUBERMAN J.A. & RIGGS A.D. (1968) On the mechanism of DNA replication in mammalian chromosomes. *J. Molec. Biol.* **32**, 327–41.

INGRAM L.O. (1970) Morphological mutations of the blue-green alga, *Agmenellum*. *J. Phycol. (Suppl.)* **6**, 13.

INGRAM L.O. & FISHER W.D. (1973a) Novel mutant impaired in division: evidence for a positive regulating factor. *J. Bacteriol.* **113**, 999–1005.

INGRAM L.O. & FISHER W.D. (1973b) Mechanism for the regulation of cell division in *Agmenellum*. *J. Bacteriol.* **113**, 1006–14.

INGRAM L.O. & THURSTON E.L. (1970) Cell division in morphological mutants of *Agmenellum quadruplicatum*, strain BG-1. *Protoplasma* **71**, 55–75.

INGRAM L.O. & VAN BAALEN C. (1970) Characterisations of a stable, filamentous mutant of a coccoid blue-green alga. *J. Bacteriol.* **102**, 784–9.

INGRAM L.O., THURSTON E.L. & VAN BAALEN C. (1972) Effects of selected inhibitors on growth and cell division in *Agmenellum*. *Arch. Mikrobiol.* **81**, 1–12.

INGRAM L.O., PIERSON D., KANE J.F., VAN BAALEN C. & JENSEN R.A. (1972) Documentation of auxotrophic mutation in blue-green bacteria: characterisation of a tryptophan auxotroph in *Agmenellum quadruplicatum*. *J. Bacteriol.* **111**, 112–18.

KALE S. & TALPASAYI E.R.S. (1969) On a spontaneous mutant strain of *Anabaena ambigua* Rao. *Phykos* **8**, 42–5.

KANEY A.R. (1973) Enzymatic characterisation of a phenylalanine auxotroph of the blue-green bacterium *Synechococcus cedrorum*. *Arch. Mikrobiol.* **92**, 139–42.

KANEY A.R. & DOLACK M.P. (1972) Temperature-sensitive mutants of *Anacystis nidulans*. *Genetics* **71**, 465–7.

KUMAR H.D. (1962) Apparent genetic recombination in a blue-green alga. *Nature, Lond.* **196**, 1121–2.

KUMAR H.D. (1963) Effects of radiations on blue-green algae. 1. The production and characterization of a strain of *Anacystis nidulans* resistant to ultraviolet radiation. *Ann. Bot., N.S.* **27**, 723–33.

KUMAR H.D. (1965) Effects of certain toxic chemicals and mutagens on the growth of the blue-green alga *Anacystis nidulans*. *Can. J. Bot.* **43**, 1523–32.

KUMAR H.D. (1968a) Action of mutagenic chemicals on *Anacystis nidulans*. II. Ethyl methanesulfonate. *Beitr. Biol. Pflanzen* **45**, 161–70.

KUMAR H.D. (1968b) Action of mutagenic chemicals on *Anacystis nidulans*. III. N-methyl-N′-nitro-N-nitrosoguanidine. *Arch. Mikrobiol.* **63**, 95–102.

KUMAR H.D. (1968c) Mutation studies in the blue-green alga *Anacystis nidulans*. *Proc. XII Intern. Congr. Genet.* **1**, 84.

KUMAR H.D. (1968d) Inhibitory action of the antibiotics mitomycin C and neomycin on the blue-green alga *Anacystis nidulans*. *Flora, Abt. A.* **159**, 437–44.

KUMAR H.D., SINGH H.N. & PRAKASH G. (1967) The effect of proflavine on different strains of the blue-green alga *Anacystis nidulans*. *Plant Cell Physiol.* **8**, 171–9.

KUNISAWA R. & COHEN-BAZIRE G. (1970) Mutations of *Anacystis nidulans* that affect cell division. *Arch. Mikrobiol.* **71**, 49–59.

LAZAROFF N. & VISHNIAC W. (1962) The participation of filamentan astomosis in the developmental cycle of *Nostoc muscorum*, a blue-green alga. *J. Gen. Microbiol.* **28**, 203–10.

LEACH C.K. & HERDMAN M. (1973) Structure and function of nucleic acids. In *The Biology of the Blue-Green Algae*, eds. Carr N.G. & Whitton B.A. pp. 186–200. Blackwell Scientific Publications, Oxford.

LEAK L.V. (1967) Studies on the preservation and organisation of DNA-containing regions in a blue-green alga: a cytochemical and ultrastructural study. *J. Ultrastruct. Res.* **20**, 190–205.

MAGNI G.E. (1963) The origin of spontaneous mutations during meiosis. *Proc. Natl. Acad. Sci. U.S.A.* **50**, 975–80.

MAKINO F. & TSUZUKI J. (1971) Absence of histone in the blue-green alga *Anabaena cylindrica*. *Nature, Lond.* **231**, 446–7.

MASTERS M. & BRODA P. (1971) Evidence for the bidirectional replication of the *Escherichia coli* chromosome. *Nature, New Biol.* **232**, 137–40.

MAUZY W., BRAUN W. & WHALLON J. (1955) Studies on DNA and DNase in *Brucella* cultures. *Bact. Proc.* 1955, 46.

MINATO S. & WERBIN H. (1972) Excitation and fluorescence spectra of the chromophore associated with the DNA-photoreactivating enzyme from the blue-green alga *Anacystis nidulans*. *Photochem. Photobiol.* **15**, 97–100.

MITRONOVA T.N., SHESTAKOV S.V. & ZHEVNER V.D. (1973) Properties of a radiosensitive filamentous mutant of the blue-green alga *Anacystis nidulans*. *Mikrobiologiya* **42**, 519–24.

ORKWISZEWSKI K.G. & KANEY A.R. (1974) Genetic transformation of the blue-green bacterium, *Anacystis nidulans*. *Arch. Mikrobiol.* **98**, 31–7.

OTTOLENGHI E. & HOTCHKISS R.D. (1960) Appearance of genetic transforming activity in *Pneumococcal* cultures. *Science* **132**, 1257–8.

PADAN E. & SHILO M. (1969) Short-trichome mutant of *Plectonema boryanum*. *J. Bacteriol.* **97**, 975–6.

PADAN E. & SHILO M. (1973) Cyanophages—viruses attacking blue-green algae. *Bact. Rev.* **37**, 343–70.

PADAN E., SHILO M. & OPPENHEIM A.B. (1972) Lysogeny of the blue-green alga *Plectonema boryanum* by LPP2-SPI cyanophage. *Virology* **47**, 525–6.

PHARES W. & CHAPMAN L.F. (1975) *Anacystis nidulans* mutants resistant to aromatic amino acid analogues. *J. Bacteriol.* **122**, 943–8.

PIKÁLEK P. (1967) Attempt to find genetic recombination in *A. nidulans*. *Nature, Lond.* **215**, 666–7.

RIMON A. & OPPENHEIM A.B. (1974) Isolation and genetic mapping of temperature-sensitive mutants of cyanophage LPP2-SPI. *Virology* **62**, 567–9.

RIS H. & SINGH R.N. (1961) Electron microscope studies on blue-green algae. *J. Biophys. Biochem. Cytol.* **9**, 63–80.

SAFFERMAN R.S. (1973) Phycoviruses. In *The Biology of Blue-Green Algae*, eds. Carr N.G. & Whitton B.A. pp. 214–37. Blackwell Scientific Publications, Oxford.

SAFFERMAN R.S., DIENER T.O., DESJARDINS P.R. & MORRIS M.E. (1972) Isolation and characterisation of AS-1, a phycovirus infecting the blue-green algae, *Anacystis nidulans* and *Synechococcus cedrorum*. *Virology* **47**, 105–13.

SAGER R. (1965) On non-chromosomal heredity in micro-organisms. In *Function and Structure in Micro-Organisms*, eds. Pollock M.R. & Richmond M.N. pp. 324–42. Fifteenth Symposium of the Society for General Microbiology, Cambridge University Press, Cambridge.

SAITO N. & WERBIN H. (1970) Purification of a blue-green algal deoxyribonucleic acid photoreactivating enzyme. An enzyme requiring light as a physical cofactor to perform its catalytic function. *Biochemistry* 9, 2610–20.

SETLOW R.B. (1964) Physical changes and mutagenesis. *J. Cell. Comp. Physiol.* 64 (Suppl. 1), 51–68.

SHESTAKOV S.V. (1972) Mutagenesis and repair systems in unicellular blue-green algae. In *Taxonomy and Biology of Blue-Green Algae*, ed. Desikachary T.V. pp. 262–3. University of Madras.

SHESTAKOV S.V. & JEVNER V.D. (1968) Study of mutagenesis in blue-green alga *Anacystis nidulans*. *Proc. XII Intern. Congr. Genet.* 1, 84.

SHESTAKOV S.V. & KHYEN N.T. (1970) Evidence for genetic transformation in blue-green alga *Anacystis nidulans*. *Molec. Gen. Genet.* 107, 372–5.

SINGH H.N. (1967) Genetic control of sporulation in the blue-green alga *Anabaena doliolum* Bharadwaja. *Planta* 75, 33–8.

SINGH H.N. (1968) Effect of acriflavine on ultra-violet sensitivity of normal, ultra-violet sensitive and ultra-violet resistant strains of a blue-green alga, *Anacystis nidulans*. *Radiation Bot.* 8, 355–61.

SINGH H.N., KUMAR H.D. & PRAKASH G. (1969) Postirradiation modification of ultra-violet-sensitivity of normal and nitrofurazone-resistant strains of the blue-green alga *Anacystis nidulans*. *Radiation Bot.* 9, 105–10.

SINGH R.N. & SINGH H.N. (1964a) Ultra-violet induced mutants of blue-green algae I. *Anabaena cycadeae* Reinke. *Arch. Mikrobiol.* 48, 109–17.

SINGH R.N. & SINGH H.N. (1964b) Ultra-violet induced mutants of blue-green algae II. *Anabaena doliolum* Bharadwaja. *Arch. Mikrobiol.* 48, 118–21.

SINGH R.N. & SINGH P.K. (1972) Ultra-violet damage, modification and repair of blue-green algae and their viruses. In *Taxonomy and Biology of Blue-Green Algae*, ed. Desikachary T.V. pp. 246–57. University of Madras.

SINGH R.N. & SINHA R. (1965) Genetic recombination in the blue-green alga *Cylindrospermum majus* Keutz. *Nature, Lond.* 207, 782–3.

SINGH R.N. & TIWARI D.N. (1969) Induction by ultra-violet irradiation of mutation in the blue-green alga *Nostoc linckia* (Roth.) Born. et Flah. *Nature, Lond.* 221, 62–4.

SINGH R.N., SINGH H.N. & SINHA R. (1966) Mutations and recombinations in blue-green algae. In *The Impact of Mendelism on Agriculture, Biology and Medicine*, pp. 405–30. International Symposium of the Indian Society of Genetics and Plant Breeding. Indian Agricultural Research Institute, New Delhi.

SINGH R.N., SINGH S.P. & SINGH P.K. (1972) Genetic regulation of nitrogen fixation in blue-green algae. In *Taxonomy and Biology of Blue-Green Algae*, ed. Desikachary T.V. pp. 264–8. University of Madras.

SINHA B.D. & KUMAR H.D. (1973) A mutational study of the nitogen-fixing blue-green alga *Anabaena doliolum*. *Ann. Bot. N.S.* 27, 673–9.

SMITH A.J. & LUCAS C. (1971) Propionate resistance in blue-green algae. *Biochem. J.* 124, 23p–24p.

SRIVASTAVA B.S. (1969) Ultra-violet induced mutations to growth factor requirement and penicillin resistance in a blue-green alga. *Arch. Mikrobiol.* 66, 234–8.

SRIVASTAVA B.S. (1970) A simple method of isolating ultraviolet resistant and sensitive strains of a blue-green alga. *Phycologia* 9, 205–8.

SRIVASTAVA B.S., KUMAR H.D. & SINGH H.N. (1971) The effect of caffeine and light on killing of the blue-green alga *Anabaena doliolum* by ultraviolet radiation. *Arch. Mikrobiol.* 78, 139–44.

STANIER R.Y. & VAN NIEL C.B. (1962) The concept of a bacterium. *Arch. Mikrobiol.* **42**, 17–35.

STEVENS S.E. & VAN BAALEN C. (1969) N-methyl-N′-nitro-N-nitrosoguanidine as a mutagen for blue-green algae: evidence for repair. *J. Phycol.* **5**, 136–9.

STEVENS S.E. & VAN BAALEN C. (1970) Growth characteristics of selected mutants of a coccoid blue-green alga. *Arch. Mikrobiol.* **72**, 1–8.

STEWART W.D.P. & SINGH H.N. (1975) Transfer of nitrogen-fixing (*nif*) genes in the blue-green alga *Nostoc muscorum. Biochem. Biophys. Res. Commun.* **62**, 62–9.

TAKAHASHI I. (1962) Genetic transformation of *Bacillus subtilis* by extracellular DNA. *Biochem. Biophys. Res. Commun.* **7**, 467–70.

TAYLOR A.L. (1963) Bacterophage-induced mutation in *Escherichia coli Proc. Natl. Acad. Sci U.S.A.* **50**, 1043–51.

TAYLOR A.L. & TROTTER C.D. (1972) Linkage map of *Escherichia coli* K-12. *Bact. Rev.* **36**, 504–24.

VAN BAALEN C. (1965a) Quantitative surface plating of coccoid blue-green algae. *J. Phycol.* **1**, 19–22.

VAN BAALEN C. (1965b) Mutation of the blue-green alga, *Anacystis nidulans. Science, N.Y.* **149**, 70.

VAN BAALEN C. (1968) The effects of ultra-violet irradiation on a coccoid blue-green alga: survival, photosynthesis and phtoreactivation. *Plant. Physiol.* **43**, 1689–95.

VAN BAALEN C. (1973) Mutagenesis and genetic recombination. In *The Biology of Blue-Green Algae*, eds. Carr N.G. & Whitton B.A. pp. 201–13. Blackwell Scientific Publications, Oxford.

VAN BAALEN C. & O'DONNELL R. (1972) Action spectra for ultra-violet killing and photo-reactivation in the blue-green alga *Agmenellum quadruplicatum. Photochem. Photobiol.* **15**, 269–74.

WERBIN H. & RUPERT C.S. (1968) Presence of photoreactivating enzyme in blue-green algal cells. *Photochem. Photobiol.* **7**, 225–30.

WHITTON B.A., CARR N.G. & CRAIG I.W. (1971) A comparison of the fine structure and nucleic acid biochemistry of chloroplasts and blue-green algae. *Protoplasma* **72**, 325–57.

WILCOX M., MITCHISON G.J. & SMITH R.J. (1975) Mutants of *Anabaena cylindrica* altered in heterocyst spacing. *Arch. Mikrobiol.* **103**, 219–23.

WITKIN E.M. (1969a) The role of DNA repair and recombination in mutagenesis. *Proc. XII Intern. Congr. Genet.* **3**, 225–45.

WITKIN E.M. (1969b) Ultraviolet-induced mutation and DNA repair. *Ann. Rev. Microbiol.* **23**, 487–514.

WU J.H., LEWIN R.A. & WERBIN H. (1967) Photoreactivation of UV-irradiated blue-green algal virus LPP-1. *Virology* **31**, 657–64.

YOSHIKAWA H. (1966) Mutations resulting from the transformation of *Bacillus subtilis. Genetics* **54**, 1201–14.

ZHEVNER V.D. & SHESTAKOV S.V. (1972) Studies on the ultraviolet-sensitive mutants of the blue-green alga *Synechocystis aquatilis* Sanv. *Arch. Mikrobiol.* **86**, 349–60.

ZHEVNER V.D., GLAZER V.M. & SHESTAKOV S.V. (1973) Mutants of *Anacystis nidulans* with modified cell division. *Mikrobiologiya* **42**, 290–7.

CHAPTER 2

FORMAL GENETICS OF
CHLAMYDOMONAS REINHARDTII

G. A. HUDOCK† AND H. ROSEN*

†Dept. of Zoology, Indiana University,
Bloomington, Indiana 47401, U.S.A.
*Dept. of Biology, California State University,
Los Angeles, California 90032, U.S.A.

1 Introduction 29

2 Mutants requiring specific metabolites 33

3 Acid-phosphatase-deficient mutants 36

4 Pigment-deficient mutants 36

5 Acetate-deficient mutants 38

6 Conditional-lethal mutants 40

7 Phototactic mutants 41

8 Mutants with conditional chlorophyll and chloroplast synthesis 42

9 Ultraviolet-sensitive mutants 42

10 General conclusions 42

11 References 44

1 INTRODUCTION

The unicellular green flagellate *Chlamydomonas reinhardtii* (Chlorophyta) is an excellent organism for genetic investigation both at the cellular and at the molecular level. Each cell contains a single cup-shaped chloroplast which occupies one third to one half of the cell volume (Bourque, Boynton & Gillham 1971; Schotz *et al.* 1972) and 10 to 15 highly branched mitochondria (Schotz *et al.* 1972) which occupy less than 5% of the cell volume (Bourque *et al.* 1971). The alga has asexual and sexual reproductive cycles with relatively short generation times and is easily maintained in pure cultures on defined media.

Wild-type cells of *C. reinhardtii* can be grown in defined mineral nutrient liquid or on solid agar media under three conditions: *phototrophic*—photosynthetic growth in light, with carbon dioxide as the sole carbon source; *heterotrophic*—growth in darkness, with acetate as the sole carbon source; and *mixotrophic*—growth in light, with both carbon dioxide and acetate as carbon sources (Sager & Granick 1953; Sueoka 1960; Gorman & Levine 1965; Bourque *et al.* 1971). *C. reinhardtii* normally reproduces asexually by two rapid successive mitotic divisions, yielding four daughter cells from a single mother cell. In liquid culture, mitotic doubling times for wild-type cells vary from a minimum of about five hours under phototrophic conditions where light and CO_2 are not limiting to about 12 hours under suitable heterotrophic conditions. Thus it is possible to obtain large populations of cells in a relatively short time. Furthermore, cultures of synchronously dividing cells can be obtained at a constant temperature, with CO_2 as the sole carbon source, by alternating twelve-hour periods of light and dark (Surzycki 1971; Kates & Jones 1964).

Sexual activity can be induced by withdrawal of nitrogen from the standard growth medium (Sager & Granick 1954). The two isogamous sexes, designated mating-type plus (*mt+*) and mating-type minus (*mt−*), are apparently determined by alternative alleles of a single Mendelian gene (see Ch 8.). Nitrogen depletion in the light generally initiates two gametogenic divisions, resulting in the production of four gametes from each vegetative parent cell (Kates *et al.* 1968). When suspensions of gametes of complementary mating types are mixed, aggregation or clumping occurs almost at once, and within minutes it is followed by the pairing and fusion of cells of opposite mating type. For genetic studies the resultant zygotes are matured in darkness for five to seven days on agar plates containing a source of nitrogen. During this period the zygotic wall thickens, making the zygotes easily distinguishable from the surrounding unmated gametes. After maturation is complete, the plates are inverted over chloroform for 30 seconds to kill these unmated cells. Germination involving synchronous meiotic divisions is induced by exposure of the zygotes to light for 16 to 20 hours. For genetic analysis the four meiotic products of each (unordered) tetrad can easily be separated by micromanipulation (cf. Levine & Ebersold 1960). The entire sexual cycle, from gametogenesis through meiosis, is completed in approximately seven days.

Mutations arise spontaneously at low frequencies (e.g. 10^{-6} for Mendelian streptomycin-resistance) and can be induced in higher proportions by mutagens such as UV irradiation (Ebersold 1956; Eversole 1956), nitrosoguanidine (Gillham 1965), and ethylmethane sulphonate (Loppes 1968). Selective techniques can be readily applied for the isolation of certain classes of mutations such as those resistant to antibiotics (Sager 1962; Gillham & Levine 1962) or deficient in photosynthetic capacity (Togasaki & Hudock 1975; Harris *et al.* 1974).

Three operationally distinct genetic systems have been recognized. A Mendelian system comprising 16 linkage groups is assumed to reside in the nucleus (see Fig. 2.1). A second genetic system, with non-Mendelian, uniparental transmission of genes through the mating-type plus (*mt+*) parent, is provisionally assigned to the DNA of the chloroplast (see Ch. 5). A third genetic system. in which genes are transmitted in a non-Mendelian, biparental fashion, is thought to reside in the mitochondrial DNA (Alexander *et al.* 1974; see Ch. 6).

In this chapter we will concentrate on isolation, characterization and general utility of various metabolic mutants of *Chlamydomonas reinhardtii*. A wide range of mutant genotypes and phenotypes is available for genetic and metabolic studies. Techniques developed for tetrad and single-strand analysis of crosses (Levine & Ebersold 1958) and for isolation of stable heterozygous diploid strains (Ebersold 1967) permit investigation not only of linkage but also of dominance, mitotic recombination, and other diploid-related phenomena. The organism is also suitable for biochemical investigations, especially because this species can be grown heterotrophically in darkness.

The mutant types that will be examined in this chapter can be roughly grouped into the following categories:

1 Mutants requiring specific metabolites
2 Acid-phosphatase-deficient mutants
3 Pigment-deficient mutants
4 Acetate-requiring mutants
5 Conditional lethal mutants of cellular metabolism
6 Phototactic mutants
7 Mutants conditionally deficient in chlorophyll synthesis
8 Ultraviolet-sensitive mutants

These mutations involve genes which primarily affect the metabolism of the alga. Almost all exhibit Mendelian segregation (2:2) when crossed with the wild type. Most of these can be assigned to known linkage groups (see Fig. 2.1); some cannot. Levine and Goodenough (1970) have questioned whether those mutant genes which cannot be so assigned are in fact located on nuclear chromosomes. The work of Sager and Lane (1972) suggested that the chloroplast DNA of only one parent is transmitted, and that therefore only uniparental inheritance can be expected from genes located on the chloroplast chromosome. The work of Chiang (1971), however, provided evidence that the chloroplast DNA of both parents is transmitted to the zygote, that chloroplast fusion and segregation are similar to nuclear fusion and gene segregation, and that recombination occurs between the two parental chloroplast DNAs during zygote development. These findings suggested that genes located on the chloroplast 'chromosome' could also exhibit Mendelian segregation patterns. For this reason, genes that cannot be assigned to existing linkage groups will be described only in terms of their segregation pattern, as Mendelian or non-Mendelian.

We will exclude some classes of mutants from this study, namely those mutants which affect flagellar structure and/or function, those which affect drug sensitivity/resistance, and those affecting cell-wall formation. Many of these mutants will be discussed in other chapters (see Ch. 3 and 4).

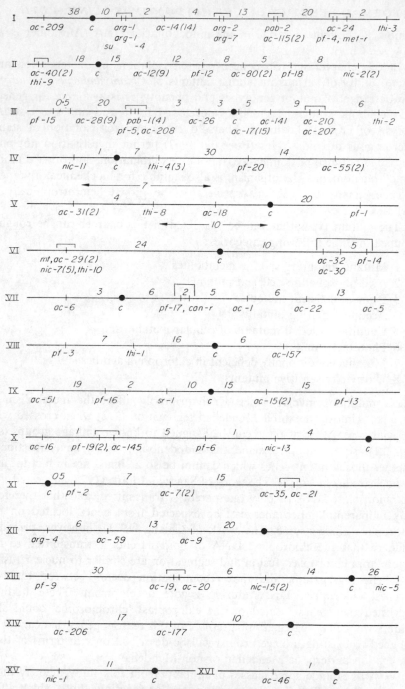

Fig. 2.1. The 16 linkage groups of *Chlamydomonas reinhardtii*. Figures in parenthesis after the name of a locus indicate the number of alleles known at that locus. The bracket above a group of markers indicates that their relative positions are uncertain or unknown. (From Levine & Goodenough 1970.)

2   MUTANTS REQUIRING SPECIFIC METABOLITES

These classes of mutants in *C. reinhardtii* can be divided into amino-acid-requiring mutants, base-requiring mutants, vitamin-requiring mutants, and mutants lacking the ability to grow on NO_3^- as their only nitrogen source. In general, isolation of these mutants simply involves growing colonies from mutagenized cells on properly supplemented agar plates, replica plating to unsupplemented agar plates, and isolating those strains which grow exclusively on the supplemented medium.

Amino-acid mutants, common in many prokaryotic and eukaryotic organisms, have been difficult to isolate in *C. reinhardtii*. Eversole (1956) reported isolation of several mutants each of which had a nutritional requirement for a single specific amino acid, but all but one were apparently lost before they could be subjected to genetic analysis. The exception was a single arginine-requiring strain designated *arg*-2, which requires arginine specifically. Stolbova *et al.* (1969) reported isolation of several other amino-acid-requiring mutant strains, but no further data on these strains are available.

Until recently the only other stable amino-acid-requiring mutants to be reported were *arg*-1 (Ebersold 1956), *arg*-4 (Gillham—see Loppes 1969b) and *arg*-7 (Gillham 1965). Each of the *arg* genes segregates 2:2 when crossed to wild type. Further crossing experiments showed that *arg*-1, *arg*-2 and *arg*-7 are located on linkage group I and *arg*-4 on linkage group XII (Ebersold *et al.* 1962; Loppes 1969b), and that *arg*-7 is closely linked to *arg*-2. Phenotypically, *arg*-2 and *arg*-7 are similar in that they both require arginine specifically (Eversole 1956; Loppes 1969b) and both lack the enzyme argininosuccinate lyase (Strijkert & Sussenbach 1969; Strijkert *et al.* 1973). Recombination and complementation analyses indicate that these mutations are located in the same cistron. By contrast, *arg*-1 can grow on arginine, citrulline or ornithine (Ebersold 1956) and lacks the enzyme acetylglutamyl-phosphate reductase (Strijkert & Sussenbach 1969), while *arg*-4 can grow only on arginine (Loppes 1969b) and lacks the enzyme ornithine trans-carbamylase (Strijkert *et al.* 1973).

Experiments utilizing these stable arginine-requiring mutant strains, and studies on the enzymes of wild-type *C. reinhardtii* (Farago & Denes 1967), have led to elucidation of much of the pathway for arginine biosynthesis, as is shown in Fig. 2.2, which also indicates the position of each mutant block.

Loppes (1969a) showed that the composition of the selective medium can have a profound effect on the experimental recovery of arginine auxotrophs. In his studies of back-mutation, he observed that unless arginine was provided in the medium immediately after mutagenesis (with the alkylating agents methylmethane sulphonate or ethylmethane sulphonate), the recovery of revertants was greatly reduced.

In a subsequent study, Loppes (1969b) studied mutation from prototrophy to arginine auxotrophy among cells first plated on a medium containing yeast extract but lacking ammonium ions (which inhibit active transport of amino acids into yeast cells; Grenson *et al.* 1966) and then replica-plated to selective and non-selective media. The *arg* mutants he obtained in this way did not grow in media containing NH_4^+. Genetic studies, in which crosses of wild type to NH_4^+-resistant *arg*-7 or *arg*-2 yielded NH_4^+-sensitive arginine mutants, indicated the concomitant presence of a mutation for NH_4^+-resistance in these four stable arginine mutants (Loppes 1970), which thus must be double mutants.

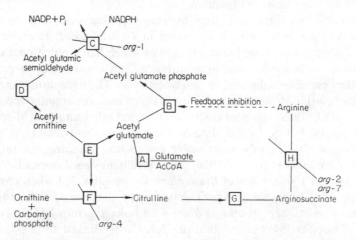

Fig. 2.2. Partial scheme of the pathways for the regulation of arginine bio-synthesis. Positions of the mutant blocks are included (modified from Sussenbach & Strijkert 1969). Enzymes in the pathway are designated by letters: *A*, acetyl-glutamate synthetase, *B*, acetyl glutamate kinase, *C*, acetylglutamyl-phosphate reductase, *D*, acetylornithine transaminase, *E*, acetyl ornithine-glutamate trans-acetylase, *F*, ornithine transcarbamylase, *G*, arginosuccinate synthetase, *H*, arginosuccinate lyase.

Biochemical studies on the phenomenon of NH_4^+-sensitive arginine mutant strains showed that arginine uptake is depressed by NH_4^+ (Loppes & Strijkert 1972). The arginine pool in the cell remains low because what little arginine does enter the cell is broken down by the action of arginine deamin-ase. It was shown that growth of arginine-requiring mutants is possible only under conditions in which a high pool level of arginine is maintained. NH_4^+-sensitive strains (such as wild-type *C. reinhardtii*) can achieve such high levels only when grown on medium with a low NH_4^+ concentration. NH_4^+-insensi-tive mutant strains can maintain high arginine pools because of their ability to take up increased amounts of arginine.

Loppes and Strijkert (1972) came to the conclusion that difficulties like those encountered in the isolation of arginine-requiring mutants may also complicate the isolation of other amino-acid-requiring mutants and that perhaps the metabolism of amino acids may make it impossible for exogenous amino acids to be used for protein synthesis. If this is true, Loppes and Strijkert speculated that the only way to isolate amino-acid-requiring strains would be to first obtain mutant strains in which the metabolism of such acids is altered (e.g. by the increased uptake or decreased breakdown of the amino acid concerned).

Chlamydomonas mutants requiring nitrogenous bases, similar to those isolated in many organisms, have been avidly sought, but without success so far. Thymine-requiring mutants in particular would be most valuable as they might facilitate radioactive labelling of DNA, permitting radioisotope-tracer studies of nucleic acid metabolism, replication and repair. Wild-type *C. reinhardtii* strains take up little exogenous thymine or thymidine. The only base-requiring mutants to have been reported are strains requiring guanine and uracil (Stolbova *et al.* 1969); but these have not been extensively studied.

Hipkiss (1968) isolated two bromouracil-resistant mutants which take up no bromouracil or exogenous thymidine into their DNA, apparently because of their inability to convert thymine to thymidylate. However, since the organism nevertheless survives and divides, it must be able to obtain thymidylate residues from another pathway, indicating that it might be necessary to isolate a double mutant in order to get a thymine-requiring strain. Such double-mutant strains would probably be difficult to obtain.

Several vitamin-requiring mutants of *C. reinhardtii* have been isolated, including strains requiring nicotinamide (*nic*), para-aminobenzoic acid (*pab*) and thiamine (*thi*) (Eversole 1956). All are attributable to single Mendelian genes with the loci well distributed in different linkage groups (Ebersold, Levine & Olmsted 1962) (see Fig. 2.1). In addition, some mutants requiring vitamin B_6 or B_{12} have been reported (Stolbova *et al.* 1969), but they have not been genetically characterized.

Analysis of the growth requirements of *thi* mutants has led to a sketchy pathway for thiamine biosynthesis (Fig. 2.3). No pathways have been determined for the biosynthesis of para-aminobenzoic acid or nicotinamide in this organism.

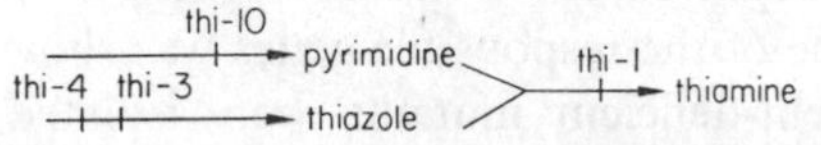

Fig. 2.3. A possible pathway for thiamine biosynthesis in *Chlamydomonas reinhardtii*. The mutant blocks are indicated on the basis of the known nutritional requirements of various mutant strains.

The wild-type 137C strains used by Sager can utilize either NO_3^- or NH_4^+ as a sole nitrogen source whereas the wild-type 137C strain used by Ebersold—though ostensibly of the same origin—cannot grow with NO_3^- as the sole nitrogen source. The difference, however it arose, is genetic; in crosses between these two 'wild-type' strains the gene shows Mendelian segregation (Ebersold personal communication). Attempts to isolate a NO_3^--utilizing strain by mutation from the Ebersold stock have been unsuccessful (Rosen unpublished).

3 ACID-PHOSPHATASE-DEFICIENT MUTANTS

Recently Loppes and Matagne (1973) isolated and characterized two classes of mutants defective for production of acid phosphatase. The mutant genes exhibit Mendelian inheritance and their loci appear to be unlinked or only loosely linked.

The isolation technique consisted of replica-plating colonies from cells mutagenized with MNNG* on sterile filter papers, which were then dried at room temperature, irreversibly fixing the colonies to the paper. When such papers were immersed in a solution of α-naphthyl phosphate and tetrazotized-o-dianisidine in phosphate buffer at pH 4·8, brown-red coloration around wild-type colonies indicated acid phosphatase activity. When the temperature of the reagent was raised to 60°C, several mutants were revealed by the absence of acid phosphatase activity indicating that there could be two types of acid-phosphatase in the cell, one thermostable and one thermolabile. A double mutant was isolated by mutagenizing one of the thermostable mutants (with MNNG) and testing surviving colonies for acid-phosphatase activity without prior heat treatment. The mutants, analysed by biochemical and electrophoretic techniques, fell into two classes. Those of one class were found to lack a heat-stable acid phosphatase normally bound to the cellular debris of a crude extract, those of the other to lack a soluble heat-sensitive enzyme.

4 PIGMENT-DEFICIENT MUTANTS

In this category we include only those strains which are unconditionally deficient, excluding such mutants as y-1 and y-2 in which formation of the normal pigment content depends on illumination (as discussed below). Sager and Palade (1954) reported the isolation of yellow, yellow-green and pale green mutants. None of the responsible genes was shown to be linked to any other. Other pigment-deficient mutants were isolated by Ebersold (1956) and some of these are now designated as acetate-requiring mutants, including ac-1, ac-5, ac-7, ac-22, ac-29, ac-30, ac-31, ac-32 and ac-55.

Goodenough, Armstrong and Levine (1969) examined in detail the photosynthetic properties and the ultrastructure of ac-31. Whereas the chlorophyll content of wild-type cells varies from about 2·0 to 4·0 μg/cell, depending on the growth conditions, that of ac-31 is reduced to about 1·0 μg/cell when grown under mixotrophic conditions and still lower, to 0·70 μg/cell, under photolithotrophic conditions. Furthermore, the chlorophyll a/b ratio, 1·8

* See p. 9.

in wild-type cells, is elevated in the mutant to about 3·0. Stacking of chloroplast lamellae is reduced or absent in *ac*-31. In the mutant the rates of CO_2-reduction and of cell division in minimal medium (supplemented with acetate) are only about half of the corresponding wild-type rates. Since *ac*-31 and certain other mutants can carry out photosynthetic electron transport, phosphorylation and carbon dioxide reduction at rates comparable to those of wild-type cells, the basis for their obligate heterotrophy or mixotrophy is unexplained.

More recently Nicholson-Guthrie (1972) isolated and characterized a mutant incapable of chlorophyll synthesis under any conditions examined to date. This mutant, designated *y.y* (yellow in the dark, yellow in the light) was derived from *y*-1 (yellow in the dark) following ultraviolet irradiation. The cells contain no detectable chlorophyll and their carotenoid complement resembles that of dark-grown *y*-1. Their ultrastructure is virtually indistinguishable from that of the naturally apochlorotic flagellate *Polytoma*, except for the presence of an outer chloroplast membrane, devoid of lamellae. The mutant is, therefore, an obligate heterotroph; but, even when supplied with a suitable organic substrate, its doubling time is three to four times that of the wild type, its growth rate and yield being independent of light intensity up to 6600 lux.

Genetically, the *y.y* strain is complex, since it is at least a double mutant. Reciprocal matings to wild type have produced complicated results. When wild-type *mt*+ cells are mated to *y.y.mt*− gametes, the progeny phenotypes segregate in a variety of patterns. Segregations are indicated by sectored colonies arising from germinated zygotes; tetrad analyses indicate that the *y.y* phenotype is characterized by two linked genes and one unlinked marker, the *y.y* gene. (From a reciprocal cross, twenty tetrads yielded only two viable clones, both *y.y* (Nicholson-Guthrie 1972). No adequate explanation has been proposed for this result.)

Mutations which affect the pigment content of *C. reinhardtii* and its photosynthetic capacity range from the chlorophyll-less mutant *y.y*, through pale green mutants with about 15% of the normal chlorophyll content, to *ac*-31 which contains about half as much chlorophyll as wild type. Hudock and Hudock (unpublished) have isolated temperature-sensitive mutants in which chlorophyll synthesis in continuous light is normal at 25° but not at 35°. Such mutants will be described in more detail below.

Since at the restrictive temperature the appearance of a pigment-deficient mutant colony is obviously and dramatically different from that of the wild type, isolation and genetic analysis of such mutants is relatively simple. The responsible genes have proved to be scattered through the genome of the organism. These mutations have been useful in studies of the control of pigment synthesis and of the roles of the pigments in the photosynthetic process.

5 ACETATE-DEFICIENT MUTANTS

The vast majority of the known mutant strains of *C. reinhardtii* are acetate-deficient: they have been divided into two general categories by Levine and Goodenough (1970). Those of the first class are capable of apparently normal CO_2-reduction, and their pigment content is similar or identical to that of the wild-type cells. The nature of the genetic lesion in these strains is unknown. The second class includes a more extensive and more intensively studied group of mutants in which the acetate requirement results from a lesion in a gene affecting photosynthesis. Levine and his co-workers have elegantly exploited such mutants to elucidate the genetic control of photosynthesis and the concomitant sequence of electron-transport reactions. In the early stages of their work, it was reasoned that a mutant incapable of photosynthesis would require a carbon source such as acetate for growth even when supplied with CO_2 and light, and this was the basis for their method of mutant selection. A wild-type culture was mutagenized, usually with UV, and the cells were first plated on acetate-supplemented medium. The resulting colonies were replica-plated to a minimal medium and to the same minimal medium supplemented with acetate. Out of several hundred putative mutants (which grew only on the acetate plates), only 10% proved to be essentially unable to reduce CO_2 (Gillham and Hudock unpublished). Many of the remainder resembled *ac*-31, being able to reduce CO_2 only at a rate considerably lower than that of wild type.

Levine (1960) developed a simple but extremely effective means for selectively identifying mutants which cannot fix carbon dioxide. A cell suspension, after treatment with a mutagen, was plated on an acetate-supplemented medium in a petri dish which was then inverted over a watch glass containing $NaH^{14}CO_3$ and a little sulphuric acid. Illumination of the dish permitted incorporation of ^{14}C by photosynthetically competent cells but not by impaired mutants. The colonies were replicated to filter paper from which a radio-autogram was prepared. By this method non-photosynthetic mutants could be identified, since only photosynthetic clones could fix $^{14}CO_2$.

When light is absorbed by chlorophyll, the energy which is not utilized in photosynthetic electron transport is dissipated as red-light fluorescence. Bennoun and Levine (1967) made use of this phenomenon to identify weakly photosynthetic mutants among a mixture of acetate-requiring strains. A plate on which such mutants were grown was illuminated with blue light and observed through a red filter, perpendicular to the illuminating beam. Mutants with impaired photosynthesis fluoresced more brightly than those with an intact photosynthetic electron-transport system.

Togasaki and Hudock (1975) developed yet another technique which efficiently enriches for photosynthetic mutants, based on the premise that mutants unable to carry out photophosphorylation could be expected to be less sensitive than wild-type cells to arsenate, a competitive inhibitor of phosphate incorporation. They demonstrated that known photosynthetic mutants survive concentrations of inorganic arsenate at least three or four times as high as those tolerated by the wild type. Experiments based on this differential sensitivity, in combination with the fluorescence technique described above, yielded a considerable proportion of mutants, up to 20% of the surviving cells. These mutant strains proved deficient in photosystem II.

Acetate-requiring photosynthetic mutants deficient in CO_2-reduction capability have been utilized to identify the components of the electron-transport system in *C. reinhardtii* and to establish their relative positions in the electron-transport pathway. Figure 2.4 summarizes the data from studies on various acetate-requiring mutant strains (Levine & Smillie 1962; Levine & Gorman 1966; Givan & Levine 1967). Deficiencies of electron-transport components in mutant strains were chemically and spectrophotometrically determined. The position of each of the pigments in the pathway was determined (1) by testing the ability of the mutant strains to carry out partial photosynthetic reactions (e.g., the Hill reaction and NADP photoreduction by the ascorbate-DPIP couple) and (2) by illumination of the mutant cells with various wave-lengths of light and the detection of oxidation and reduction patterns of the cytochromes in the mutant cells. F-54, another acetate-requiring mutant which cannot fix CO_2, was found to possess an intact electron-transport pathway, but to be impaired in the ATP-synthesizing apparatus coupled to electron transport (Sato *et al.* 1971).

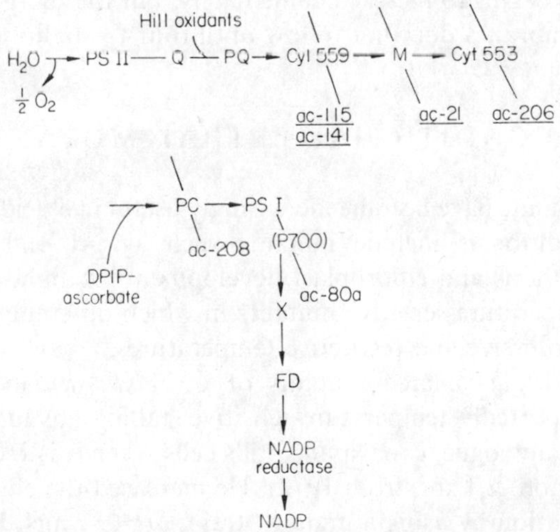

Fig. 2.4. The photosynthetic electron-transport chain in *C. reinhardtii* (modified from Levine 1968). The dashed lines refer to the components affected in the different mutant strains. Sites of action of the Hill-reaction oxidants and the ascorbate-DPIP couple are also indicated. *Q*, the quencher of fluorescence of system II, *Cyt*, cytochrome, *PC*, plastocyanin, *PQ*, plastoquinone, *P700*, reaction centre of system I, *X*, postulated electron acceptor of system I, and FD, ferredoxin.

The mutant *ac*-20 is an acetate-requiring strain which can grow only slowly on minimal medium in the light. It appears that the mutation affects a single nuclear gene located on linkage group XII, about 30 map units from the centromere. The mutant is thought to possess primary defects in the chloroplast protein-synthesizing machinery, and therefore produces pleiotropic effects on gene products translated in the chloroplast. The ribulose diphosphate (RuDP) carboxylase activity of this strain is more than normally sensitive to the environment in which the cells are grown, being highest during autotrophic growth and lowest (about 50% of wild-type activity) during organotrophy or mixotrophy (Levine & Togasaki 1965). Correspondingly, when organotrophically cultured this mutant reduces CO_2 at only about one-half of the wild-type rate; when returned to photolithotrophic conditions, carbon-dioxide reduction and RuDP carboxylase activity recover (Togasaki & Levine 1970). The recovery of both activities is light-dependent and evidently necessitates protein synthesis, since the increase in carboxylase activity in the double mutant *ac*-20, *arg*-2 is dependent on exogenous arginine.

Levine and Paszewski (1970), who examined photosynthetic electron transport in *ac*-20, revealed that this strain lacks not only cytochrome 559 but also a 'quencher' of fluorescence in photosystem II. Goodenough and Levine (1970) observed that in mixotrophic cultures its chloroplasts usually lack pyrenoids, have disorganized chloroplast membranes, and contain only 5 to 10% as many chloroplast ribosomes as wild-type cells. When mutant cells are transferred to photolithotrophic conditions, the number of chloroplast ribosomes begins to increase immediately, but the reorganization of the chloroplast membranes does not follow until four to six hours later.

6 CONDITIONAL-LETHAL MUTANTS

Conditional mutants have become increasingly useful in elucidating biological processes. Conditionals include mutants such as *y*-1 and *y*-2, in which chlorophyll synthesis and chloroplast development are light-dependent, and a variety of temperature-sensitive mutants in which different phenotypes are expressed at permissive and restrictive temperatures.

McMahon (1971) isolated mutants of *C. reinhardtii* in which protein synthesis is purportedly temperature-sensitive, taking advantage of the fact that an arginine analogue, canavanine, kills cells when it is incorporated into protein (McMahon & Langstroth 1972). He increased the efficiency of canavanine incorporation by using a mutant strain, *arg*-2, which has an absolute requirement for arginine. Cells of mutants with impaired protein synthesis could be expected not to incorporate canavanine, and thus would be more resistant to its killing effect than wild-type cells. Washed *arg*-2 cells were mutagenized for a period with nitrosoguanidine, after which the survivors were washed, permitted to divide for several generations, transferred to a growth medium lacking arginine, and incubated at 33° for 30 min to deplete endogenous arginine. Canavanine was then added, and the survivors (1% to 10%) were resuspended in arginine. In all, three similar rounds of growth and enrichment were performed at three-day intervals. The cells were then cultured on agar, and after three days the resulting small colonies were replica-plated and respectively incubated at 25° and 33°. Using these procedures, about 0·3% of the colonies proved to be mutants with temperature-sensitive protein synthesis, the extent of inhibition at the restrictive temperature ranging from 15 to 90%.

Howell and Naliboff (1973) isolated temperature-sensitive mutants blocked at various stages in the vegetative cell cycle. Mutation was induced with ultraviolet light, and the survivors, after incubation in the dark to prevent photoreactivation, were grown mixotrophically for four or five days, and were then replica-plated and incubated at the chosen permissive (25°) and restrictive (33°) temperatures. 'Tight' mutants, which ceased cell division soon after the temperature was raised, were specifically selected. The initial selection was followed by a second step in which liquid suspensions were spread on agar plates and incubated at 33° for 72 hours. Non-mutants formed macroscopic colonies during this time, whereas mutants divided to produce microcolonies of no more than four to eight cells each. About 20% of the clones isolated after initial selection of this sort satisfied their criterion, being blocked in division: the overall fractional yield was 2 to 10×10^{-5}. Howell and Naliboff defined as a 'block point' the last time in a cell division cycle in which a shift to the restrictive temperature can block subsequent cell divisions. In different mutants such block points have been found to occur at different stages, ranging through most of the cell cycle.

7 PHOTOTACTIC MUTANTS

Smyth and Ebersold (1970) and Hudock and Hudock (1973) have described methods for the isolation of mutant strains with altered phototactic responses. Since the mutant cells respond negatively to a light source to which wild-type cells respond positively, isolation of these mutants was quite simple. A vessel containing a suspension of wild-type cells was placed in a light beam, which attracted most of the cells towards its source. A sample from the side of the vessel away from the light source was used to inoculate a second culture, and the process was repeated through six rounds of selection (Hudock & Hudock 1973). The selected cells were then plated on agar. Out of 80 clones isolated and analysed, 15 were found to have cells which were motile but nonresponsive to light, while those in three others responded negatively under all conditions studied. No obvious structural or physiological anomalies have been noted. In each case the mutant character segregated 2:2 in crosses with wild-type. Smyth and Ebersold (1970) demonstrated that one mutation is located on linkage group I, close to the centromere. Though the other mutant genes have not been assigned to linkage groups, two of the three isolated by Hudock and Hudock are linked; the third is on a separate chromosome. From the frequency of tetratype tetrads it was concluded that at least two of the three loci are relatively far from their centromere(s).

More recently, Hudock and Hudock (1973) isolated, and Wood and Hudock (unpublished) have begun to characterize, mutant strains in which the phototactic response is conditional, being negative in cells grown at 25° but positive in cells grown at 35°. Initial genetic analysis of one such mutant suggests that the mutation leading to this phenotype may be inherited in a uniparental manner.

8 MUTANTS WITH CONDITIONAL CHLOROPHYLL AND CHLOROPLAST SYNTHESIS

Although wild-type *C. reinhardtii* is capable of sustained chlorophyll synthesis in continuous darkness, spontaneous mutants lacking this ability occur with a high frequency; the most intensively studied of these are the mutants *y*-1 and *y*-2. Growth of either mutant in the dark results in extensive chlorosis and dedifferentiation of the chloroplast: transfer of such chlorotic cultures to the light results in rapid development of the chloroplast and synthesis of chlorophyll. Both genes segregate in a Mendelian pattern, but it has not been possible to assign either to any known linkage group.

Nicholson-Guthrie and Hudock (unpublished) isolated and characterized a mutant with temperature-sensitive chlorophyll synthesis. Grown at 25° the cells are pale green, having roughly one-third of the chlorophyll content of wild-type; grown at temperatures above 30° they contain no chlorophyll at all. Hudock and Fuller (1965) found that changes in the activity of NADP-dependent triose-phosphate dehydrogenase precisely parallel changes in chlorophyll content, being otherwise independent of illumination. Studies of such mutants make it possible to separate the effects of illumination from those of chlorophyll synthesis in the regulation of the biogenesis of the chloroplast and its components.

9 ULTRAVIOLET-SENSITIVE MUTANTS

Several ultraviolet-sensitive mutant strains have been isolated in *C. reinhardtii* by a relatively simple method (Davies 1967a; Rosen & Ebersold 1972). Colonies from cells mutagenized by UV, nitrosoguanidine, or ICR-170 (a monofunctional quinacrine mustard) are replica-plated in duplicate, and each replicate is irradiated with UV light. One replicate is placed in the light, the other in darkness for 18 hours (to prevent photoreactivation) and then in the light. Colonies which grow on the plate placed immediately in the light but which do not develop on the other replicate plate are classified as UV-sensitive. The UV-sensitivity of these strains is confirmed by comparing their survival, after various UV doses, with that of wild type. Most of such mutants are at least ten times as sensitive as the wild-type strain.

It has been determined in other organisms that UV sensitivity is the result of a mutation in the DNA-repair mechanism. Two classes of such mutants have been isolated, *uvr* mutants (Howard-Flanders *et al.* 1966) and *rec*⁻ mutants (Clark & Margulies 1965). The *uvr* mutants appear to be deficient in an enzyme system that replaces a damaged region of one DNA strand with DNA copied from the intact complementary strand (Boyce & Howard-Flanders 1964; Pettijohn & Hanawalt 1964). In *rec*⁻ mutants, UV sensitivity appears to be a direct consequence of a deficiency in the recombination mechanism (Howard-Flanders *et al.* 1969).

All UV-sensitive mutations studied genetically, in crosses with wild type, have been found to segregate 2:2, and although none of the genes have been mapped, crosses between UV-sensitive strains have indicated that few of them are linked to each other. UV-sensitivity of several double-mutant strains was greater than that of single-mutant strains, indicating that more than a single process is affected (Davies 1967b; Rosen unpublished). Diploid strains homozygous for the UV-sensitive gene *uvse-1* have been found to be more sensitive to UV than diploid strains homozygous for the wild-type (UV-sensitive) allele (Martinek, personal communication).

Davies (1967b) irradiated cells undergoing a synchronized meiotic cycle and found that UV sensitivity varied at different stages, being most sensitive at the time when recombination is presumed to take place. This was interpreted to indicate that repair enzymes do not operate during this period; and Davies hypothesized that this might permit freer recombination (assuming that breakage and rejoining of non-sister DNA strands occur during the process).

Most UV-sensitive mutants have been found to have normal meiotic recombination. However, Rosen and Ebersold (1972) isolated one UV-sensitive mutant strain which appears analogous to *rec*⁻ mutants in other organisms, consistently reducing the meiotic recombination frequency when homozygous. Investigations of such a mutant could ultimately lead to a better understanding of the mechanism of recombination in *C. reinhardtii*. The effect of UV-sensitive genes on mitotic recombination has yet to be studied in this organism.

At present, due to our lack of thymine-requiring mutants and to the poor incorporation of exogenous thymine into the DNA of wild-type cells, only a few biochemical experiments have been performed which could elucidate the mechanism of repair in this species. Swinton and Hanawalt (1973a, b), who developed a technique for separation of ³²P-labelled pyrimidine dimers, found that UV irradiation induces dimers in DNA; but they adduced no evidence for dimer excision or repair replication in either chloroplast or nuclear DNA, and they therefore concluded that there is no excision-repair mechanism in this alga. Experiments by White (personal communication), however, indicate that some portion of the DNA is excised. If Swinton and Hanawalt are correct, repair must occur by the recombination-repair mode; but this does not explain the fact that most UV-sensitive mutations tested do not alter recombination rates (Rosen & Ebersold 1972).

10 GENERAL CONCLUSIONS

Several different general classes of mutants of *C. reinhardtii* have been discussed here and elsewhere in this book. The organism is already well characterized genetically, and its genotype is rapidly becoming even better known. Since it is unicellular, it can be cultured easily either on agar plates, in bulk or as clones, or in liquid, in batch, synchronous or continuous culture. It can be propagated indefinitely by vegetative cell division. Sexuality can be induced easily, and mating efficiency is generally high. Zygote dormancy, long an obstacle to genetic studies of certain green algae, need not last for more than a few days, and the progeny are normally viable. Meiotic products can be examined by either single-strand or tetrad analysis, making possible the accumulation of genetic information of kinds not available for most other organisms. In addition, stable diploids can now be obtained in this species (Ebersold 1967), permitting us to distinguish between dominant and recessive alleles and to study mitotic recombination, a system so far unique among algae although comparatively well studied among ascomycetes. This alga also presents systems of non-Mendelian inheritance; present evidence indicates that a significant fraction of the genetically relevant DNA is contained in the chloroplast. The cells can be cultured photolithotrophically, mixotrophically or organotrophically, making it possible to separate photosynthetic from oxidative phosphorylation functions in its metabolism and to study their respective genetic control mechanisms. The organism is sensitive to a variety of mutagens, and the isolation of mutant strains (with the exception of amino-acid auxotrophs) seems limited only by the ability of investigators to design suitable selective and enrichment techniques. Mutant strains can be stored for years under liquid nitrogen, with reasonable viability and demonstrated genetic stability (Hwang & Hudock 1971). Finally, in spite of all that is known about the organism, *C. reinhardtii* remains as delightfully vexing as are all living things when we stop to examine them closely, and promises to provide enough problems to keep many investigators busy for years to come.

11 REFERENCES

ALEXANDER N.J., GILLHAM N.W. & BOYNTON J.E. (1974) The mitochondrial genome of Chlamydomonas: Induction of minute colony mutations by acriflavin and their inheritance. *Molec. gen. Genet.* (in press).

BENNOUN P. & LEVINE R.P. (1967) Detecting mutants that have impaired photosynthesis by their increased level of fluorescence. *Plant. Physiol.* **42,** 1284–7.

BOURQUE D.P., BOYNTON J.E. & GILHAM N.W. (1971) Studies on the structure and cellular location of various ribosome and ribosomal RNA species in the green alga *Chlamydomonas reinhardi. J. Cell Sci.* **8,** 153–83.

BOYCE R.P. & HOWARD-FLANDERS P. (1964) Release of ultra-violet light induced thymine dimers from DNA in *E. coli* K-12. *Proc. Nat. Acad. Sci. U.S.A.,* **51,** 293–300.

CHIANG K.S. (1971) Replication, transmission and recombination of cytoplasmic DNA's in *Chlamydomonas reinhardi.* In *Autonomy and Biogenesis of Mitochondria and Chloroplasts,* eds. Boardman N.K., Linnane A.W. & Smillie R.W. pp. 235–59, North-Holland, Amsterdam.

CLARK A.J. & MARGULIES A.D. (1965) Isolation and characterization of recombination-deficient mutants of *Escherichia coli* K-12. *Proc. Nat. Acad. Sci. U.S.A.,* **53,** 451–9.

DAVIES D.R. (1967a) UV-sensitive mutants of *Chlamydomonas reinhardi. Mutation Res.* **4,** 765–70.

DAVIES D.R. (1967b) The control of dark repair mechanisms in meiotic cells. *Molec. Gen. Genetics,* **100,** 140–9.

EBERSOLD W.T. (1956) Crossing over in *Chlamydomonas reinhardi Am. J. Botany,* **43,** 408–10.

EBERSOLD W.T. (1967) *Chlamydomonas reinhardi*: heterozygous diploid strains. *Science,* **157,** 447–9.

EBERSOLD W.T., LEVINE R.P. & OLMSTED M.A. (1962) Linkage maps in *Chlamydomonas reinhardi. Genetics* **47,** 531–43.

EVERSOLE R.A. (1956) Biochemical mutants of *Chlamydomonas reinhardi Am. J. Botany,* **43,** 404–7.

FARAGO A. & DENES G. (1967) Mechanism of arginine biosynthesis in *Chlamydomonas reinhardti.* II. Purification and properties of H. Acetylglutamate 5-phosphotransferase, the allosteric enzyme of the pathway. *Biochim. Biophys. Acta,* **136,** 6–18.

GILLHAM N.W. (1965) Induction of chromosomal and non-chromosomal mutations in *Chlamydomonas reinhardi* with N-methyl-N'-nitro-N-nitrosoguanidine. *Genetics,* **52,** 529–37.

GILLHAM N.W. & LEVINE R.P. (1962) Studies on the origin of streptomycin resistant mutants in *Chlamydomonas reinhardi. Genetics,* **47,** 1463–74.

GIVAN A.L. & LEVINE R.P. (1967) The photosynthetic electron transport chain of *Chlamydomonas reinhardi.* VII. Photosynthetic phosphorylation by a mutant strain of *Chlamydomonas reinhardi* deficient in active P700. *Plant Physiol.* **42,** 1264–8.

GOODENOUGH U.W., ARMSTRONG J.J. & LEVINE R.P. (1969) Photosynthetic properties of *ac*-31, a mutant strain of *Chlamydomonas reinhardi* devoid of chloroplast membrane stacking. *Plant. Physiol.* **44,** 1001–12.

GOODENOUGH U.W. & LEVINE R.P. (1970) Chloroplast structure and function in *ac*-20, a mutant strain of *Chlamydomonas reinhardi.* III. Chloroplast ribosomes and membrane organization. *J. Cell. Biol.* **44,** 547–62.

GORMAN D.S. & LEVINE R.P. (1965) Cytochrome F and plastocyanin: Their sequence in the photosynthetic electron transport chain of *Chlamydomonas reinhardi. Proc. Nat. Acad. Sci. U.S.A.* **54,** 1665–9.

GRENSON M., MOUSSET M., WIAME J.M. & BECHET J. (1966) Multiplicity of the amino acid permeases in *Saccharomyces cerevisiae*. 1. Evidence for a specific arginine transporting system. *Biochim. Biophys. Acta (Amst.)* **127**, 325–38.

HARRIS E.H., BOYNTON J.E. & GILLHAM N.W. (1974) Chloroplast ribosome biogenesis in *Chlamydomonas*: selection and characterization of mutants blocked in ribosome formation. *J. Cell Biol.* **63**, 160–79.

HIPKISS A.R. (1968) Bromouracil resistant mutant of *Chlamydomonas*. *Canad. J. Biochem.* **46**, 421–3.

HOWARD-FLANDERS P., BOYCE R.P. & THERIOT L. (1966) Three loci in *Escherichia coli* K-12 that control the excision of pyrimidine dimer and contain other mutagen products from DNA. *Genetics*, **53**, 1119–36.

HOWARD-FLANDERS P., THERIOT L. & STEDEFORD B. (1969) Some properties of excision-defective recombination-deficient mutants of *Escherichia coli* K-12. *J. Bacteriol.* **97**, 1134–41.

HOWELL S.H. & NALIBOFF J.A. (1973) Conditional mutants in *Chlamydomonas reinhardi* blocked in the vegetative cell cycle. *J. Cell Biol.* **57**, 760–72.

HUDOCK G.A. & FULLER R.C. (1965) Control of triose phosphate dehydrogenases in photosynthesis. *Plant. Physiol.* **40**, 1201–11.

HUDOCK G.A. & HUDOCK M.O. (1973) Phototaxis: Isolation of mutant strains of *Chlamydomonas reinhardi* with reversed sign of response. *J. Protozool.* **20**, 139–40.

HWANG SHUH-WEI & HUDOCK G.A. (1971) Stability of *Chlamydomonas reinhardi* in liquid nitrogen storage. *J. Phycol.* **7**, 300–3.

KATES J.R., CHIANG K.S. & JONES R.F. (1968) Studies on DNA replication during synchronized vegetative growth in gametic differentiation in *Chlamydomonas reinhardtii*. *Exp. Cell Res.* **49**, 121–35.

KATES J.R. & JONES R.F. (1964). The control of gametic differentiation in liquid cultures of *Chlamydomonas*. *J. Cell. Comp. Physiol.* **63**, 157–64.

LEVINE R.P. (1960) A screening technique for photosynthetic mutants in unicellular algae. *Nature*, **188**, 339–40.

LEVINE R.P. (1968) Genetic dissection of photosynthesis. *Science*, **162**, 768–71.

LEVINE R.P. & EBERSOLD W.T. (1958) Gene recombination in *Chlamydomonas reinhardi*. *Cold Spring Harbor Symp. Quant. Biol.* **23**, 101–9.

LEVINE R.P. & EBERSOLD W.T. (1960) The genetics and cytology of *Chlamydomonas*. *Ann. Rev. Microbiol.* **14**, 197–216.

LEVINE R.P. & GOODENOUGH U.W. (1970) The genetics of photosynthesis and of the chloroplast in *Chlamydomonas reinhardi*. *Ann. Rev. Gen.* **4**, 397–408.

LEVINE R.P. & GORMAN D.S. (1966) Photosynthetic electron transport chain of *Chlamydomonas reinhardi*. III Light-induced changes in chloroplast fragments of the wild type and mutant strains. *Plant Physiol.* **41**, 1293–1300.

LEVINE R.P. & PASZEWSKI A. (1970) Chloroplast structure and function in *ac*-20, a mutant strain of *Chlamydomonas reinhardi* II. Photosynthetic electron transport. *J. Cell. Biol.* **44**, 540–6.

LEVINE R.P. & SMILLIE R.M. (1962) The pathway of triphosphopyridine nucleotide photoreduction in *Chlamydomonas reinhardi*. *Proc. Nat. Acad. Sci. U.S.A.*, **48**, 417–21.

LEVINE R.P. & TOGASAKI R.K. (1965) A mutant strain of *Chlamydomonas reinhardi* lacking RuDP carboxylase activity. *Proc. Natl. Acad. Sci.* **53**, 987–90.

LOPPES R. (1969a) Effect of the selective medium on the manifestation of mutations induced with mono-alkylating agents in *Chlamydomonas reinhardi*. *Mutation Res.* **7**, 25–34.

LOPPES R. (1968) Ethyl methanesulfonate; an effective mutagen in *Chlamydomonas reinhardi*. *Molec. Gen. Genet.* **102**, 229–31.

LOPPES R. (1969b) A new class of arginine requiring mutants in *Chlamydomonas reinhardi*. *Molec. Gen. Genet.* **104**, 172–7.

LOPPES R. (1970) Growth inhibition by NH_4^+ ions in arginine requiring mutants of *Chlamydomonas reinhardi. Molec. Gen. Genet.* **109**, 233–40.

LOPPES R. & MATANGE R.F. (1973) Acid phosphatase mutants in *Chlamydomonas.* Isolation and characterization by biochemical, electrophoretic, and genetic analysis. *Genetics,* **75**, 593–604.

LOPPES R. & STRIJKERT P.J. (1972) Arginine metabolism in *Chlamydomonas reinhardi.* Conditional expression of arginine-requiring mutants. *Molec. Gen. Genet.* **116**, 248–57.

McMAHON D. (1971) The isolation of mutants conditionally defective in protein synthesis in *Chlamydomonas reinhardi. Molec. Gen. Genet.* **112**, 80–6.

McMAHON D. & LANGSTROTH P. (1972) The effects of canavine and of arginine starvation on macromolecular synthesis in *Chlamydomonas reinhardi. J. Gen. Microbiol.* **73**, 239–50.

NICHOLSON-GUTHRIE C.S. (1972) Isolation and characterization of a mutant of *Chlamydomonas reinhardi* lacking chlorophyll. Doctoral dissertation submitted to Department of Zoology. Indiana University.

PETTIJOHN D. & HANAWALT P. (1964) Evidence for repair-replication in ultra-violet damaged DNA in bacteria. *J. Mol. Biol.* **9**, 395–410.

ROSEN H. & EBERSOLD W.T. (1972) Recombination in relation to ultraviolet sensitivity in *Chlamydomonas reinhardi. Genetics,* **71**, 247–53.

SAGER R. (1962) Streptomycin as a mutagen for non chromosomal genes. *Proc. Nat. Acad. Sci. U.S.A.* **48**, 2018–26.

SAGER R. & GRANICK S. (1953) Nutritional studies with *Chlamydomonas reinhardi. Ann. N.Y. Acad. Sci.* **56**, 831–8.

SAGER R. & GRANICK S. (1954) Nutritional control of sexuality in *Chlamydomonas reinhardi J. Gen. Physiol.* **37**, 729–42.

SAGER R. & LANE D. (1972) Molecular basis of maternal inheritance. *Proc. Nat. Acad. Sci. U.S.A.* **69**, 2410–13.

SAGER R. & PALADE G.E. (1954) Chloroplast structure in green and yellow strands of *Chlamydomonas. Exp. Cell. Res.* **7**, 584–8.

SATO V.L., LEVINE R.P. & NEUMANN J. (1971) Photosynthetic phosphorylation in *Chlamydomonas reinhardi.* Effects of a mutation altering an ATP-synthesizing enzyme. *Biochim. Biophys. Acta.*, **253**, 437–48.

SCHOTZ F., BATHELT H., ARNOLD C.G. & SCHIMMER O. (1972) Die architektur und organisation der *Chlamydomonas*-Zelle. Ergebnisse der Elektron-mikroskopie von Serienschnitten und der daraus resultierenden dreidimensionalen Rekonstruktion. *Protoplasma,* **75**, 229–54.

SMYTH R.D. & EBERSOLD W.T. (1970) A *Chlamydomonas* mutant with altered phototactic response. *Genetics,* **64**, s62.

STOLBOVA A.V., MURASHKO G.N., LaVRENCHOUK V.Ya. & KVITKO K.V. (1969) Genetic analysis of the mutants from *Chlamydomonas reinhardi. Prog. in Protozool.* pp. 120–121.

STRIJKERT P.J., LOPPES R. & SUSSENBACH J.S. (1973) The actual biochemical block in the *arg*-2 mutant of *Chlamydomonas reinhardi. Biochem. Genetics,* **8**, 239–48.

STRIJKERT P.J. & SUSSENBACH J.S. (1969) Arginine metabolism in *Chlamydomonas reinhardi.* Evidence for a specific regulatory mechanism of the biosynthesis. *Europ. J. Biochem.* **8**, 408–12.

SUEOKA N. (1960) Mitotic replication of deoxyribonucleic acid in *Chlamydomonas reinhardi. Proc. Nat. Acad. Sci. U.S.A.* **46**, 83–91.

SURZYCKI S. (1971) Synchronously grown cultures of *Chlamydomonas reinhardi.* In *Methods in Enzymology* Vol. xxiii (part A), pp. 67–73, ed. A. San Pietro, Academic Press, New York.

SUSSENBACH J.S. & STRIJKERT P.J. (1969) Arginine metabolism in *Chlamydomonas reinhardi.* On the regulation of the arginine biosynthesis. *Europ. J. Biochem.* **8**, 403–7.

SWINTON D.C. & HANAWALT P.C. (1973a) Absence of ultra-violet-stimulated repair replication in the nuclear and chloroplast genomes of *Chlamydomonas reinhardi*. *Biochim. Biophys. Acta*, **294**, 385–95.

SWINTON D.C. & HANAWALT P.C. (1973b) The fate of pyrimidine dimers in ultra-violet-irradiated *Chlamydomonas*. *Photochem. Photobiol.* **17**, 361–75.

TOGASAKI R. & HUDOCK M.O. (1975) Effect of inorganic arsenate on photosynthetic algae I. Response of wild type and four photosynthetic mutant strains of *Chlamydomonas reinhardi* to inorganic arsenate. *Plant Physiol.* (in press).

TOGASAKI R.K. & LEVINE R.P. (1970) Chloroplast structure and function in *ac*-20, a mutant strain of *Chlamydomonas reinhardi*. I. CO_2 fixation and ribulose-1,5-diphosphate carboxylase synthesis. *J. Cell. Biol.* **44**, 531–9.

CHAPTER 3

GENETIC DETERMINANTS OF FLAGELLUM PHENOTYPE IN *CHLAMYDOMONAS REINHARDTII*

SIR JOHN RANDALL AND DAVID STARLING

Department of Zoology, University of Edinburgh,
King's Buildings, West Mains Road, Edinburgh, U.K.

1   Introduction 49

2   Straight-flagella strains 52

3   Complementation tests on closely linked flagellar genes 57

4   Information derived from the study of temporary dikaryons 58

   4.1   Motility in heterokaryons 58

4.2   Motility of regenerated flagella in heterokaryons 59

4.3   The flagella of dikaryons containing the long-flagella mutant *Lf*1 59

5   Summary of genetic results 59

6   Conclusions 60

7   References 61

(In 1971, Sir John Randall and David Starling contributed, to an international symposium on the genetics of the spermatozoon, a comprehensive article in which they reviewed much of the work that they and their colleagues had done on the genetic control of flagellar form and function in *Chlamydomonas reinhardtii*. Since that article contained just the kind of information that we needed for the compilation of this book on algal genetics, but since the volume in which it was published is not generally available in the private or institutional libraries of phycologists, we asked the authors to allow us to republish those sections which are specifically pertinent to our subject. They and their publisher kindly consented. Sections on the dynamics and fine-structure of wild-type flagella have been deleted—for these the reader is referred to the original publication (Randall & Starling 1971)—but the parts relating to algal genetics are printed below almost *verbatim*.)

1 INTRODUCTION

The suitability of *Chlamydomonas* for various types of genetic study has been evident for some time. Of the three species, *C. moewusii*, *C. eugametos* and *C. reinhardtii* chiefly used for this purpose, investigations have largely been concentrated on the last-named organism. Because of the more detailed linkage-group data on *C. reinhardtii* this species has also been used in the present studies. The earliest isolation of paralysed flagella mutants was however carried out by Lewin (1954) on *C. moewusii* and detailed investigation of *C. reinhardtii* was carried out by Levine and Ebersold (1960) and Ebersold *et al.* (1962). Smith and Regnery (1950) were the first to show that mating type was determined by a single pair of alleles. The methods of tetrad analysis of mutant crosses with wild-type—involving the transfer of single zygotes to fresh agar medium, the subsequent germination of the zygotes and the separation and characterization of the meiotic products—have been fully worked out by previous investigators and have been used in our laboratory with only minor variations.

Until recently no definitive study of chromosome number in *C. reinhardtii* has emerged. Thus the haploid number quoted by Buffaloe (1958), Levine and Folsome (1959) is eight, while that given by Wetherell and Krauss (1956) and Schaechter and De Lamater (1955) is 16. The problem has been complicated by the small size of the chromosomes and by the induction of polyploidy under certain conditions of light intensity. Meanwhile it is noticeable that the number of linkage groups obtained by tetrad analysis has gradually increased from nine (Levine & Ebersold 1960) to 11 (Ebersold *et al.* 1962) and now to 16 (Levine & Goodenough 1970, modified from Hastings *et al.* 1965). The recent work of McVittie and Davies (1971) gives the haploid number of chromosomes as 16, a figure in agreement with the known number of linkage groups.

One of the first objectives was to search for strains with structural abnormalities in the flagellar apparatus, and a mutant non-motile strain of *C. reinhardtii* which occurred in the presence of proflavine was isolated and seen to have straight flagella. In the electron microscope the flagella showed a distinctive structural abnormality. The central pair of microtubules of the axoneme assembly was replaced by an irregular but approximately axial core of apparently disorganized material (Randall *et al.* 1964). Because of the absence of the central pair of tubules in their normal form such strains have been designated as 9 + 0 mutants. Since that time many more non-motile mutants with differing phenotypic characteristics have been isolated (Warr *et al.* 1966; McVittie 1969 a, b; 1971 a, b). Some of the '*pf*' mutants first isolated by Lewin have also been examined. The techniques of mapping are time-consuming and only about one-third of the roughly 120 mutants available to us have so far been mapped. To show the distribution of these mapped strains on the 16 linkage groups we have simplified the map recently published by Levine and Goodenough (1970) and added information about some of our strains not available to them (Fig. 3.1).

From a genetic standpoint it is notable that the sites on the linkage groups of those genes and their alleles that have been mapped are well distributed. No mutant-flagella strains have yet been assigned to linkage groups XIV, XV

and XVI, but the mapping even of known strains is at a comparatively early
stage. The distribution of [genes of] paralysed strains which seem likely to
have a bearing on the morphopoiesis of the organelle compares sharply with
the situation that has been found to apply to even-number bacteriophage
(Edgar & Wood 1966), where marked clustering of genes associated with
particular structures occurs.

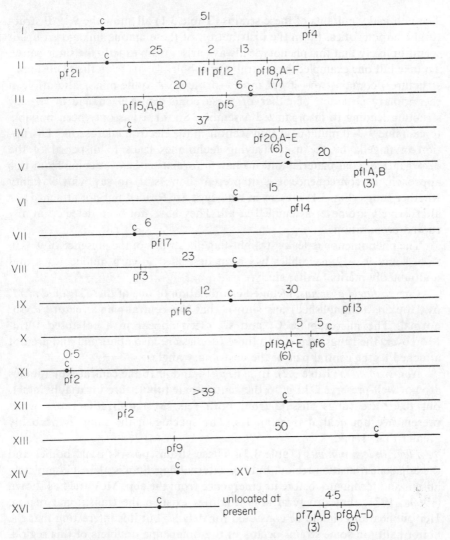

Fig. 3.1. Simplified linkage map of *C. reinhardtii* showing only mutations
affecting the flagellar apparatus, based on the data of Hastings *et al.* (1965),
Warr *et al.* (1966) and McVittie (1971).

The information in Tables 3.1 to 3.4 about fine-structural abnormalities so far discovered in paralysed flagella is necessarily summarized. It will not be possible to go into greater detail here; new information is continually becoming available and some features, especially those relating to Table 3.3, are in course of publication (McVittie 1971 a, b).

2 STRAIGHT FLAGELLA STRAINS

The non-leaky mutants of these strains (Table 3.1) all show the 9 + 0 structural characteristics. From the distribution of these among linkage groups it seems unlikely that this phenotype always arises from exactly the same cause. To take but one example: the central pair tubules (apart from their subsidiary structures) contain two—if not more—proteins. A quite minor alteration in the primary structure of either of these could cause a change in tertiary structure leading to disorganized assembly. So far it has only been possible to examine 9 + 0 mutants in thin section in the electron microscope. Disruption of flagella by the usual fraying techniques leads to dispersal of the disorganized axial material and no new information has yet come from this approach. In consequence it is at present impossible to say with certainty whether rungs (Fig. 3.1) present in wild-type are formed but not attached; or alternatively are never assembled at all. They have not been detected in disrupted preparations.

The phenomenon in leaky straight-flagella strains of the presence of 9 + 0, 9 + 1 and 9 + 2 assemblies has been described in early publications and undoubtedly merits further study.

Curved-flagella strains. Detailed examination of one of these strains, *pf* 17, by Hopkins (unpublished) has shown that the central-pair structure is abnormal. The microtubules C_1 and C_2 often appear in a collapsed form. Moreover, the rungs attached to these tubules are also abnormal and are not attached to the central pair at the customary angles.

Non-swimmers (Table 3.2). In *pf* 6, *pf* 14 and *pf* 16 the central-pair tubules are not well-preserved. In *pf* 16 the central-pair tubules are invariably intact, but there are rungs missing from both sets. Moreover, the rungs when present are not held at right angles. The spacing of the rungs is probably greater than 160A.

Flagella-less mutants (Table 3.3). These strains possess basal bodies and the adjacent transition region, but the external flagellum is almost completely absent and terminates before its emergence from the cup. McVittie has shown (1969a, 1971) that structural abnormalities exist in the transitional region. Her publications should be consulted for details; but it is interesting that the abnormality in some strains occurs in the outer nine doublets of this region. In wild-type these doublets are continuous with the AB tubules of the external flagellum. No abnormalities of structure of these have yet been found in mutant strains.

Table 3.1. Genetic and fine-structural data on flagellar mutants of *Chlamydomonas reinhardtii*: chiefly straight, curved and swollen flagella.

| | | Genetic data | | Fine structure where known* | | | | | |
| | | Non-leaky (NL) or leaky (L) | Linkage group | Tubules A, B | Side arms | Spokes | C_1 | C_2 | Rungs |
Phenotype	Strains			Outer tubules and attachments			Central pair and attachments		
Tendency to palmelloid 9 + 2	pf 4		I						
Straight flagella 9 + 0	pf 18A–F	NL	II	N	N		None	None	None
Straight flagella 9 + 0	pf 19, pf 19A–D	NL	X	N	—	—	None	None	None
Straight flagella 9 + 0	pf 15, pf 15A–B	NL	III	N	N	—	None	None	None
Straight flagella 9 + 0, 9 + 1,9 + 2	pf 20, pf 20A–E	L	IV	N	—	—		See col. 1	
Curved flagella 9 + 2	pf 1, pf 1A–B		V	P,M,N,	P	P	P,M	P,M	P,M
Curved flagella 9 + 2	pf 17		VII	P,M,N,	P	P	P,M	P,M	P
Swollen flagella	Spon 1, NG 11, NG 24, pf 21, S 13	—	—						

* In the last six columns the symbols P, M, N refer to poor, moderate and normal preservation of structure.

D

Table 3.2. Genetic and fine-structural data on flagellar mutants of *Chlamydomonas reinhardtii*: non-swimmers with flagella subtending abnormal angles, and abnormal swimmers.

| | | Genetic data | | Fine structure where known* | | | | | |
| | | | | Outer tubules and attachments | | | Central pair and attachments | | |
Phenotype	Strains	Non-leaky (NL) or leaky (L)	Linkage group	Tubules A, B	Side arms	Spokes	C_1	C_2	Rungs
Non-swimmers with Flagella subtending Abnormal angles	*pf* 2 9 + 2		XI	P,M	P,M	P,M	P,M	P,M	P
	pf 6 9 + 2		X	M	P	P,M	P,M	P,M	P
	pf 12 9 + 2		II	—	—	—	—	—	—
	pf 14 9 + 2		VI	—	—	—	—	—	—
	pf 16 9 + 2		IX						
	AO 4 9 + 2								
	HA 3 9 + 2		—	—	—	—	—	—	—
	HA 28 —	L							
Abnormal swimmers	*pf* 3 9 + 2					P			
	pf 9 9 + 2		VIII						
	pf 10 9 + 2		XIII						
	AO 3 9 + 2			—	—	—	—	—	—
	HA 7, NG 10								
	HG 14, NG 17		—						
Also long flagella See Table 3	*Lf* 1 9 + 2								
	Lf 2		II						

Table 3.3. Genetic and fine-structural data on flagellar mutants of *Chlamydomonas reinhardtii*: flagellaless, stumpy, short- and long-flagella mutants.

Phenotype	Genetic data			Fine structure where known*					
	Strains	Non-leaky (NL) or leaky (L)	Linkage group	Outer tubules and attachments			Central pair and attachments		
				Tubules A, B	Side arms	Spokes	C_1	C_2	Rungs
Flagella-less mutants	NG 6, NG 8, NG 9, NG 27, NG 28, NG 29, NG 30, S 3	—	—	Transitional-region abnormalities; see text and *McVittie* (1969, 1971)			None		
Stumpy-flagella mutants	HA 1, HA 13, NG 16, NG 25, NG 35, NG 36	—	—	Abnormalities in the external stump; see text and *McVittie* (1969, 1971)					
Short-flagella mutants	*pf* 7, *pf* 7A and probably *pf* 7B, *pf* 8A, *pf* 8B, *pf* 8C, *pf* 8D, *pf* 21	See text and *McVittie* 1971		Abnormalities in the external axoneme; see text and *McVittie* (1969, 1971)					
Long flagella See also Table 4	*Lf* 1 *Lf* 2		II XII	Normal except for swollen tips Normal except for swollen tips					

* For explanation of nomenclature, see footnote to Table 3.1.

Table 3.4. Genetic and fine-structural data on flagellar mutants of *Chlamydomonas reinhardtii*: miscellaneous strains.

| | | Genetic data | | Fine structure where known | | | | | |
| | | | | Outer tubules and attachments | | | Central pair and attachments | | |
Phenotype	Strains	Non-leaky (NL) or leaky (L)	Linkage group	Tubules A, B	Side arms	Spokes	C_1	C_2	Rungs
Impaired motility									
9 + 2	*pf* 3		VIII						
9 + 2	*pf* 9								
9 + 2	*pf* 10								
9 + 2	AO 3, AO 6, HA 7, NG 10, NG 14, NG 17	—	—	—	—	—	—	—	—
9 + 2	Lf 1		II		See Table 3.3				
9 + 2	Lf 2		XII						
Mutants with abnormal numbers of flagella	cyt 1		VI						
Suppressor mutant	SU 1								

Stumpy-flagella mutants (Table 3.3). In these strains the external flagella are just visible in the light microscope and generally less than 1 μm in length. The basal body and transition region are usually normal in structure, but a variety of abnormalities was found in the external stumps. Phenotypic heterogeneity suggests that more than one genetic locus may be involved.

Short-flagella mutants (Table 3.3). The length of the external flagella in these strains varies from ∼ 1·5 μm to 7 μm with a mean value of about 3 μm and fine-structural differences from wild-type were restricted to the axoneme of the external flagella. Poorly formed abnormally placed tubules were found together with abnormal spokes. The disorganization of the short axoneme was most evident in the *pf*7 strains, less in the *pf*8 strains and least in *pf*21.

Long-flagella mutants (Tables 3.3 and 3.4). These two mutants isolated by McVittie (1969a, b) are characterized by having flagella of considerably greater length, which however varies more widely between the flagella of the same cell and also more widely than in wild-type. These strains are distinguished not only by the greater length but by the slightly bulbous tips of the flagella, by their slow flagellar growth, and by their tendency to swim abnormally.

3 COMPLEMENTATION TESTS ON CLOSELY LINKED FLAGELLAR GENES

Starling (1969, 1970) has carried out complementation tests on a group of seven closely linked genes, the *pf*18 series, affecting the formation and organization of the central pair of tubules in the flagella of *C. reinhardtii*. It has previously been demonstrated (Warr *et al.* 1966) that recombination did not take place between *pf*18 and *pf*18A, *pf*18B, *pf*18C and *pf*18D. This early work however did not answer the question whether some or all of these mutations are located in the same cistron; or alternatively whether the mutations are in several closely linked cistrons. By means of a technique similar to that developed by Ebersold (1967) diploid cells were isolated. Having shown that the mutations under investigation were recessive to their wild-type alleles, two mutations were introduced into the same diploid nucleus and the progeny investigated. If the mutations are in different cistrons the progeny should be phenotypically wild-type. Conversely, if the two mutations are in the same cistron, the progeny should exhibit paralysed flagella. By this means Starling was able to show that the *pf*18 alleles all lie within the same cistron.

4 INFORMATION DERIVED FROM THE STUDY
OF TEMPORARY DIKARYONS

If we imagine a heterokaryon to be formed between wild-type and flagellar
mutant phenotypes of *C. reinhardtii*, it is clear that in principle the heterokaryon could be used to examine the interaction of cytoplasmic and nuclear
factors in the control of the flagellar apparatus. The adaptation of the techniques devised by Harris and Watkins (1965) to algal cells requires the use of
gametes—from which part of the cell wall has become detached in the process
of differentiation—and a suitable cohesive agent. Harris and Watkins (1965)
employed inactivated Sendai influenza virus for this purpose for hybridizing a
wide variety of genetically unrelated animal cells.

We have not so far succeeded in developing a suitable technique along
these lines. However, we have successfully made use of the normal mating
process of this organism (Levine & Ebersold 1960). The mating of mutant
phenotypes of *C. reinhardtii* has proved to be generally a good deal less
efficient than that of mt^+ and mt^- wild-types, but has been useful. As is
well known, conjugation takes place initially by flagellar contact at the tips
of the mating cells. Thus, while it has proved possible to cross short-flagella
strains with wild-type, the crossing of stumpy and flagellaless strains with
those possessing flagella has been impossible. The present scope of this
approach is correspondingly restricted.

Mating of *Chlamydomonas* cells results first in the formation of a prozygote which, for a somewhat variable period up to about two hours, remains a
quadriflagellate dikaryon possessing two nuclei and a common cytoplasm.
Subsequent to this the external flagella regress, the basal bodies disappear and
the nuclei fuse; this is the beginning of zygote maturation. In the prozygote
phase, the flagella, if removed by physical, chemical or micrurgical means,
subsequently regenerate. For the period of up to two hours already mentioned
it is therefore possible to examine the effect of the separate nuclei and
the common cytoplasm on (a) the two pairs of flagella from the cells forming the heterokaryon and (b) the regenerating flagella which form when
the originals are removed. Our first results on this system have now been
published (Starling & Randall 1971) and only a summary will be presented
here.

4.1 *Motility in Heterokaryons*

Wild-type × *pf16*. About ten minutes after cell fusion has taken place motility
of paralysed flagella of *pf*16 has been restored to all four flagella. There is
no stage during which only two flagella are used. Restoration of motility to
previously immotile flagella has therefore been shown to take place rapidly
in this heterokaryon.

Wild-type×pf18C. After a lapse of about ten minutes the heterokaryon begins to swim with two flagella only. After some 40 minutes the motility is still not restored to the (presumptively) mutant flagella. Reference to Tables 1 and 2 shows that the structural lesions in the paralysed flagella of *pf*16 and *pf*18C are different. Superficially it would appear that the 9 + 0 structure of *pf*18C cannot be repaired in this cross; this conclusion requires verification in the electron microscope.

Wild-type×pf19B. Motility of the heterokaryon occurs in two stages. The heterokaryon begins to swim in about ten minutes using the (presumptively) normal pair of flagella. Twenty minutes after cell-fusion, the second pair of flagella comes into use. Strain *pf*19B is also a 9 + 0 mutant; the necessity for electron microscope study becomes imperative with its implications for the importance or otherwise of the central pair of tubules and its accessory rung structures.

4.2   *Motility of Regenerated Flagella in Heterokaryons*

*Wild-type×pf*16. All four regenerated flagella take part in swimming.

Wild-type×pf18C. All four regenerated flagella are motile and restoration of function thus takes place as part of the regeneration process.

4.3.   *The Flagella of Dikaryons containing the Long-Flagella mutant Lf1*

The time required for the flagella of *Lf*1 to regenerate is of an order of magnitude greater than for wild-type (McVittie 1969a, b) and offers useful material for the examination of the effect of wild-type on mutant. Prior to deflagellation there is no difficulty in distinguishing between wild-type and mutant by their flagellar lengths and there is no obvious adjustment of either as a consequence of cell fusion. After deflagellation, regeneration of all four flagella takes place at approximately the same rate as for wild-type. After full regeneration the lengths of all four flagella are approximately the same and rather less than those of wild-type. The wild-type genome thus, as in other instances above, dominates the phenotype of the heterokaryon flagella.

5   SUMMARY OF GENETIC RESULTS

1. Although in *C. reinhardtii* there is non-chromosomal inheritance for streptomycin resistance, the inheritance of all flagellar genes so far investigated is Mendelian. The gamete contains about $1 \cdot 2 \times 10^{-13}$ g of DNA of which nearly 7% is in the chloroplast. The nuclear DNA has a density of $1 \cdot 723$ g cc^{-1}; the chloroplast DNA a density of $1 \cdot 695$ g cc^{-1}. A third minor peak, presumptively identifiable with mitochondrial DNA, has a density of $1 \cdot 715$ g cc^{-1}. No 'basal-body' DNA has yet been found; even if it is present it would probably be

beyond present technical resources to prove its existence and cytological location.

2. Examination of numerous strains by tetrad analysis has shown that in each case the phenotypic change is associated with mutation in a single gene.

3. Similar phenotypic changes may arise from several different causes. Thus the change from normal, curved motile flagella to straight non-motile flagella with a 9 + 0 structure has been observed in 22 strains with mutant genes occurring on four different linkage groups (II, III, IV and X).

4. Since flagellar contact is apparently an essential early step in the mating process, the straightforward genetic analysis of stumpy and flagellaless mutants has not been possible. The electron microscope therefore provides the only present means for examination of these strains.

5. The natural cell-fusion process of mating has been exploited to produce and examine early prozygotes—termed temporary dikaryons—in which a common cytoplasm is associated with separate nuclei for a limited number of hours. The wild-type genes are dominant and restoration of motility has been demonstrated in existing defective flagella and in regenerated ones. Rates of growth and final values of length are also affected.

6 CONCLUSIONS

These studies we have outlined are perhaps more remarkable for the questions left unanswered than for the positive results they provide. While much diversity of phenotype has been achieved, a still wider range is now required. Not only new phenotypes, but also new or improved methods of isolation are required. The use of temperature-sensitive strains may prove profitable, although many are markedly leaky. Examination of the phenotype in terms of fine structure is in part straightforward; no one can fail to distinguish between a 9 + 2 and a 9 + 0 strain. But the assessment of small differences in the existence, location and stability of minor features requires great care. Undoubtedly the use of optical diffraction techniques—rather than purely visual scrutiny—is now an obligatory aspect of the classification of the phenotype.

The comparatively small scale of our investigations has limited the number of strains it has been possible to isolate and characterize. No detectable lesions in the basal body have yet been produced, in terms of either number of tubules or structure. Although structural defects in the transition region have been observed, these have not been accompanied by any corresponding change in the number of outer-pair doublets, and in this sense mutagenesis has not yet changed the basic architecture of this part of the axoneme. The disorganization of the central-pair tubules has so far shown no effect on the outer tubules.

The search for new strains is therefore imperative; but this is not enough, since normal mating—as we have already shown—may be impossible. Stable, temperature-sensitive mutants would be a great help; but present indications are not very promising. Our work on naturally produced heterokaryons also suggests the need to develop new techniques of cell-fusion to aid genetic analysis within and outwith the species. *A priori*, however, it is uncertain whether the heterokaryons produced from an enforced cell fusion will go through the normal processes of zygote maturation and germination, and thus whether tetrad analysis would be possible.

Future work must also contrive to penetrate more deeply into control processes. As we have observed, the explanation of regeneration and regression must be carried still further. This implies the study of the influence of each flagellum on the other and whether this is direct—e.g., via the twin basal-body system—or through the medium of the cytoplasm or membrane.

This far from complete analysis must suffice for now. The way ahead is not simple; but the problems posed by the development and properties of flagella present a challenge which we shall continue to accept. In the field of morphopoiesis, a definitive organelle such as the flagellum stands high in the list of priorities. *Chlamydomonas reinhardtii*—with all its attendant problems—remains a very suitable organism for the purpose.

We wish to acknowledge permission from Dr A. C. McVittie and J. M. Hopkins to refer to various items from their unpublished data and for helpful comments on the manuscript.

7 REFERENCES

BUFFALOE N.D. (1958) A comparative cytological study of four species of *Chlamydomonas*. *Bull. Torrey Bot. Club* **85**, 157–78.

EBERSOLD W.T., LEVINE R.P., LEVINE E.E. & OLMSTED M.A. (1962) Linkage maps in *Chlamydomonas reinhardtii*. *Genetics* **47**, 531–54.

EBERSOLD W.T. (1967) *Chlamydomonas reinhardtii*; heterozygous diploid strains. *Science* **157**, 447–8.

EDGAR R.S. & WOOD W.B. (1966) Morphogenesis of bacteriophage T4 in extracts of mutant-infected cells. *Proc. Nat. Acad. Sci.* **55**, 498–505.

HARRIS H. & WATKINS J.F. (1965) Hybrid cells derived from mouse and man: artificial heterokaryons of mammalian cells from different species. *Nature, Lond.* **205**, 640–6.

HASTINGS P.J., LEVINE E.E., COSBEY E., HUDDOCK M.O., GILLHAM N.W., SURZYCKI, S.J., LOPPES R. & LEVINE R.P. (1965) The linkage groups of *Chlamydomonas reinhardtii*. *Micr. Genet. Bull.* **23**, 17–19.

LEVINE R.P. & FOLSOME C.E. (1959) The nuclear cycle in *Chlamydomonas reinhardtii*. *Zeitsch. für Vererbung* **90**, 215–22.

LEVINE R.P. & EBERSOLD W.T. (1960) Genetics and cytology of *Chlamydomonas*. *Ann. Rev. Microbiol.* **14**, 197–216.

LEVINE R.P. & GOODENOUGH V.W. (1970) The genetics of photosynthesis and of the chloroplast in *Chlamydomonas reinhardtii*. *Ann. Rev. Genet.* **4**, 397–408.

LEWIN R.A. (1954) Mutants of *C. moewusii* with impaired motility. *J. Gen. Microbiol.* **11**, 358–63.

62 CHAPTER 3

McVittie A.C. (1969a) Flagellum mutants of *Chlamydomonas reinhardtii*: genetic and electron microscope studies. Ph.D. Thesis. University of London.

McVittie A.C. (1969b) Studies on flagella-less, stumpy and short flagellum mutants of *Chlamydomonas reinhardtii*. *Proc. Roy. Soc. B.* **173**, 59–60.

McVittie A.C. (1972) Flagellum mutants of *Chlamydomonas reinhardtii*. I. Isolation and characterisation. *J. Gen. Microbiol.* **71**, 525–40.

McVittie A. & Davies D.R. (1971) The location of the mendelian linkage groups in *Chlamydomonas reinhardtii*. *Molec. Gen. Genetics* **112**, 225–8.

Randall J.T., Warr J.R., Hopkins J.M. & McVittie A. (1964) A single gene mutation of *C. reinhardtii* affecting motility: a genetic and electron microscope study. *Nature, Lond.* **203**, 912–4.

Schaechter M. & De Lamater E.D. (1955) Mitosis of *Chlamydomonas*. *Am. J. Bot.* **42**, 417–22.

Smith G.M. & Regnery D.C. (1950) Inheritance of sexuality in *Chlamydomonas reinhardtii*. *Proc. Nat. Acad. Sci.* **36**, 246–8.

Starling D. (1969) Complementation tests on closely linked flagellar genes in *Chlamydomonas reinhardtii*. *Genet. Res. Camb.* **14**, 343–347.

Starling D. (1970) Genetic, morphopoietic and microbeam irradiation studies on the flagella of *Chlamydomonas reinhardtii*. Ph.D. Thesis, University of London.

Starling D. & Randall Sir John (1971) The flagella of temporary dikaryons of *Chlamydomonas reinhardtii*. *Genet. Res. Camb.* **18**, 107–113.

Warr J.R., McVittie A., Randall Sir John & Hopkins J.M. (1966) Genetic control of flagellar structure in *C. reinhardtii*. *Genet. Res. Camb.* **7**, 335–51.

Wetherell D.F. & Krauss R.W. (1956) Colchicine induced polyploidy in *Chlamydomonas*. *Science* **124**, 25–6.

Original article from which this chapter was derived

Randall J. & Starling D. (1971) In *The Genetics of the Spermatozoon*, pp. 13–36, eds R. A. Beatty and S. Gluecksohn-Waelsch, Proceedings of International Symposium, 16–20 August, 1971, Edinburgh.

CHAPTER 4

GENETICS OF CELL WALL SYNTHESIS IN *CHLAMYDOMONAS REINHARDTII*

D. R. DAVIES AND K. ROBERTS

John Innes Institute,
Colney Lane, Norwich, NR4 7UH, England

1 Introduction 63

2 Mutation induction 63

3 Genetic analyses 64

4 Cell wall structure and composition 65

5 References 67

1 INTRODUCTION

Genetic studies on the development of the flagella and of the cell wall in *Chlamydomonas reinhardtii* are among the few analyses of morphogenetic processes that have been undertaken in eukaryotic systems. A necessary prerequisite for such studies is the availability of mutants which show a defect in some aspect of the structure being investigated. We have accumulated 79 mutants in *C. reinhardtii* defective in wall synthesis; the induction, isolation, genetic analysis and characterization of these mutants have been described (Davies & Plaskitt 1971). In addition to their intrinsic value as a means of elucidating the nature and mode of assembly of the cell wall, some of these mutants have proved valuable as experimental systems in which to isolate certain organelles and subcellular structures, without our having to resort to the destructive physical and chemical techniques that normally have to be used for removing the cell wall (Miller & McMahon 1974).

2 MUTATION INDUCTION

Mutants were recovered following treatment of gametes with N-methyl-N′-nitro-N-nitrosoguanidine (NG), UV (253·7nm) and ethylmethane sulphonate (Davies & Plaskitt 1971; Hyams & Davies 1972). Colonies of mutant

cells with defective walls were found to exhibit a distinctive physical configuration on agar. Instead of the normal ovoid or spherical shape of the
wild-type (Cw) cells, mutant cells having a defective wall (cw) become somewhat amoeboid and flat when plated on an agar surface because of their lack
of structural rigidity. Mutant colonies are therefore readily distinguishable
under a low-power microscope both by their lighter appearance, due to the
greater transparency of the flat colonies, and by the irregular form of the cells
at their edges (see Davies & Plaskitt 1971, plate 1).

No spontaneous mutants were recovered from a sample of over 2×10^5
colonies examined (Hyams & Davies 1972), but following exposure to high
doses of NG or UV up to 3% of the surviving cells were cw mutants.
When grown on a medium containing yeast extract, acetate and peptone, the
growth rate of many of the mutants was comparable to that of Cw cells
(Davies & Plaskitt 1971). No osmotic correction of the medium was necessary, even for cells lacking all traces of wall, because the osmotic pressure of
the cells is regulated by the contractile vacuole. Cw and cw cells grew at comparable rates in a wide range of concentrations of sucrose, glucose or sorbitol.

Reverse mutation rates could not be determined, as no selective technique
has been found for readily distinguishing the rare revertant (wild-type) from
mutant cells. Attempts to discriminate between them on the basis of a
differential sensitivity either to antibiotics which differentially affect coccal
and L forms of *Staphylococcus aureus*, or to polycationic substances such as
DEAE dextran and poly-1-ornithine, were unsuccessful (Hyams 1972).

3 GENETIC ANALYSES

All mutants were crossed with a Cw strain, and analyses of the progeny
indicated that most of them were distinguished from the wild type by a single
gene. Some of the mutants were intercrossed in pairs in two separate groups,
the first involving 15cw mutants (Davies & Plaskitt 1971), the second 20
additional mutants together with 12 of the original group (Hyams & Davies
1972). From genetic analyses of their progeny it was concluded that the 32
mutants so analysed could probably best be classified into 19 groups, each
group being characterized by the low frequency or complete absence of wild-
type recombinant cells when individual members of the group were intercrossed.

However there were exceptions to this pattern of behaviour, and some of
these have been studied in considerable detail. One example, $cw18$, will be
considered further here. This mutant, $cw18.mt^+$ (mt = mating type), which
was first isolated following NG treatment, frequently segregated wild types
when selfed, and gave anomalous ratios in other crosses. From a cross with
wild-type, all the 25 tetrads analysed were found to segregate regularly,
$2Cw:2cw$. However, from one of these tetrads a $cw18.mt^-$ strain was isola-

ted which, when crossed with mt^+, gave tetrads which segregated (66%) $2Cw:2cw$, (29%) $1Cw:3cw$ and (5%) $0Cw:4cw$. Similar aberrant results were obtained from the reciprocal cross, $Cw.mt^+ \times cw.mt^-$. When selfed, $cw18 \times cw18$ gave tetrads 70% of which segregated $1Cw:3cw$, the remainder segregating $0Cw:4cw$. (Remarkably, diploid cells produced by the fusion of $cw18.mt^+$ and $cw18.mt^-$ cells were wild-type in phenotype.) Those mutants which gave aberrant tetrad ratios on crossing were stable in somatic cultures and no post-meiotic segregation was observed. No evidence of the presence of a readily transmissible agent could be found. Later generation crosses of derivatives of these tetrads (Davies 1972a) indicated the involvement of both non-Mendelian and Mendelian systems of genetic control. Other markers known to be located in the nucleus segregated normally in these crosses. Conversion events were apparently not involved in the production of the anomalous ratios, and the excess of cw strains from some of the crosses was far too high to be attributable to mutation (Davies 1972a). UV irradiation of gametes prior to mating had no influence on the genetic results; this is in contrast to another non-Mendelian system in this species (Sager & Ramanis 1967).

Several interpretations have been considered to account for these phenomena (Davies 1972a), and the one which most readily accounts for all the data involves an extra-nuclear control of cell-wall synthesis (Davies & Lyall 1973). If an extra-nuclear system is involved in the $cw18$ mutants then it must be capable also of giving $2Cw:2cw$ segregations, as observed in many of the pedigrees from crosses of mutant and wild-type. If the extra-nuclear system can segregate 2:2 then patently it must be a possibility that all our mutations affecting cell wall structure are regulated by an extra-nuclear rather than a nuclear system! Segregating 2:2 at meiosis need not be the prerogative of nuclear alleles. By mapping the loci involved, in preliminary experiments we have been unable to show linkage between any cw gene and nuclear genes, but we have interesting observations on the relation between two cw loci and the 'yellow' locus $y1$, which is apparently not located on any of the 16 chromosomes of this species (Ebersold personal communication). We found that crosses of either $cw2$ or $cw17$ with a Cw strain carrying $y1$ yield only ditype tetrads, both parental and non-parental. So far we have analysed only 15 tetrads involving $cw2$ and 22 involving $cw17$, but the available data suggest that in these mutants affecting cell-wall assembly we may be dealing with another extra-nuclear system.

4 CELL WALL STRUCTURE AND COMPOSITION

In order to understand the nature of defects in the various cell-wall mutants it is essential to know something about the structure and composition of the normal cell wall. Our approach to this problem has been twofold. We have used various electron-microscopical techniques with a resolution of about 2·5nm, to elucidate the structure of the wild-type cell wall, and various biochemical techniques to study its composition. On the basis of information from these studies we have then examined the structure and composition of some of the *cw* mutants. Lastly we have found a system in which the isolated cell wall may be dissociated and reassembled *in vitro*, and this in turn has indicated how cell-wall biogenesis may occur *in vivo*.

The wild-type cell wall is composed of seven basic layers (W1–W7). The inner layer (W1) is amorphous and of variable thickness, and accounts for about 7·5% of the wall by weight. It contains about 30% protein, the rest being largely made up of a polymer of arabinose, galactose and mannose. The protein has a high proportion of hydroxyproline (Roberts *et al.* 1972; Roberts 1974). The next five layers (W2–W6) together account for most of the remainder of the wall. They form a structure of constant thickness; negative staining has shown this to be composed of a crystalline lattice. The detailed structure of this lattice has been determined by using both negative staining and rotary shadowing, and by complementing these with optical diffraction analysis, linear integration, and computer reconstruction (Hills *et al.* 1973; Roberts 1974). On the basis of hydrodynamic studies we have concluded that the lattice contains one basic homogeneous glycoprotein (Hills *et al.* 1975), which can be separated into several discrete species of glycoproteins by electrophoresis on SDS* polyacrylamide gel (Davies 1972b; Roberts 1974). This glycoprotein has a bulk chemical composition very similar to that of the inner amorphous wall layer, W1. A very thin amorphous wall layer (W7), of very variable appearance, often appears outside the crystalline wall component.

The crystalline part of the cell wall and the inner amorphous layer, W1, may be separated in the presence of certain chaotropic agents such as lithium chloride or sodium perchlorate. The W1 layer remains insoluble while the crystalline glycoprotein lattice is solubilized. Removal of the salt by dialysis results in the reassembly *in vitro* of a product which, on the basis of shape, size, electron microscopy, optical and electron diffraction analysis, chemical composition and antigenicity, is identical to the original cell wall (Hills 1973; Roberts 1974). In the absence of the W1 layer the salt-soluble glycoproteins alone can self-assemble on dialysis to form small fragments (Hills *et al.* 1975), which have the crystalline structure typical of the original lattice but are not capable of forming a spherical wall-shaped shell. An analogous situation occurs in many of the *cw* mutants. Some (e.g. *cw2*) are able to make and excrete the requisite glycoproteins for the assembly of the crystalline lattice; but, presumably since they do not possess an intact inner amorphous wall layer to which these can be attached, they merely assemble small crystalline fragments free to diffuse away from the cell. This results in a wall-less

* Sodium dodecyl sulphate.

mutant. As judged by optical diffraction, the crystalline structure of the lattice is invariant in all those mutants able to form it; no abnormal lattice has been found. In those mutants unable to form a lattice, the pattern of glycoproteins, as revealed by SDS gel electrophoresis, is distinctly abnormal (Hyams 1972). No mutant has been found with only an inner layer.

The conditions under which wall assembly can take place, both *in vitro* and *in vivo*, are not simple and are not yet fully understood. *In vitro* the salt removal must be swift and must be across a suitable surface (agar, millipore, dialysis sac), and both the salt-soluble and salt-insoluble components must be present. In those mutants unable to assemble a complete wall a suitable surface appears to be also necessary for lattice formation *in vivo* (Davies & Lyall 1973).

From the available evidence it is hard to choose between two possible models of wall assembly *in vivo*. According to one model (Davies & Lyall 1973) the crystalline lattice layer of the wall is laid down first around the naked daughter protoplasts within the mother cell wall. Upon completion of this layer the inner layer is then accreted at its inner surface. However the evidence from the *in vitro* reassembly experiments suggests that the opposite could well be true, i.e. that the inner layer is formed first and that the glycoproteins responsible for the crystalline lattice layer are then excreted and assemble on the outer surface of the wall. Loss of the ability to produce the glycoprotein lattice, and its reacquisition on passing through the diploid zygospore stage, have been observed in particular mutants (Davies & Lyall 1973); they could be related to the need for a 'template', such as the W1 layer.

A firm basis for the ultrastructural, biochemical and genetical characterization of cell-wall mutants is not yet available, and many details of the normal assembly of the various layers of the wild type cell wall remain to be elucidated. However these initial studies of cell wall formation in *C. reinhardtii* inidicate the scope of the system for studying problems of synthesis and assembly of complex structures in a eukaryotic cell.

5 REFERENCES

Davies D.R. (1972a) Cell wall organization in *Chlamydomonas reinhardi*: The role of extra nuclear systems. *Mol. Gen. Genet.* **115**, 334–48.

Davies D.R. (1972b) Electrophoretic analysis of wall glycoproteins in normal and mutant cells. *J. Exptl. Cell. Res.* **73**, 512–6.

Davies D.R. & Lyall V. (1973) The assembly of a highly ordered component of the cell wall: The role of heritable factors and of physical structure. *Molec. Gen. Genet.* **124**, 21–34.

Davies D.R. & Plaskitt A. (1971) Genetical and structural analyses of cell wall formation in *Chlamydomonas reinhardi*. *Genet. Res. Camb.* **17**, 33–43.

Hills G.J. (1973) Cell wall assembly *in vitro* from *Chlamydomonas reinhardi*. *Planta*, **115**, 17–23.

HILLS G.H., GURNEY-SMITH M. & ROBERTS K. (1973) Structure composition and morphogenesis of the cell wall of *Chlamydomonas reinhardi*. II Electron microscopy and optical diffraction analysis. *J. Ultrastruct. Res.* **43**, 179–92.

HILLS G.H., PHILLIPS J.M., GAY M.R. & ROBERTS K. (1975) The *in vitro* assembly of a plant cell wall. *J. Molec. Biol.* (in press).

HYAMS J.S. (1972) Genetical and ultrastructural analysis of cell wall mutants in *Chlamydomonas reinhardi*. Ph.D. Thesis, University of East Anglia, 1972.

HYAMS J.S. & DAVIES D.R. (1972) The induction and characterization of cell wall mutants of *Chlamydomonas reinhardi*. *Mutation Research* **14**, 381–9.

MILLER M.J. & McMAHON D. (1974) Synthesis and maturation of chloroplast and cytoplasmic ribosomal RNA in *Chlamydomonas reinhardi*. *Biochem. Biophys. Acta.* **366**, 35–44.

ROBERTS K. (1974) Crystalline glycoprotein cell walls of algae: their structure, composition and assembly. *Phil. Trans. Roy. Soc. Lond. B.* **268**, 129–46.

ROBERTS K., GURNEY-SMITH M. & HILLS G.J. (1972) Structure composition and morphogenesis of the cell wall of *Chlamydomonas reinhardi* I. Ultrastructure and preliminary chemical analysis. *J. Ultrastruct. Res.* **40**, 599–613.

SAGER R. & RAMANIS Z. (1967) Biparental inheritance of nonchromosomal genes induced by ultraviolet irradiation. *Proc. Nat. Acad. Sci. (Wash.)* **58**, 931–7.

CHAPTER 5

PLASTID INHERITANCE
IN *CHLAMYDOMONAS REINHARDTII*

G. MICHAEL W. ADAMS†, KAREN P. VANWINKLE-SWIFT†
NICHOLAS W. GILLHAM† AND JOHN E. BOYNTON*
Departments of Botany* and Zoology†
Duke University, Durham, North Carolina 27706, U.S.A.

1   Introduction 69

2   Chloroplast DNA 71
    2.1   Discovery 71
    2.2   Size and composition 74
    2.3   Kinetic complexity 76
    2.4   Replication during the vegetative cycle 78
    2.5   Replication during the sexual cycle 81
    2.6   Transmission in crosses 81

3   Evidence that uniparentally inherited genes reside in the chloroplast 86

4   Methods for altering the inheritance pattern of chloroplast genes 88

5   Segregation and recombination of chloroplast genes 90

6   Linkage and mapping of chloroplast genes 98

7   Models for inheritance of chloroplast genes 110

8   Conclusions 114

9   References 115

1   INTRODUCTION

As shown by Sager (1954), mutations to streptomycin resistance in *C. reinhardtii* may be inherited in either a Mendelian fashion (*sr-1*, Ch. 2) or in a distinctive non-Mendelian, uniparental fashion (*sr-2*). Thus whereas the *sr-1* mutation regularly segregates 2:2 in a cross regardless of the parental mating type carrying the mutation, the *sr-2* mutation is transmitted to almost all of the offspring when present in the mt^+ parent but is only rarely transmitted when donated by the mt^- parent (Fig. 5.1). Since 1954, many other mutations have been found exhibiting the same uniparental pattern of inheritance as *sr-2* (Table 5.1; Sager 1972; Gillham 1969).

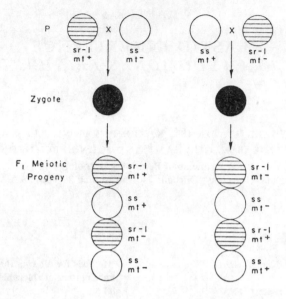

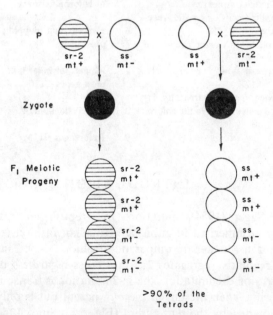

Fig. 5.1. Patterns of inheritance of the Mendelian *sr-1* and chloroplast *sr-2* mutants in *Chlamydomonas reinhardtii*. Segregation of the Mendelian gene *mt* (mating type) is also shown. (Gillham 1969.)

The maternal transmission of various plant characteristics, especially chlorophyll deficiencies, has been recognized for the past 60 years (Baur 1909; Correns 1909). Such uniparental transmission of genes has long been thought to imply their location in some cytoplasmic genetic element rather than in the nuclear chromosomes. The discovery of significant amounts of DNA in mitochondria (Luck & Reich 1964) and chloroplasts (Chun *et al*. 1963; Sager & Ishida 1963) of eukaryotic cells has made these organelles the prime targets for speculation on the location of non-Mendelian genes. Although the uniparental pattern of inheritance in *C. reinhardtii* is similar to that observed in higher plants (Rhoades 1955; Kirk & Tilney-Bassett 1967), a fundamental difference exists between the two cases. The cytoplasmic contributions of the two gametes of higher plants are patently unequal, with virtually all of the cytoplasm being donated by the maternal parent. Under these conditions it is fairly easy to envisage that any cytoplasmic genes are more likely to be transmitted to the progeny by the maternal rather than by the paternal parent. *C. reinhardtii*, however, is isogamous and both parents contribute equal nuclear and cytoplasmic contents to the formation of the zygote. Chloroplasts have been shown to fuse four to six hours after mating, at approximately the same time that the parental nuclei fuse (Cavalier-Smith 1970). Thus in *Chlamydomonas* some mechanism other than cytoplasmic exclusion must account for the observed preferential transmission of maternal cytoplasmic genes.

In the following pages, we will deal with the discovery and characterization of chloroplast DNA in *C. reinhardtii* and the physical transmission of this DNA in crosses. Secondly, we will examine the genetic evidence for assignment of uniparental mutants to chloroplast DNA, and the transmission, segregation, and recombination of these mutations during the sexual cycle. Third, we will consider the published maps of chloroplast genes. In conclusion we will compare various models which correlate the genetic and physical properties of chloroplast DNA in an attempt to explain the observed patterns of transmission, recombination and segregation in terms of a molecular mechanism.

2   CHLOROPLAST DNA

2.1 *Discovery*

In 1962 the existence of chloroplast DNA was established in *Chlamydomonas* by Ris and Plaut (1962). Studies on three separate species, *C. reinhardtii*, *C. eugametos*, and *C. moewusii*, demonstrated the presence of Feulgen-positive regions within the chloroplast which stained as intensely as the nucleus. Treatment of cells with deoxyribonuclease abolished the Feulgen reaction and also susceptibility to staining with acridine orange both in the chloroplast

Table 5.1. Summary of known chloroplast gene mutations in *Chlamydomonas reinhardtii*. Modified from Sager (1972) with nomenclature revised so that each mutant is given a symbol descriptive of its phenotype, followed by *-u-* for uniparentally inherited, and by its previously published designation or its isolation number. In the case of mutants with altered sensitivity to antibiotics, this symbol indicates the name of the antibiotic and whether the mutant is resistant (*r*) or dependent (*d*). Conditional mutants are indicated with a subscript (*d*). Reversions are designated by the original mutant symbol followed by *R* and a subscript indicating the new phenotype.

Abbreviations: CARB = carbomycin, CLE = cleocin, EMS = ethyl methane sulphonate, ERY = erythromycin, KAN = kanamycin, NEA = neamine, NG = nitrosoguanidine, OLE = oleandomycin, SPEC = spectinomycin, SPI = spiramycin, STREP = streptomycin.

References: 1, Behn & Arnold, 1972; 2, Boschetti & Bogdanov, 1973; 3, Boynton et al., 1973; 4, Boynton et al., 1976; 5, Burton, 1972; 6, Conde et al., 1975; 7, Gillham, 1965; 8, Gillham et al., 1970; 9, Gillham et al., 1974; 10, Gillham & Fifer, 1968; 11, Gillham & Levine, 1962; 12, Lee et al., 1973; 13, Mets & Bogorad, 1971; 14, Sager, 1954; 15, Sager & Ramanis, 1963; 16, Sager & Ramanis, 1965; 17, Sager & Ramanis, 1970; 18, Sager & Ramanis, 1971c; 19, Sager & Ramanis, 1972; 20, Surzycki & Gillham, 1971.

Gene	Previous designation	Induced by	References	Phenotype	Mapped (Ref.)
ac-u-1	ac1	STREP	15	Requires acetate (leaky)	17, 19
ac-u-2	ac2	STREP	15	Requires acetate (stringent)	17, 19
ac-u-3	ac3	STREP	18	Requires acetate (stringent)	17, 19
ac-u-4	ac4	STREP	18	Requires acetate (leaky)	17, 19
car-u-1	car1	STREP	19	Resistant to 50 µg/ml CARB	17, 19
clr-u-1	cle1	STREP	19	Resistant to 50 µg/ml CLE	17, 19
er-u-1[a]	ery1	STREP	19	Resistant to 200 µg/ml ERY	4, 17, 19
er-u-1a	ery-U1a	EMS	13	Resistant to 200 µg/ml ERY	4
er-u-37[b]	ery-u-37	Spontaneous; selected on ERY	9	Resistant to 200 µg/ml ERY	4
er-u-3	ery3	STREP	18	Resistant to ERY, CARB, OLE, SPI	—
knr-u-1	kan1	STREP	18	Resistant to 100 µg/ml KAN	—
nd-u (4 mutants)	nd	NG	7	Dependent on at least 10 µg/ml NEA	—
nr-u-2-1	nr-2, nr-2-1; = nea-2-1 in Sager, 1972	NG	7	Resistant to 100 µg/ml NEA	4, 10, 17, 19
olr-u-1, olr-u-2, olr-u-3	ole1 through ole3	STREP	19	Resistant to 50 µg/ml OLE	17, 19
sd-u-sm4	sm4	STREP	16, 17	STREP dependent	17
sd-u-3-18	sd-3-18	NG; selected on STREP	7, 8	Dependent on at least 20 µg/ml STREP	—
sdc-u-tm2	tm2	STREP	18	Requires STREP when grown at 35°C	—
sdR-u-sm4-D-310, -D-371	D-310, D-371	Growth of sm4 without STREP	18	Resistant to 500 µg/ml STREP	—
sdR-u-sm4-D- (4 mutants)[c]	D	Growth of sm4 without STREP	19	Resistant to 500 µg/ml STREP	—

Strain	Synonym/origin	Method	Ref.	Phenotype	Ref.
... (11 mutants)			18	Resistant to various low levels of STREP: 20 µg/ml; 50 µg/ml; 100 µg/ml.	—
sdR_{ca}-u-sm4-D-769	D-769	Growth of sm4 without STREP	17	Conditional STREP dependent	17, 19
sdR_{ca}-u-sm4-D^e (3 mutants)	D	Growth of sm4 without STREP	18	Conditional STREP dependent	—
sdR_s-u-3-18R19		Growth of sd-3-18 without STREP	1, 8	STREP-sensitive	—
sdR_s-u-3-18R35		Growth of sd-3-18 without STREP	1, 8	STREP-sensitive	—
sdR-u-3-18R20		Growth of sd-3-18 without STREP	1, 8	STREP-resistant	—
sdR_s-u-UV-16, -UV-17		UV (from sm4)	18	Resistant to 500 µg/ml STREP	—
sdR_s-u-UV- (4 mutants)		UV (from sm4)	18	Resistant to 20 µg/ml STREP	—
sdR_s-u-UV- (3 mutants)^d		UV (from sm4)	18	Resistant to 20 µg/ml STREP	—
spir-u-1 through spir-u-5	spi1 through spi5	STREP	19	Resistant to 100 µg/ml SPI	17, 19
spr-u-1-27-3	spr-1,^{f} spr-1-27, sp-2-73, (refs. 10, 20, 5, 3)	NG; selected on SPEC	10	Resistant to 100 µg/ml SPEC only in the presence of acetate	4
spr-u-1-6-2	spr-1^t	NG; selected on SPEC	9, 10	Resistant to 100 µg/ml SPEC	4
spr-u-sp-23	spc1, sp-23 (Sager, pers. comm.)	STREP	19	Resistant to 50 µg/ml SPEC	4, 17, 19
spr-u-1-5-2-4	spr-1^t	NG; selected on SPEC	10	Resistant to 100 µg/ml SPEC	—
sr-u-2-60	sr-2-60	Spontaneous; selected on STREP	8, 9, 11	Resistant to 500 µg/ml STREP	4
sr-u-2-23^g	sr-u-2-NG-2-23	NG; selected on STREP	6	Resistant to 500 µg/ml STREP only in the presence of acetate	4
sr-u-sm2^g	sm2	STREP	14	Resistant to 500 µg/ml STREP	4, 17, 19
sr-u-sm3	sm3	STREP	16	Resistant to 50 µg/ml STREP	17
sr-u-sm5^h	sm5	STREP	18	Resistant to 500 µg/ml STREP	17
sr-u-sm-3a, sr-u-sm-3b, sr-u-sm-3c, sr-u-sm-3d	sm-3a, sm-3b, sm-3c, sm-3d	NG; selected on STREP	12	Resistant to 100 µg/ml STREP	—
sr-u-35	sr$_{35}$	STREP	2	Resistant to 100 µg/ml STREP	—
sr-u-2-1, sr-u-2-280, sr-u-2-281	sr-2-1, sr-2-280, sr-2-281	Spontaneous; selected on STREP	8, 11	Resistant to 500 µg/ml STREP	—
ti-u-1 through ti-u-5	ti1 through ti5	NG	18	Tiny colonies on all media	—
tm-u-1	tm1	STREP	19	Cannot grow at 35°C	17, 19
tm-u- (7 mutants)	tm	STREP	18	Cannot grow at 35°C	—

[a] ery1 = ery11 cited in Sager references (Sager, personal communication); [b] Recombines with er-u-1a; [c] Segregate like persistent hets (sd/low sr); [d] Segregate like persistent hets (sd/sr); [e] Segregate like persistent hets (sd/cond. sd); [f] Three independently isolated spr-1 mutants (10); [g] Recombines with sr-u-2-60; [h] Recombines with sr-u-sm2.

and in the nucleus. Electron micrographs also revealed the presence of 25Å DNase-sensitive fibrils, presumed to be chloroplast DNA molecules, within the Feulgen-positive areas of the chloroplast.

2.2 Size and composition

Following the cytological discovery of chloroplast DNA in *Chlamydomonas*, investigators have proceeded to establish the biochemical properties of plastid DNAs from a variety of higher plants and algae. By centrifugation of whole-cell preparations of plant DNA in cesium chloride density gradients, different classes of DNAs from a single organism can be distinguished on the basis of their buoyant density (Kirk 1971; Sager 1972; Borst & Kroon 1969). Studies with *C. reinhardtii* have shown a major DNA species (α) comprising about 85% of the total cell DNA, with a buoyant density of 1·721 g/cm³ (Chun *et al.* 1963; Leff, Mandel, Epstein & Schiff 1963; Chiang & Sueoka 1967a, b; Sager & Ishida 1963), which is assumed to be nuclear DNA (Fig. 5.2). There is also a minor satellite DNA (β) with a buoyant density of 1·695 g/cm³ and comprising about 14% of the total DNA of vegetative cells (Chiang & Sueoka 1967b). This satellite can be greatly enriched in the chloroplast fractions of whole-cell preparations (Sager & Ishida 1963) and is now recognized as chloroplast DNA.

A second satellite DNA (γ) in *C. reinhardtii* with a buoyant density of 1·712 g/cm³, although originally thought to be mitochondrial DNA (Chiang & Sueoka 1967b), has now been reported to hybridize with cytoplasmic ribosomal RNA (Bastia *et al.* 1971). Studies of isolated mitochondrial preparations from *C. reinhardtii* show a DNA species, δ, with a buoyant density of 1·706 g/cm³ (Ryan *et al.* 1973), a kinetic complexity of $1·6 \times 10^7$ daltons, and a circular contour length of 4·1–5·1 μm (Ryan *et al.* 1974). Such purified mitochondrial preparations contain less than 25% α and no detectable β or γ DNA. δ DNA comprises less than 1% of the total cell DNA.

Chiang and co-workers (though not Sager & Lane 1972) consistently report still another DNA species in *C. reinhardtii*, designated 'M-Band' DNA, which has a buoyant density equivalent to that of γ DNA (1·712 g/cm³), and which is formed in large amounts during zygote maturation (Chiang & Sueoka 1967b) but which does not hybridize with ribosomal RNA (Chiang, personal communication). The cellular location and function of M-band DNA remain unknown.

Many of the physical and biochemical properties of chloroplast DNA are now well defined (Table 5.2, p. 87). The buoyant density of 1·695 g/cm³ in CsCl reflects a G-C content of 35·7% as compared to 1·723 g/cm³, corresponding to a 64·3% G-C content, for nuclear DNA. Chloroplast DNA has been shown to be double-stranded by its increase in buoyant density and hyperchromicity upon heat denaturation (Wells & Sager 1971; Bastia *et al.* 1971) and by the quantitative equivalence of complementary bases (Sager & Ishida 1963). Analysis of whole-cell DNA by cesium chloride density-gradient centrifugation reveals that although β DNA may represent as much as 15% of the total DNA of vegetative cells, gametes contain only about half this amount (Chiang & Sueoka 1967b). The amount of chloroplast DNA measured on a per-cell basis may well depend not only upon the reproductive state of the cell, but also upon conditions of growth prior to DNA extraction, the particular strain being studied, and, in addition, the methods of DNA extraction and purification (Chiang & Sueoka 1967a).

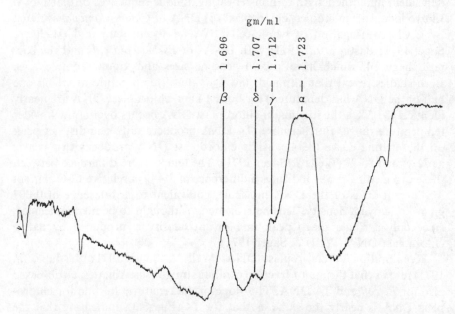

gm/ml

Fig. 5.2. Densitometer tracing of a CsCl equilibrium gradient containing DNA extracted from whole cells of *Chlamydomonas reinhardtii*. The gradient was purposely overloaded to display the minor DNA species, thereby sacrificing resolution of the major component. (R. W. Lee, unpublished). Chloroplast (β) = 1·696; Mitochondrial (δ) = 1·707; Cytoplasmic r-DNA (γ) = 1·712; Nuclear (α) = 1·723 g/cm³.

Attempts to determine the precise amount of chloroplast DNA in partially purified cell fractions have produced somewhat ambiguous results. Chloroplast preparations appear to contain at least 3·3% of the original whole-cell DNA (Sager & Ishida 1963), but upon further analysis no more than 40% of this is of the buoyant density characteristic of β DNA. A considerable amount of chloroplast DNA could be lost during the purification procedure, and the contaminating DNA of higher buoyant density may either be nuclear DNA or, alternatively, a second component of chloroplast DNA which coincidentally has a buoyant density equal to that of nuclear DNA. Assuming that chloroplast DNA represents a minimum of 6% of the total cell DNA, the weight of the chloroplast genome of a gamete should be approximately $8 \cdot 6 \times 10^{-15}$ grams, or 5×10^9 daltons (Wells & Sager 1971; Bastia *et al.* 1971).

2.3 *Kinetic complexity*

The most direct method for determining the degree of repetitiveness of a given DNA is by analysis of its renaturation kinetics (Wetmur & Davidson 1968; Britten & Kohne 1968), in which the observed rate of renaturation is dependent upon the length of non-repeating, unique sequences of that DNA. In two independent studies, chloroplast (β) DNA of *Chlamydomonas* isolated by CsCl centrifugation of whole-cell DNA has been compared (Wells & Sager 1971; Bastia *et al.* 1971) with DNAs of *Escherichia coli* and the bacteriophage T4, which have known genome sizes and kinetic complexities. Both studies reveal that although the T_m values (temperature at which one half of the DNA has denatured) for both T4 and chloroplast DNA are nearly identical ($83 \cdot 5°C$), the melting of chloroplast DNA occurs over a much wider temperature range. Furthermore, T4 DNA produces only one distinct peak on the melting curve ($83 \cdot 4°$), while chloroplast DNA produces two peaks, at $82 \cdot 1°$ and $85 \cdot 2°$ (Wells & Sager 1971). The temperature difference between these two melting peaks indicates a difference of $7 \cdot 4\%$ in relative G-C content in different areas of the DNA molecule, equivalent to a difference of $0 \cdot 007$ g/cm^3 in buoyant density. This heterogeneity is thought to be intramolecular, since only a single sharp peak in buoyant density is produced by native chloroplast DNA (Wells & Sager 1971).

Renaturation of chloroplast DNA (Wells & Sager 1971; Bastia *et al.* 1971) reveals that the major fraction exhibits straight-line kinetics, as observed also for *E. coli* and T4 DNA. The slope of the renaturation line for chloroplast DNA is nearly the same as that for T4 (Fig. 5.3), indicating that the unique-sequence sizes of T4 virus and of *C. reinhardtii* chloroplast DNA are approximately the same, about 2×10^8 daltons. That is, each unique sequence has the potential to code for about 600 proteins of 20,000 molecular weight. One of the studies (Wells & Sager 1971), however, reports that the initial period of non-linearity (a characteristic of renaturation studies) is more prolonged in the case of chloroplast DNA and may represent the existence of a very fast-renaturing fraction within the chloroplast DNA sample. Since DNA of high G-C content tends to renature more rapidly than predicted (Wetmur & Davidson 1968), the fast-renaturing fraction could represent intramolecular areas of high G-C content rather than long sequences that are highly repetitive. The fast-renaturing fraction is estimated to comprise as much as 10% of the chloroplast DNA and to have a 2 to 20 times higher level of redundancy than the major fraction of chloroplast DNA. Since the actual amount of chloroplast DNA in a gamete has been estimated at 5×10^9 daltons (Bastia *et al.* 1971), the unique sequence of 2×10^8 daltons, the major fraction of DNA, must be present in 20 to 30 copies per gamete chloroplast, and in 40 to 60 copies in the chloroplast of a vegetative cell. However, one does not know whether this genome is represented by one continuous molecule of 2,000 μm—twice the length of the *E. coli* genome—or by many smaller DNA molecules. Chloroplast DNA of *Euglena gracilis*, which has a

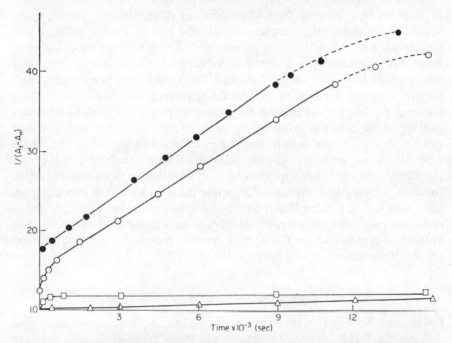

Fig. 5.3. Comparison of renaturation kinetics of chloroplast DNA and nuclear DNA from *Chlamydomonas reinhardtii* with DNA from *Escherichia coli* and from a bacteriophage, T4. Reciprocal k_2 plot of *E. coli* DNA, 15 μg/ml (—△—△—); *Chlamydomonas* nuclear DNA, 15 μg/ml (—□—□—); T4 DNA, 10 μg/ml (—●—●—); and chloroplast DNA, 12·5 μg/ml (—○—○—). The ordinate represents the OD_{260} relative to native DNA corrected for water expansion. (Wells & Sager 1971).

kinetic complexity of $1·8 \times 10^8$ daltons (Stutz 1970), occurs as 40-μm circumference circles (Manning *et al.* 1971). In order to reconcile the higher kinetic complexity of *Euglena* chloroplast DNA, which would yield 87-μm circles with the smaller physical size observed, Manning *et al.* (1971) point out that the low G-C content of this DNA leads to an over-estimate of its complexity from renaturation kinetics (Wetmur & Davidson 1968). Chloroplast DNA of *C. reinhardtii* also has a low G-C content which could likewise lead to an overestimate of its kinetic complexity and thus to an underestimate of the number of copies of the chloroplast genome per cell. If we assume the chloroplast DNA of *C. reinhardtii* to be arranged in 40-μm circles, the single chloroplast of each gamete would contain 50 of these circles, compared to the 20 to 30 copies estimated from the kinetic complexity. A comparable redundancy of 40–50 copies has been found for the mitochondrial DNA of a yeast cell (Grimes *et al.* 1974) in which one observes 26-μm circles (Borst 1972).

A contradiction arises when one attempts to correlate results of these physical studies, showing the *Chlamydomonas* chloroplast to contain many copies of the same unique sequences, with the genetic studies of Sager and Ramanis (1968), who claim a 'copy number' of only two for the chloroplast genome of vegetative cells. To explain this contradiction, Sager has suggested that perhaps only two large chloroplast DNA molecules are present, each carrying tandem repeats of nucleotide sequences, or, alternatively, that although the majority of chloroplast genes may be present in 40 to 60 copies per vegetative cell, the cytoplasmic genes so far discovered and analysed genetically are present in only two copies (Wells & Sager 1971). The latter explanation may be based upon the assumption that, as in higher organisms, the highly repeated DNA sequences have an unorthodox, non-protein-coding function (Davidson & Britten 1973), while those genes which code for proteins (and are thus recognizable in mutant form) are more often present in one, or in only a few copies, per haploid genome. On the basis of genetic data Gillham, Boynton and Lee (1974) have recently proposed a multicopy model which will be examined in a later section (p. 112).

2.4   *Replication during the vegetative cycle*

Chiang and Sueoka (1967b) have investigated the replication of chloroplast and nuclear DNAs during the mitotic cell cycle in synchronously dividing cultures of vegetative cells of *C. reinhardtii*. Under such conditions, each cell divides into four daughter cells near the end of the 12-hour dark period. By determining the total amount of DNA per cell at two-hour intervals throughout the cell cycle, they have shown the course of DNA synthesis during vegetative growth to be biphasic, suggesting that chloroplast DNA and nuclear DNA might replicate at different times. The precise time and mode of chloroplast DNA replication have been determined by a modification of the original Meselson and Stahl (1958) experiment which demonstrated the semiconservative replication of *E. coli* DNA. *Chlamydomonas* cells were grown synchronously, in medium containing ^{15}N-labelled ammonium chloride as the sole nitrogen source, until all cellular DNAs became labelled and the culture attained the desired cell density. Cells were then transferred to normal ^{14}N-containing medium and allowed to complete one division cycle, during which samples were removed periodically. DNA was then extracted, and the density of the nuclear and chloroplast DNA species determined by centrifugation in CsCl gradients (Fig. 5.4). In such experiments replication of a given DNA is recognized by a shift in buoyant density from the higher value produced by the ^{15}N label to intermediate values as newly synthesized ^{14}N strands are added.

The first replication of chloroplast DNA occurs at about 4·5 hours into the light period (Chiang & Sueoka 1967a, b; Chiang 1971) and leads to the production of hybrid ^{15}N–^{14}N molecules with the predicted intermediate buoyant density. Approximately two hours later, a second round of chloroplast DNA replication results in the production of pure light β DNA with the characteristic buoyant density of 1·695 g/cm^3. The relative amounts of 'light' and 'hybrid' chloroplast DNAs are approximately equal, and no change in their buoyant densities or amounts occurs during the remainder of the cell cycle, indicating that no further replication of chloroplast DNA occurs.

The nuclear DNA of *C. reinhardtii*, in contrast, shows no shift in density during the 12-hour light period and the first six hours of the dark period (Chiang & Sueoka 1967a, b). During the last half of the dark period, just prior to cell division, two rounds of nuclear DNA replication occur. Thus both chloroplast and nuclear DNA of *C. reinhardtii* appear to replicate in a semiconservative manner, these events being temporally separated by about 12 hours. This separation could be explained by the utilization of a DNA polymerase specific for replication of each DNA species, or, alternatively, may merely reflect limiting amounts of nucleotide precursors in the chloroplast and the nucleus at different points in the cell cycle. The conclusion that β DNA replication is semi-conservative depends on (1) the clear-cut production first of hybrid, then of fully light DNA, and (2) the buoyant density of a hybrid DNA species being precisely intermediate between that of pure heavy and pure light DNAs. Isotopic transfer studies on synchronous cultures of *Chlamydomonas* by Lee and Jones (1973) have demonstrated only approximately hybrid and light molecules of β DNA, and a gradual shift towards lighter density. This may be explained by either the presence of a pool of ^{15}N precursor molecules remaining after transfer, or by extensive recombination occurring between ^{15}N and ^{14}N strands. The two studies concur in the clear demonstration of the temporal separation of nuclear and chloroplast DNA replication.

2.5 *Replication during the sexual cycle*

Development of methods for studying the replication of chloroplast DNA during the vegetative growth of *C. reinhardtii* has led directly to attempts to follow the replication and fate of β DNA during the sexual, meiotic cycle. Since non-Mendelian, uniparental genes donated by the *mt*$^-$ parent to the zygote are not transmitted to the meiotic progeny (Gillham 1969; Sager 1972), one might predict the physical disappearance of *mt*$^-$ chloroplast DNA from the zygote.

The physical fate of *mt*$^-$ chloroplast DNA in the zygote of *C. reinhardtii* has been examined by both Sager and Lane (1972) and Chiang (1968, 1971), who have employed somewhat similar methods but have drawn dramatically different conclusions. To distinguish *mt*⁻ and *mt*⁺ chloroplast DNA in the zygote, gametes from each parent must be differentially marked using density and/or radioactive labels. Initially the times of β DNA and nuclear DNA

Fig. 5.4. Demonstration of the time and nature of nuclear (α) and chloroplast (β) DNA replication during vegetative cell division in synchronous cultures of *Chlamydomonas reinhardtii* after transfer from ¹⁵N to ¹⁴N medium. (Chiang & Sueoka 1967a.)

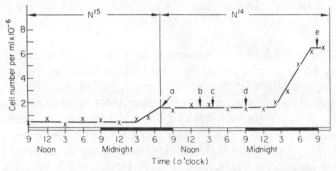

Fig. 5.4.1. The density transfer experiment: A suspension containing *mt*⁺ cells at 10⁴/ml of ¹⁵N 3/10 HSM* was subjected to a light-dark cycle to maintain synchronous cell division. As soon as the concentration of cells reached 1·64 × 10⁶/ml, they were harvested, suspended in a preconditioned and adenine-supplemented ¹⁴N 3/10 HSM medium, and quickly returned to the light-dark cycle. Samples were taken at different times (indicated by a, b, c, d, and e) and the DNA was extracted.

replication during the sexual cycle were determined by prelabelling gametes of both mating types with ¹⁵N and transferring the young zygotes to ¹⁴N medium for maturation (Chiang & Sueoka 1967b). β DNA was shown to replicate soon after the fusion of gametes, whereas nuclear DNA did not replicate until just prior to germination of the zygote (Chiang & Sueoka 1967a, b). Thus, as is true during vegetative growth, plastid and nuclear DNA syntheses are temporally separate throughout the sexual cycle (Fig. 5.5).

2.6   Transmission in crosses

To follow the behaviour of chloroplast DNAs from the two parents in the zygote, Chiang (1968, 1971) labelled gametes of opposite mating types with ³H adenine or ¹⁴C adenine. DNA was extracted and analysed from newly fused gamete pairs, from young zygotes after 30 hours of light maturation (prior to incubation in the dark), and from the zoospores resulting from germination of the zygotes. From the relative amounts of the two radioactive labels observed, Chiang concluded that both parental chloroplast DNAs

* HSM=high-salt medium.

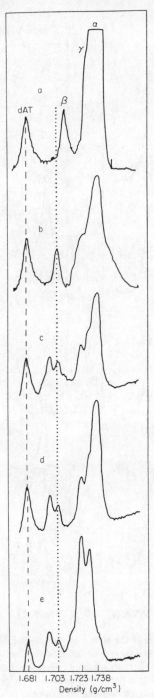

Fig. 5.4.2. Analyses by CsCl density gradient centrifugation showing shifts of the buoyant densities of α and β DNA, indicating their respective times of replication.

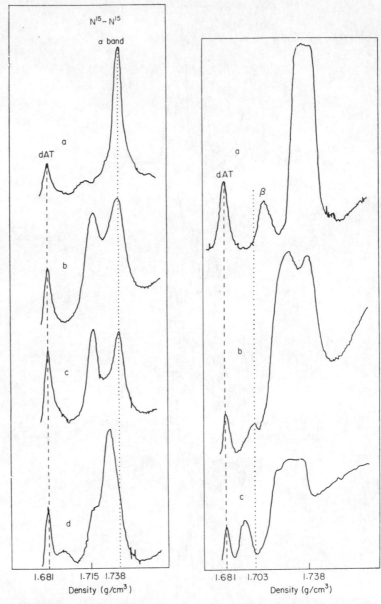

Fig. 5.5.1. Nuclear DNA (α)

Fig. 5.5.2. Chloroplast DNA (β)

Fig. 5.5. Demonstration of the time and nature of DNA replication during the sexual cycle of *Chlamydomonas reinhardtii*. The technique used was similar to that described in Fig. 5.4.1. (a) Freshly mated ^{15}N-labelled gamete pairs; (b) developing zygotes at termination of light period of maturation in ^{14}N medium; (c) zygospores at end of maturation; (d) germinating zygospores prior to mitosis. (Chiang & Sueoka 1967b.)

were conserved in the zygote, from mating through germination. However, the presence of both labels fails to rule out the possibility that the labelled mt^- nucleic acids were degraded and the label reincorporated into newly synthesized DNA. Alternatively, extensive recombination between mt^+ and mt^- chloroplast genomes could lead to the preservation of both labels without the preservation of either of the intact plastid genomes (Chiang 1971). The latter alternative, however, necessitates an explanation for the failure of expression of the mt^- genes in the progeny of the zygote.

The possibility of separate intracellular pools of labelled nucleic acid precursors in the chloroplast has not been ruled out. Chiang (1968) tried to preclude this possibility in the original experiments by 'chasing' labelled cells with cold adenine and analysing the labelling of M-band DNA, which replicates extensively during zygote maturation. Since M-band DNA was found to contain no radioactivity, Chiang concluded that the chase had been effective. However, Siersma and Chiang (1971) later demonstrated that both cytoplasmic and chloroplast ribosomes are degraded during gametogenesis and that labelled nucleotides from degraded ribosomal RNA are used for subsequent gametogenic DNA synthesis. Further studies by Hinckley, Chiang and Goodenough (1974) demonstrated that synthesis of rRNA and ribosomal proteins, as well as the assembly of new ribosomes in both chloroplast and cytoplasm, occurred concomitantly with ribosomal degradation throughout the period of gamete differentiation. Chloroplast ribosomes turned over more rapidly than cytoplasmic ribosomes. Clearly considerable transfer of label between various cellular components can and does occur during gametogenesis. Thus, label incorporated into ribosomal RNA by cells of both mating types could be transferred to newly synthesized DNA in the zygote. If this were to occur following chloroplast fusion in the zygote, radioactivity originally present in the ribosomes of both the mt^+ and mt^- parents could appear in newly synthesized β-DNA and be indistinguishable from radioactivity originally associated specifically with mt^- chloroplast DNA.

The absence of label in newly synthesized M-band DNA, as reported in the original study by Chiang (Chiang & Sueoka 1967b), is especially puzzling in light of the more recent report of extensive nucleotide precursor pools within the gametic cytoplasm and chloroplast (Siersma & Chiang 1971). Since chloroplast DNA replicates soon after the fusion of gametes (Sager & Lane 1972), chloroplast DNA may be more subject to such pool problems than other DNA species which replicate later during maturation. Chiang showed that both chloroplast and M-band DNA replicate within 24 hours after mating (Chiang & Sueoka 1967b) and that ribosomal deficiencies are remedied within 48 hours (Siersma & Chiang 1971). However, the precise times of any of these events were not determined, and the possibility of a temporal separation has not been ruled out.

Alternatively, M-band DNA may be physically isolated from the precursor pool utilized for chloroplast and/or nuclear DNA synthesis, although it is admittedly difficult to imagine what cellular location might be inaccessible to degradation products of both cytoplasmic and chloroplast ribosomes. Before one can accurately estimate the effect of precursor pools on the experiments aimed at determining the fate of mt^- β DNA during the sexual cycle, it will be necessary to determine more precisely the time of initial synthesis of each DNA species after mating, the times of resynthesis of cytoplasmic and chloroplast ribosomes, and the mobility within the cell of precursors derived from chloroplast or cytoplasmic ribosomes.

Chiang (1971) also combined density labelling with radioactive labelling to investigate the extent of β DNA recombination during zygote maturation. mt^+ gametes carried $^{14}N^3H$ labels, while the mt^- parent carried $^{15}N^{14}C$ labels. DNA extracted at the gamete-pair and young zygote stages and analysed by CsCl centrifugation showed the radioactive labels still associated with the original density labels. In the zoospores produced, however, both radioactive labels appeared in the same area of the gradient, indicating considerable scrambling of density and radioactive labels. Chiang interpreted this to be a consequence of recombination between both parental nuclear genomes and plastid genomes, and suggested that it further supported the hypothesis that there is no selective destruction of chloroplast DNA from either parent. These studies are, of course, subject to the same precursor pool problems as have been discussed above.

Sager has challenged Chiang's conclusions regarding the fate of chloroplast DNA during the sexual cycle (Sager 1972; Sager & Lane 1972). In these experiments optical density profiles were followed directly using density labels and radioactive isotopic tags were unnecessary since M-band synthesis did not occur. If the cross mt^+ (^{14}N) × mt^- (^{15}N) was made, the chloroplast DNA of the young zygotes was light, but if the reciprocal cross was made, it was heavy (Fig. 5.6c, d). These results were interpreted as meaning that chloroplast DNA from the mt^+ but not the mt^- parent was conserved in the zygote and that chloroplast DNA from the mt^- parent was destroyed. In all crosses, including $^{14}N × ^{14}N$ controls (Fig. 5.6a), the chloroplast DNA showed a slightly lower buoyant density than expected. Sager and Lane suggested that this shift could be accounted for by methylation.

Using radioactive labels, Sager (1972) initially supported Chiang's earlier conclusion that β DNA from both parents is present in the zygote. However, the presence of both parental labels in this case was interpreted to be a consequence of metabolic turnover and reutilization of DNA precursors during repair synthesis and/or repeated replication.

More recently Schlanger and Sager (1974) reported the exclusive recovery of mt^+ chloroplast DNA from zygotes derived from ^3H- and ^{14}C- adenine-labelled parental gametes. Furthermore, UV treatment of the mt^+ gametes prior to mating, which permits transmission of both maternal and paternal chloroplast genes (see p. 88), reportedly results in recovery of the paternal chloroplast DNA from the zygote, but not the maternal chloroplast DNA.

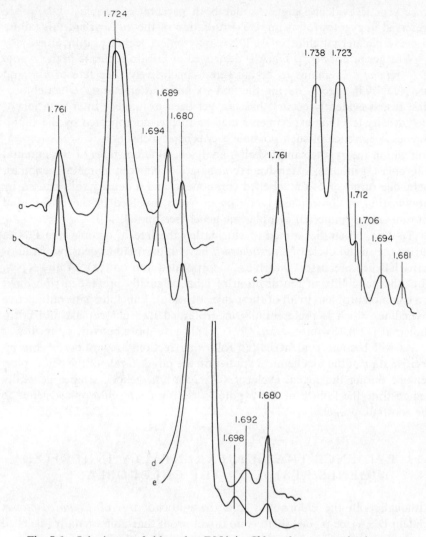

Fig. 5.6. Inheritance of chloroplast DNA in *Chlamydomonas reinhardtii* crosses, as indicated by cesium chloride density-gradient centrifugation analyses. (a) DNA from 24-hr zygotes from cross of ^{14}N $\times$ ^{14}N (unlabelled) gametes, showing shift of chloroplast DNA to a lighter density (1·689 g/cm^3) than is normally found in vegetative cells and gametes. (b) DNA from one of the ^{14}N gamete types used as parents in the cross shown in *a*. The chloroplast DNA has a density of 1·694 g/cm^3. (c) DNA from a 1:1 mixture of ^{14}N and ^{15}N gametes, showing resolution of both light and heavy nuclear DNAs (1·723 and 1·738 g/cm^3) and chloroplast DNAs (1·694 and 1·706 g/cm^3). (d) DNA from 6-hr zygotes from a cross of ^{14}N $\times$ ^{15}N gametes, showing a single chloroplast DNA peak at 1·692 g/cm^3. (e) DNA from 6-hr zygotes from a cross of ^{15}N $\times$ ^{14}N gametes, showing a single chloroplast peak at 1·698 g/cm^3. DNAs used as density markers: SP-15 DNA at 1·761 g/cm^3 and poly (dAT) at 1·680 or 1·681 g/cm^3. (Sager & Lane 1972.)

E

Since genetic evidence suggests that both parental chloroplast DNAs are preserved in zygotes following UV irradiation of the mt^+ parent, this failure to recover maternal chloroplast DNA may reflect technical difficulties.

Whether or not mt^- β DNA is destroyed or fails to replicate in the zygote of *C. reinhardtii* remains to be answered conclusively. The fate of mt^+ and mt^- β DNA in zygotes during the first six hours after mating, when chloroplast fusion evidently occurs, has not yet been examined. Interpretation of such density-labelling experiments may be further complicated by the simultaneous occurrence of such possible events as asynchronous β DNA replication within the zygote sample being analysed, different fates of the parental chloroplast genomes, extensive recombination between parental genomes, metabolic turnover of the labelled component, and a density shift caused by modification of chloroplast DNA by enzymes involved in the preferential transmission of maternal cytoplasmic genes (see pp. 84, 110).

To date, all studies aimed at elucidating the fate of cytoplasmic DNAs during the sexual cycle of *C. reinhardtii* have utilized either density or radioactive labelled precursors which are incorporated into both DNA and RNA. Thus the possibility of reutilization of label originally present in ribosomes is a common problem to all of these investigations. Labelling with radioactive thymidine, which is preferentially incorporated into chloroplast and mitochondrial DNA (Swinton *et al.* 1974), could give more convincing results.

As will become evident in the following sections, regardless of the indecisive state of the biochemical studies on the physical fate of the chloroplast genome during the sexual cycle of *C. reinhardtii*, genetic studies generally demonstrate the failure of the zygote to transmit mt^- chloroplast genes to the zoospore progeny.

3   EVIDENCE THAT UNIPARENTALLY INHERITED GENES RESIDE IN THE CHLOROPLAST

Although both the chloroplast and the mitochondria of *Chlamydomonas* contain DNA (see p. 74), there is no direct proof that uniparentally inherited genes are localized in either genome. However, several pieces of circumstantial evidence make the chloroplast the most likely site.

First, most uniparentally inherited mutations (Table 5.1) either require acetate, and are thought to be impaired in photosynthesis (Sager 1972), or confer antiobiotic resistance on chloroplast ribosomes (see Ch. 6). Second, the mutagen N-methyl-N'-nitro-N-nitrosoguanidine (NG), which is thought to act at the replication fork of DNA (Cerdá-Olmedo *et al.* 1968), increases the frequency of Mendelian (*sr-1*) and uniparentally inherited (*sr-2*) mutations to streptomycin resistance at different times in the cell division cycle (Lee & Jones 1973). During the period of nuclear DNA replication in synchronous cultures NG increases the frequency of *sr-1* mutants 15–30 fold, while that of *sr-2* mutants is unaffected. When NG is added during the time of chloroplast DNA replication, however, both *sr-1* and *sr-2* mutations show a small but significant (two-fold) increase. The increase in *sr-1* mutants can be attributed to mutation in the small percentage of cells which are out of synchrony and are replicating their nuclear DNA at a time when the rest of the cells are

Table 5.2. DNA species of *Chlamydomonas reinhardtii*. From Chiang 1971, modified according to Ryan *et al.* 1973, and others.

Growth stage	μg DNA/ cell × 10⁷	Com- ponent	Origin	Buoyant density in CsCl			% GCᵈ	Relative amount of total DNAᵉ
				Native	Dena- turedᵃ	Rena- turedᵇ		
Vegetative cells	1·24 ± 0·081ᶜ	α	Nuclear	1·723	1·738	1·738	64·3	85
		β	Chloroplast	1·695	1·710	1·695	35·7	14
		γ	Nuclear	1·715	1·730	1·715	56·1	1
		δ	Mito- chondrial	1·706	1·720	—	46·9	<1
Gametes	1·23 ± 0·064ᶜ	α	Nuclear	1·723	1·738	1·738	64·3	89
		β	Chloroplast	1·695	1·710	1·695	35·7	7
		γ	Nuclear	1·715	1·730	1·715	56·1	4

ᵃ Heating at 100° for 10 min in 0·015 M NaCl, plus 0·015 M Na citrate pH 7, then quickly cooled.

ᵇ Thermal denatured DNA in 0·3 M NaCl, 0·03 M Na citrate pH 7 was heated to 70° for 5 hr, then slowly cooled.

ᶜ Standard deviation.

ᵈ Composition is obtained from the density, assuming no unusual bases; density (ρ) = $1·660 + 0·098 \times (G+C$ mole percent).

ᵉ Estimated by integrating area under each DNA peak in the microdensitometer tracings of UV-absorption photographs.

replicating their chloroplast DNA (Lee & Jones 1973; Lee, personal communication). The small increase in *sr-2* mutants probably reflects the presence of many copies of the chloroplast genome, so that few of the induced mutations are expressed at the time the cells are challenged with streptomycin, or the reduced susceptibility of the chloroplast DNA to NG. Third, Sager and Lane (1972) found a correlation between the loss of the expression of uniparental genes from the *mt⁻* parent and the disappearance of the *mt⁻* DNA during zygote maturation, although this has not yet been firmly established (see p. 84). Fourth, Alexander *et al.* (1974) demonstrated that acridine induces, with nearly 100% efficiency, 'minute' colony mutants in *Chlamydomonas*, apparently analogous to the mitochondrial 'petite' mutations that this dye induces in yeast (cf. Mahler 1973). The 'minute' mutations of *Chlamydomonas* are inherited in a non-Mendelian but biparental fashion distinct from the uniparentally inherited chloroplast mutations, and they are therefore thought to reside in mitochondrial DNA. Fifth, streptomycin, a drug known to affect chloroplast structure and function, was reported to be an effective mutagen for uniparental genes (Sager 1962, 1972). Although individually none of these pieces of evidence is compelling, collectively they support the hypothesis that uniparentally inherited genes are located on the chloroplast DNA, and we will provisionally refer to them hereafter as chloroplast genes.

4 METHODS FOR ALTERING THE INHERITANCE PATTERN OF CHLOROPLAST GENES

Analysis of segregation and recombination of chloroplast genes requires transmission from both the mt^+ (*maternal*) and the mt^- (*paternal*) parent to the zygote and its progeny. In 90% or more of all zygotes (*maternal zygotes*) only the mt^+ parent transmits its chloroplast genes to all four meiotic products. Chloroplast genes from the mt^- parent are transmitted spontaneously in less than 10% of the zygotes (*exceptional zygotes*) (Gillham 1969; Sager 1972). Two kinds of such exceptional zygotes occur. *Biparental zygotes* transmit chloroplast genes from both parents to the progeny. In the progeny of such zygotes chloroplast genes, unlike Mendelian genes, continue to segregate not only in the initial meiotic divisions but also in postmeiotic, mitotic divisions (Gillham 1969; Sager 1972). *Paternal zygotes*, the rarest class of all, transmit only chloroplast genes carried by the mt^- parent. Some recombination analysis has been done with rare spontaneous biparental zygotes (Gillham 1965; Gillham & Fifer 1968). Sager and Ramanis (1967) showed that the frequency of biparental zygotes could be increased manyfold by treating the mt^+ gametes with UV prior to mating them (Fig. 5.7). The proportion of each type of zygote depends on the UV dosage used and may be further influenced by photoreactivation of the mated pairs. At low UV dosages biparental zygotes are more numerous than paternal zygotes, whereas increasing the dosage raises the frequency of paternal zygotes. Furthermore, potentially paternal zygotes are converted to biparental zygotes by photoreactivation of the mated pairs more readily than biparental zygotes are converted to maternal zygotes.

The effect of UV is most pronounced when the mt^+ gametes are irradiated immediately before mating (Sager & Ramanis 1967). If the gametes are irradiated and then held, in light or darkness, prior to mating, the production of exceptional zygotes is reduced. A similar reduction in the frequency of exceptional zygotes results from treating the cells with UV early in gametogenesis, before they are fully differentiated as gametes. It is not clear whether the cells are less sensitive to UV at this stage or whether recovery occurs during the time needed to complete gametogenesis. UV irradiation of the mt^- gametes does not increase the proportion of exceptional zygotes; irradiating the zygotes themselves is reported to be highly lethal (Sager & Ramanis 1967). By adjusting the UV dosage to the *mt*[+] parent, and by photoreactivating the mated pairs, a high proportion of biparental zygotes can be routinely produced for genetic analysis.

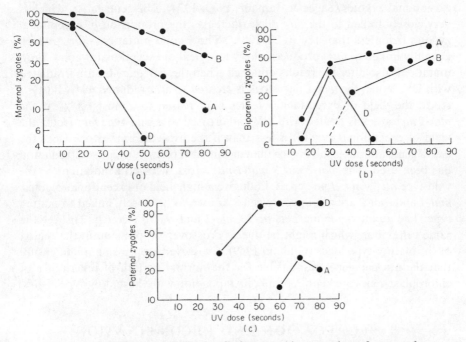

Fig. 5.7. Effectiveness of UV irradiation on: (a) conversion of maternal to "exceptional" zygotes; (b) yield of biparental zygotes; (c) yield of paternal zygotes. Female (*mt*[+]) gametes were irradiated, mated with unirradiated males (*mt*[-]), and kept in darkness until zygote formation was completed. After 2 hours, the zygotes were diluted, plated, and incubated in the dark (D), or in the light (A) for photoreactivation. In addition, mating gametes (B) were exposed to light during and after zygote formation. After 24 hours in the light, the A and B series were transferred to the dark and incubated with the D series for one week. All plates were then exposed to light to induce germination of the zygotes. (Sager & Ramanis 1967.)

Inhibitors of protein synthesis and of other metabolic functions have also been examined for their effect on increasing the frequency of occurrence of exceptional zygotes (Sager & Ramanis 1973). Ethidium bromide was found to be the most potent, although other agents, such as erythromycin, DL-ethionine, rifamycin SV, spiramycin and cycloheximide, all increase the proportion of exceptional zygotes recovered when applied to fully differentiated gametes. Some of these inhibitors are at least as effective when applied to the *mt*[-] gametes, in contrast to UV, which does not act on *mt*[-] gametes in this way. Unlike UV, many inhibitors are also effective when applied during gametogenesis. Cycloheximide, which inhibits cytoplasmic protein synthesis, was found to be particularly effective when applied to mt^- cells during gametogenesis, although zygote viability was greatly reduced.

Sager has recently described two mutations which affect the frequency of exceptional zygotes (Sager & Ramanis 1973, 1974). The gene *mat-1*, which is very closely linked to the mt^- allele, increases the proportion of exceptional zygotes from less than 1% to 20–100%. The second mutation, *mat-2*, which is tightly linked to mt^+, produces a very low yield of exceptional zygotes. Both mutations also alter the results obtained when the mt^+ gametes are irradiated with UV. When the *mat-1* mt^- stock is crossed to an irradiated mt^+ wild-type stock, the yield of exceptional zygotes is increased to almost 100%. If the *mat-2* mt^+ stock is treated with UV and crossed to a wild-type mt^- stock, the yield of exceptional zygotes is less than if the irradiated mt^+ stock lacks the *mat-2* gene, although still higher than in similar crosses in which no irradiation has been used. The *mat-2* mt^+ × *mat-1* mt^- cross, with no irradiation, or the wild-type mt^+ × *mat-1* mt^- cross, both give a high yield of exceptional zygotes. *mat-1* and *mat-2* are not the only genes known to be closely linked to mating type. The auxotrophic markers *ac-29*, *nic-7* and *thi-10* (see Ch. 2) show the same behaviour, which might be due to crossover suppression in the region of the mating-type locus (Gillham 1969). Crossover suppression might ensure that the nuclear genes responsible for the normal pattern of inheritance of chloroplast genes are kept in proper juxtaposition with the mating-type alleles.

5   SEGREGATION AND RECOMBINATION OF CHLOROPLAST GENES

Somatic segregation of non-Mendelian genes has been demonstrated in other organisms (cf. Jinks 1964; Mahler 1973) and in *Chlamydomonas*, where chloroplast genes segregate both in the meiotic and in postmeiotic mitotic divisions (Gillham 1969; Sager 1972).

Two principal methods have been used to study segregation and recombination. Sager has used pedigree analysis of individual biparental zygote progeny (Fig. 5.8; Sager & Ramanis 1970). After one postmeiotic mitotic division the eight daughter cells are replated and allowed to divide once more; the 16 resultant daughter cells are then separated; and each cell, after it has given rise to a colony, is then tested for genotype. By using three pairs of unlinked Mendelian genes the progeny can usually be scored according to the meiotic division from which they were derived.

Gillham has used zygote clone analysis (Fig. 5.9; Conde *et al.* 1975; Gillham 1965; Gillham & Fifer 1968). Zygotes are allowed to germinate on non-selective medium and form colonies which are then replica-plated on selective media to distinguish the biparental zygotes. Cells from the biparental

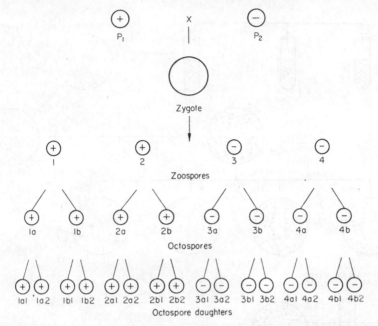

Fig. 5.8. Procedure for pedigree analysis. After zygospore germination, the four zoospores are allowed to undergo one mitotic doubling, and the eight cells (octospores) are then transferred to a fresh Petri plate and respread. After one further doubling, each pair of octospore daughters is separated and each cell is allowed to form a colony. The sixteen colonies, derived from the first two doublings of each zoospore, are then classified for all segregating markers. (Sager & Ramanis 1970.)

zygote colonies formed on the non-selective plates are replated on non-selective medium and later replica-plated in order to determine the frequency of each of the alleles in the zygote clone. Since biparental zygotes contain chloroplast genes from both parents, some of the homozygous progeny will carry the maternal allele for a marker, while others will carry the paternal allele. The allelic ratio of a biparental zygote is defined as the fraction of progeny clones in the total progeny bearing a given allele of a chloroplast gene (Gillham *et al.* 1974).

Zygotes heterozygous for a given chloroplast gene can exhibit three segregation patterns (Fig. 5.10a–c; Sager 1972). Type-I segregation, the commonest, produces two daughter cells, each of which is still heterozygous for the pair of alleles (e.g. a/a^+). Type-II segregation yields one daughter cell still heterozygous (e.g. a/a^+) and the other homozygous for one of the alleles (e.g. a/a or a^+/a^+). Type-III segregation, the rarest, results in the production of two homozygous cells, one pure for the allele donated by the mt^+ parent (e.g. a/a), the other pure for the allele donated by the mt^- parent (e.g. a^+/a^+).

As mentioned, type-II segregations give one homozygous daughter cell

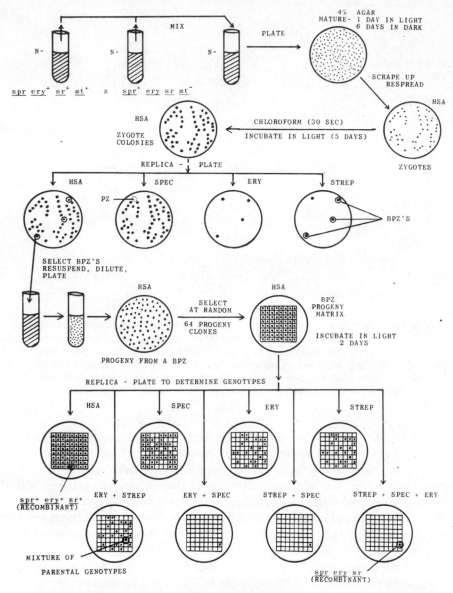

Fig. 5.9. Diagram showing methods used for zygote clone analysis. N⁻, nitrogen-free medium for gametogenesis; HSA, high-salt acetate medium; SPEC, HSA + 100 μg/ml spectinomycin base; ERY, HSA + 200 μg/ml erythromycin base; STREP, HSA + 100 μg/ml streptomycin base; PZ, paternal zygote; BPZ, bi-parental zygote. Multiple antibiotic plates contain individual antibiotics, specified in concentrations given above, in HSA. (Conde *et al.* 1975.)

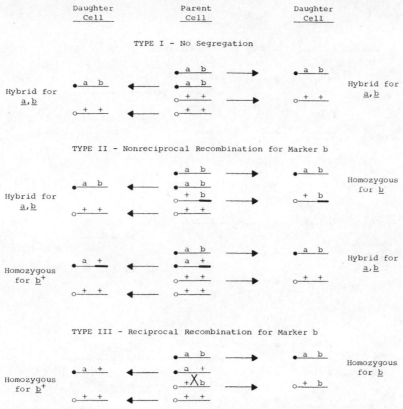

Fig. 5.10a. Chloroplast gene segregation patterns in Sager's two-copy model. Recombination is diagrammed as taking place after the two copies have replicated. For the sake of clarity the chloroplast genomes are shown as linear, although they may really be circular. (Gillham *et al.* 1974.)

and one heterozygous. The mechanism by which this is achieved is in dispute (cf. Sager 1972; Gillham *et al.* 1974). Since, by definition, type-III segregations give rise to daughter cells each homozygous for a given allele of the pair, and type-I segregations only to daughter cells heterozygous for a given pair of alleles, the relative frequency of type-II segregations for each member of a pair of alleles *must* determine the allelic ratio among the progeny of a biparental zygote. Sager (Sager & Ramanis 1968; Sager 1972) has consistently reported a ratio of 1:1 for cells homozygous for a given pair of alleles; i.e. among type-II segregations both alleles appear as homozygotes with equal frequency. This contrasts sharply with results reported by Gillham *et al.* (1974) in which the ratios of alleles in individual biparental zygotes were found to deviate markedly from 1:1, the deviation nearly always favouring the chloroplast allele carried by the maternal (*mt*⁺) parent. This implies that type-II segregations for a given pair of alleles are not equal in frequency but

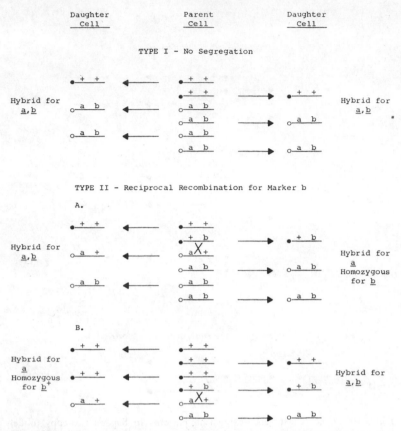

Fig. 5.10b. Origin of type-I and -II segregants from hybrid cells containing four copies of one genome and two of the other. (Gillham *et al.* 1974.)

favour homozygosity for the chloroplast allele carried by the maternal parent. The discrepancy is of considerable importance in comparisons between theoretical models for uniparental inheritance (see pp. 109 ff). The model proposed by Sager (Sager & Ramanis 1968; Sager 1972) indicates that there are only two copies of the chloroplast genome per vegetative cell, whereas that of Gillham *et al.* (1974) suggests a larger and variable copy number. According to either model, type-I segregation is the result of oriented distribution of the genomes in the absence of any crossing over, and type-III segregation is due to reciprocal recombination prior to the segregation in cells having only two copies. The mechanism responsible for type-II segregation, however, is explained differently by the two hypotheses. Sager's model requires non-reciprocal recombination between the two copies, in a process akin to gene conversion, while Gillham *et al.* propose reciprocal recombination between two of the several copies. As seen in Fig. 5.10a, the two-copy model proposed

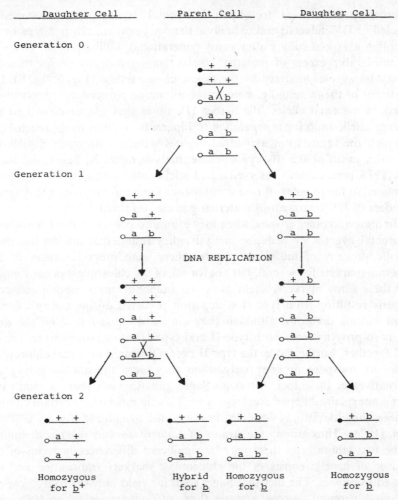

Fig. 5.10c. Pedigree showing skewing of marker ratio by type-II segregation among the progeny of a hybrid cell which originally contained four copies of one genome and two of the other. Notice that type-II segregation for marker *b*, resulting from reciprocal recombination at the first division, changes the *b⁺*:*b* ratio from 1:2 in generation 0 to 2:1 in the resulting hybrid cells of generation 1. By generation 2 a cell homozygous for *b⁺* can be segregated by type-II segregation. However, if this event occurs, the allelic ratio of *b⁺*/total will only be 0.33 since two of the four cells are homozygous for *b*, one homozygous for *b⁺* and the other hybrid for *b* and *b⁺*. (Gillham *et al.* 1974.)

by Sager predicts a 1:1 ratio, while the model of Gillham *et al.* (Fig. 5.10b, c) would lead to a skewed ratio, favouring the parent with the highest input fraction of chloroplast genomes.

Such differences in allelic input, as observed both by Gillham and by Sager, could be due either to genetic differences in the strains being studied or to the different methods used to obtain biparental zygotes (i.e. spontaneous, selected or UV-induced) and to analyse their progeny (i.e. by pedigree or by sampling a zygote colony after many generations). Gillham *et al.* observed a considerable excess of maternal alleles among progeny of spontaneous biparental zygotes analysed by the zygote-clone method (Fig. 5.11a, b). UV treatment of the maternal gametes prior to mating reduced the proportional excess of maternal alleles: the higher UV doses they employed caused the average allelic ratio for a *population* of biparental zygotes to approach 1 : 1, although the ratios from individual zygotes were not normally distributed around a mean of 0.5. By zygote-clone analysis methods, Sager and Ramanis (1975, pers. comm.) observed a 1 : 1 allelic ratio for all chloroplast genes examined in the progeny of rare spontaneous biparental zygotes and detected no effect of UV treatment of maternal gametes on this 1 : 1 ratio.

In these and other crosses, when they examined progeny from UV-induced biparental zygotes by pedigree analysis, they found that among the post-meiotic progeny of individual zygotes there were preponderances of the maternal markers for several, but not for all, of the chloroplast genes studied. For these same markers, slight maternal preponderances among zoospore progeny resulting from type-II segregation persisted during the two subsequent mitotic divisions, although they almost disappeared when the *total* zoospore progeny from both type-II and type-III segregations were considered together. Furthermore, the type-II events favoured maternal chloroplast alleles in zoospore progeny only when they were the antibiotic-resistant alternatives at these loci. Although Sager and Ramanis (pers. comm.) observed unexplained departures from a 1 : 1 allelic ratio under certain circumstances, these deviations were clearly not of the magnitude seen by Gillham *et al.* (1974). Thus among the strains of *Chlamydomonas reinhardtii* studied in the two laboratories there may be significant differences in terms of the number of genetic copies of the chloroplast markers transmitted and expressed in crosses. The fact that Sager's strains yield only 0·1 to 0·5% spontaneous biparental zygotes, whereas those of Gillham yield 1 to 10%, may well be another manifestion of these genetic differences.

Although typically the progeny of biparental zygotes segregate during vegetative cell divisions, rare cases of persistent heterozygotes, or 'cytohets', were reported by Sager (1972). These produced mainly heterozygous offspring in each mitotic division, with only a low percentage of homozygotes. When such cells were crossed, however, the majority of the resulting zoospores were homozygous, and those that were heterozygous tended to segregate normally. The behaviour of these cytohets, which may represent a breakdown in the normal mechanism for segregation, has not been satisfactorily explained.

Another unusual segregation pattern was observed by Schimmer and Arnold (1969, 1970a, b, c). A uniparental streptomycin-dependent mutant was found to produce streptomycin-sensitive offspring at a low frequency. These sensitive cells in turn gave rise to more dependent offspring, at a rate which varied from one clone to another; certain clones segregated dependent cells with rates as high as 10^{-2} per cell division. The new dependent cells in turn segregated more sensitive cells at variable rates. Because Sager asserted that chloroplast genes are present only in duplicate, Schimmer and Arnold suggested that the genes in question may be mitochondrial in origin since they

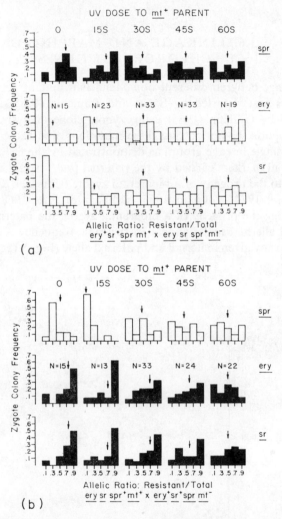

(a)

(b)

Fig. 5.11. Allelic ratios among biparental zygotes from the reciprocal crosses (a) *ery-u-37+ sr-u-2-60+ spr-u-1-6-2 mt+* × *ery-u-37 sr-u-2-60 spr-u-1-6-2+ mt−*, and (b) *ery-u-37 sr-u-2-60 spr-u-1-6-2+ mt+* × *ery-u-37+ sr-u-2-60+ spr-u-1-6-2 mt−*. The *mt+* parent was treated with various doses (seconds) of UV prior to mating. Zygote clones have been grouped into classes based on the fraction of antibiotic-resistant cells per clone. The number of progeny counted per zygote clone varied between 150 and 1000 on each of the diagnostic media. (Gillham *et al.* 1974).

appear to be present in multiple copies. However, Sager (1972) explained the results of Schimmer and Arnold by assuming that the original cells were cytohets in chloroplast DNA, and not mitochondrial mutations. Later Gillham (1974) argued that the *sd* gene is located in the chloroplast genome since it shows linkage to other chloroplast markers, and attributed Schimmer and Arnold's results to the presence of multiple copies of the chloroplast genome.

6 LINKAGE AND MAPPING OF CHLOROPLAST GENES

Genetic linkage between different non-Mendelian mutant genes in *Chlamydomonas* was first reported by Sager and Ramanis (1965) using pedigree analysis and by Gillham (1965) using zygote clone analysis. All available evidence supports the notion that these mutations, and those isolated since, constitute a single linkage group, as demonstrated by the following observations. First, all markers carried by the paternal (mt^-) parent are usually cotransmitted to the progeny of a biparental zygote (Chu-Der & Chiang 1974; Gillham *et al*. 1974; Gillham 1965; Gillham & Fifer 1968; Sager 1972). Second, among the progeny of a biparental zygote the paternal alleles, like the maternal alleles, usually appear in the same frequency regardless of the ratio between any given maternal and paternal allele (Fig. 5.12; Boynton *et al*.

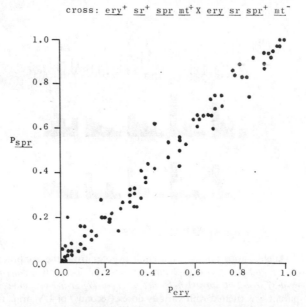

Fig. 5.12. Coordinate plot of the paternal allelic ratios (P) for the chloroplast genes *spr-u-1-6-2* and *ery-u-37* observed in progeny from individual biparental zygotes derived from the above cross. (Boynton *et al*. 1976.)

1976). Third, Sager's experiments (Sager 1972; Sager & Ramanis 1971b) showed that segregation rates for different chloroplast genes indicate that they can be arranged in a linear order with respect to a hypothetical attachment point (*ap*) (Fig. 5.13). Fourth, data from recombinational analysis for most, though possibly not all (Tables 5.6, 5.7), of the markers lend additional support to this hypothesis.

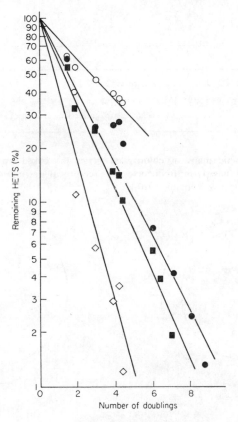

Fig. 5.13. Comparison of segregation rates in liquid culture of four chloroplast genes in the progeny of a single cross. Symbols: ○, temperature sensitivity (*tm1*); ■, streptomycin resistance (*sm2*); ● spectinomycin resistance (*spc*); ◇, erythromycin resistance (*ery*). (Sager 1972.)

Several linear maps of chloroplast genes in *Chlamydomonas*, based principally on data contained in a single paper (Sager & Ramanis 1970), have been published by Sager and her colleagues (Fig. 5.14). On the basis of unpublished data, she claimed that the chloroplast genetic map in *Chlamydomonas* is circular and not linear (Fig. 5.15a, b) (Sager & Ramanis 1971b; Sager 1972). Since the published map distances and orders are based on a

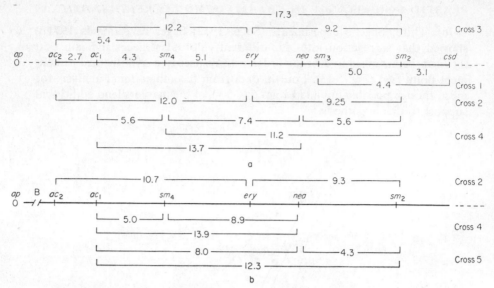

Fig. 5.14. Genetic maps of chloroplast genes (a) based on recombination frequencies; (b) based on frequencies of reciprocal recombination (type-III segregation). (Sager & Ramanis 1970.)

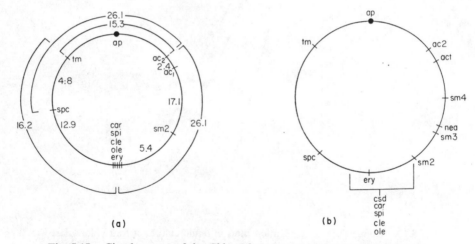

(a)

(b)

Fig. 5.15. Circular maps of the *Chlamydomonas reinhardtii* chloroplast genome. (a) Gene order and intergenic distances determined from recombination and cosegregation frequencies in progeny of four crosses, each segregating for one of the markers *car*, *cle*, *ole*, and *spi*, in addition to all six markers *ac2*, *ac1*, *sm2*, *ery*, *spc*, and *tm*. (b) Map incorporating data from several crosses. The bracketed genes all map close to *ery* and no linear order has been established. Symbols: *ap*, attachment point; *ac2* and *ac1*, acetate requirement; *sm4*, streptomycin dependence; *nea*, neamine resistance; *sm3*, low-level streptomycin resistance; *sm2*, high-level streptomycin resistance; *ery*, erythromycin resistance; *csd*, conditional streptomycin dependence; *car*, carbomycin resistance; *spi*, spiramycin resistance; *cle*, cleocin resistance; *ole*, oleandomycin resistance; *spc*, spectinomycin resistance; and *tm*, temperature sensitivity. (Sager 1972.)

combination of methods, Sager's mapping methods are reviewed here to indicate which are likely to be more useful for future mapping of chloroplast genes in *Chlamydomonas*.

Most of the earlier data published by Sager and her colleagues on recombination between chloroplast genes were obtained by pedigree analysis of clones from the octospore daughter stage (Fig. 5.8), although more recently she also used the segregation rate of different chloroplast genes in biparental zygote progeny growing in liquid culture (Fig. 5.13). At the octospore daughter stage, cells can be classified either as homozygous parental (P) or recombinant (R) for a given pair of markers, or as heterozygous (H) for one or both markers. Using such data, Sager and Ramanis (1971a) calculated recombination frequencies in three ways, none of which are independent. In the R_A method heterozygotes are ignored ($R_A = \dfrac{R}{P+R} \times 100$), whereas in the R_B method heterozygotes are included in the denominator ($R_B = \dfrac{R}{P+R+H} \times 100$). By using the R_A method one assumes that H does not contribute differentially to the R or P classes at later generations, whereas by the R_B method one assumes a preferential contribution of H to the P class. Since none of the published data suggest such a preferential contribution, we consider the conventional R_A method preferable. In the R_C or co-segregation method, Sager's assumption seems to be that closely linked markers should co-segregate to form homozygotes more frequently than markers far apart. However, as stated mathematically this method is dictated solely by the frequency of remaining heterozygotes in the population $R_C = \dfrac{P+R+H}{P+R}$, i.e. $R_C = 1/(1-H)$ and yields numbers from one to infinity rather than a fraction or percentage.

Values obtained by the R_A, R_B and R_C mapping methods, when applied to a single set of data, do not correspond closely (Table 5.3; Sager & Ramanis 1971a), as the following argument shows. If, say, two genes are ten map units apart, as calculated by the R_A method, R_B and R_C will differ according to the remaining percentage of heterozygotes (as shown in Fig. 5.16). Thus the R_B and R_C methods not only fail to yield values comparable to R_A, but also fail to yield values comparable to each other (except at one point) because

$$R_A = R_B \times R_C, \text{ i.e. } \frac{R}{P+R} = \frac{R}{P+R+H} \times \frac{P+R+H}{P+R}$$

Clearly the three methods are not independent. The R_B and R_C methods cannot be giving correct map distances, since R_B will only equal R_C when $R_B = R_C = \sqrt{R_A}$. The reason that the values determined by Sager and Ramanis (1971a) for R_B and R_C appear approximately equal, ignoring the fact that one is a percentage and the other a number > 1, is that the proportion

Table 5.3. Analysis of recombination in the cross ac_2^+ ery sm_2^+ $mt^+ \times ac_2$ ery^+ sm_2 mt^- (Modified from Sager & Ramanis 1971a). The recombination frequencies have been recalculated from the raw data.

Marker pair	Cell type and number			%R$_A$	%R$_B$	%R$_C$	%R$_D$
	Parental	Recombinant	Heterozygote				
ac_2–ery	100	36	415	26·4	6·5	4·05	4·6 (ac$_2$)
ery–sm_2	140	27	384	16·2	4·9	3·30	7·9 (ery)
ac_2–sm_2	62	35	454	36·1	6·4	5·68	10·7 (sm$_2$)

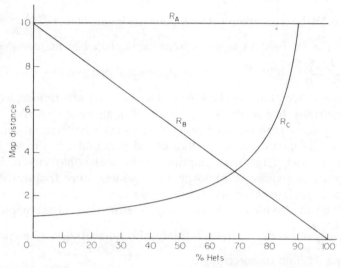

Fig. 5.16. Theoretical map distances estimated by the R_B and R_C methods, for a pair of genes 10 map units apart by the R_A method, plotted as a function of percentage heterozygotes (Hets).

of heterozygotes (H) is high at the octospore daughter stage. For a given map distance R_A there is a point where $R_B = R_C$, at a specific percentage of remaining heterozygotes which depends on the value of R_A (Fig. 5.17). Thus, if $R_A = 10\%$, then R_B will equal R_C when $R_B = 3\cdot2\%$, i.e., when there are about 70% heterozygotes (Fig. 5.16). The percentages of heterozygotes needed for $R_B = R_C$ in the case of different map distances are shown in Fig. 5.17. If R_A is $> 10\%$ and the remaining heterozygotes are $> 60\%$, then there will be relatively little difference between R_B and R_C. Since most of Sager's values for R_B are greater than four (i.e. $R_A > 16$) and the percentage of heterozygotes in her experiments is greater than 60%, R_B always approximates R_C at the octospore daughter stage.

A fourth method used by Sager (R_D) was based on her observation that in a cross the percentage of type-III segregations at the octospore daughter stage was different for each chloroplast gene (Sager & Ramanis 1970, 1971b).

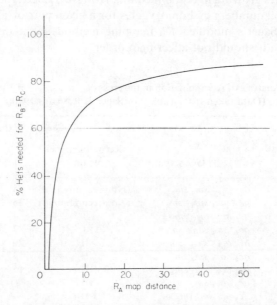

Fig. 5.17. Theoretical percentages of heterozygotes (Hets) needed for $R_B = R_C$, plotted as functions of map distance.

This characteristic variation in segregation rates among different chloroplast genes led Sager to postulate that the chloroplast genome has a membrane-attachment point (*ap*) functionally equivalent to a mitotic centromere and that the chloroplast genes are arranged sequentially (i.e. show a polarity) with respect to *ap*. Homozygosity for a given gene results from a recombinational event between *ap* and the gene, akin to mitotic recombination. The map distance between *ap* and that gene can then be computed from the percentage of type III segregations for that gene

$$R_D = \frac{\text{No. type-III segregations}}{\text{Total progeny}} \times 100.$$

Mapping using the R_D method was later modified when Sager observed that, during growth of zoospore clones from biparental zygotes, segregation and recombination proceeded at a constant rate per cell doubling, and that type-II events likewise occurred with a constant frequency and yielded equal frequencies of the two homozygous types (Fig. 5.13; Sager & Ramanis 1971b; Sager 1972). In this case the disappearance of heterozygotes from the population should be a function of the frequency of type-III segregations. Since the frequency of type-III segregations varies from marker to marker, the rate of disappearance of heterozygotes for each marker should be constant. This permitted Sager to compute gene distances with respect to ap by observing zoospore clones growing in liquid medium. However, since type-II events may not yield equal numbers of homozygotes for a given pair of alleles (Gillham *et al.* 1974), Sager's modified R_D mapping method might distort map distances, though it should not affect map order.

Table 5.4. Analysis of recombination in the cross $ac_2^+ ac_1 sm_3^+ sm_2^+ mt^+ \times ac_2 ac_1^+ sm_3 sm_2 mt^-$. (Data taken from Table 2 of Sager & Ramanis 1970.)

Progeny class	Genotype	Recombinant in region [b]	Experiment 1	Experiment 2
1	$ac_2 ac_1^+ sm_3 sm_2$	(Non-recombinant)	64	36
2[a]	$ac_2^+ ac_1 sm_3^+ sm_2^+$	(Non-recombinant)	75	32
3	$ac_2 ac_1 sm_3^+ sm_2^+$	I	– [c]	–
4	$ac_2^+ ac_1^+ sm_3 sm_2$	I	7	2
5	$ac_2 ac_1^+ sm_3^+ sm_2^+$	II	20	15
6	$ac_2^+ ac_1 sm_3 sm_2$	II	13	7
7	$ac_2 ac_1^+ sm_3 sm_2^+$	III	15	6
8	$ac_2^+ ac_1 sm_3^+ sm_2$	III	–	–
9	$ac_2 ac_1 sm_3 sm_2$	I-II	–	–
10[a]	$ac_2^+ ac_1^+ sm_3^+ sm_2^+$	I-II	5	3
11	$ac_2 ac_1 sm_3^+ sm_2$	I-III	–	–
12[a]	$ac_2^+ ac_1^+ sm_3 sm_2^+$	I-III	2	0
13	$ac_2 ac_1^+ sm_3^+ sm_2$	II-III	–	–
14[a]	$ac_2^+ ac_1 sm_3 sm_2^+$	II-III	13	0
15	$ac_2 ac_1 sm_3 sm_2^+$	I-II-III	–	–
16	$ac_2^+ ac_1^+ sm_3^+ sm_2$	I-II-III	–	–
		Total	214	101

[a] Progeny classes whose phenotype should unambiguously indicate the genotype.

[b] The recombination intervals are I $= ac_1\text{-}ac_2$; II $= ac_1\text{-}sm_3$; III $= sm_3\text{-}sm_2$.

[c] No progeny listed in original table.

Having considered Sager's mapping methods, let us turn to the chloroplast gene map itself. The construction of the original linear map by Sager and Ramanis (1970; Fig. 5.14) is open to criticism, not only because of difficulties inherent in the mapping methods used but also because most of the crosses were made in such a way that they yielded progeny classes in which the genotype for certain marker pairs could not accurately be determined from the phenotype. The only complete set of data in that paper were obtained from the cross $ac_2^+ \, ac_1 \, sm_3^+ \, sm_2^+ \times ac_2 \, ac_1^+ \, sm_3 \, sm_2$ (Table 5.4). Only nine of the 16 possible genotype combinations for these four genes were listed in the original published table. Among the classes not listed were two which should have resulted from single recombinational events (progeny classes 3 and 8). The reciprocal combinations of these were reported from each cross, but it is not clear whether the missing genotypes were not found or whether they could not be scored definitively. Sager and Ramanis stated that the $ac_1 \, ac_2$ double mutant is phenotypically similar to $ac_1^+ \, ac_2$, and must be distinguished genetically, as is also true for $sm_2 \, sm_3$ and $sm_2 \, sm_3^+$, which show an identical streptomycin-resistant phenotype. If this is the case, only four out of the 16 possible genotypes from this cross can be distinguished unambiguously in the absence of progeny testing. An examination of the other crosses made (Table 5.5), for which complete data were not given, reveals that only in one case (cross 2) would scoring be totally unambiguous.

The recombination frequencies were used by Sager to construct the genetic map, employing the R_B method instead of the R_A method. As mentioned earlier, this method is questionable not only because heterozygotes are included in the denominator but also because map distances are influenced by a factor depending on the frequency of heterozygotes (Fig. 5.16). If the heterozygote frequency for either one or both markers is high, the genes will appear closely linked, whereas if the frequency is low, the markers will appear more distantly linked. For several map intervals, recombination frequencies can be computed from the data of Sager and Ramanis (1971a) by the R_A method (Table 5.3). As expected, the apparent map distances between markers are much greater when calculated by the R_A method than by the R_B method, although the marker order is the same.

The map obtained by the R_B method can then be compared to that obtained by the R_D method, using for such mapping the polarity with respect to ap in type-III segregations of individual markers. Although type-III segregation appears polarized, problems arise in trying to assign an order to markers by this method (Table 5.6). Data from crosses 2 and 5 involving ery and sm_2 give different orders for these markers with respect to ap. In cross 3, sm_4 appears closer to ap than the closest known marker, ac_2 (see crosses 1 and 2), whereas in cross 4, which uses ac_1 instead of ac_2, the marker order appears to be ap-ac_1-sm_4. This means that of the six markers used in these crosses, ambiguities arise in the ordering of four (i.e. ery, sm_2, sm_4, ac_2). In order to make comparisons between values from different crosses, evaluated by the R_D method, it is necessary to normalize the data to account for differing amounts of marker segregation at the octospore daughter stage.

Although normalization is a perfectly legitimate procedure if a consistent method is used, Sager and Ramanis used both additive and multiplicative methods of normalization, which affect map distances differently (Table 5.6). In crosses 1 and 2, the adjustment was achieved by assuming the map

Table 5.5. Analysis of recombination of chloroplast genes in four crosses. The genotypes of the two parents are separated by a solid line. Antibiotic-resistant mutants are indicated by '-r' and their sensitive wild-type alleles by '-s'. (From Sager & Ramanis 1970.)

Cross	Markers	Number of progeny	Recombinants	(%)*	Relative map distance†
1	$ac_2{}^+ac_1\ sm_3\text{-}s\ sm_2\text{-}s\ csd\text{-}s$ $ac_2\ ac_1{}^+\ sm_3\text{-}r\ sm_2\text{-}r\ csd\text{-}r$	1978‡ 559‡	$ac_2\text{-}ac_1$ $sm_3\text{-}sm_2$ $sm_2\text{-}csd$ $sm_3\text{-}csd$	2·1 3·9 4·8 6·8	2·7§ 5·0§ 3·1‖ 4·4‖
2	$ac_2{}^+\ ery\text{-}r\ sm_2\text{-}s$ $ac_2\ ery\text{-}s\ sm_2\text{-}r$	551	$ac_2\text{-}ery$ $ery\text{-}sm_2$ $ac_2\text{-}sm_2$	6·4 4·9 6·7	12·0 9·25 12·6♯
3	$ac_2{}^+\ ac_1\ sm_4\text{-}d\ ery\text{-}s\ sm_2\text{-}s$ $ac_2\ ac_1{}^+\ sm_4\text{-}s\ ery\text{-}r\ sm_2\text{-}r$	370	$ac_2\text{-}ac_1$ $ac_1\text{-}sm_4$ $ac_2\text{-}sm_4$ $sm_4\text{-}ery$ $ery\text{-}sm_2$ $ac_1\text{-}ery$ $sm_4\text{-}sm_2$ $ac_1\text{-}sm_2$	1·35 4·3 5·4 5·1 9·2 12·2 8·65 9·5	2·7§ 4·3 5·4 5·1 9·2 12·2 17·3 9·5¶
4	$ac_1\ sm_4\text{-}d\ nea\text{-}s\ sm_2\text{-}s$ $ac_1{}^+\ sm_4\text{-}s\ nea\text{-}r\ sm_2\text{-}r$	360	$ac_1\text{-}sm_4$ $sm_4\text{-}nea$ $nea\text{-}sm_2$ $ac_1\text{-}nea$ $sm_4\text{-}sm_2$	2·5 3·3 2·5 6·1 2·5	5·6 7·4 5·6 13·7 11·2§

* Number of recombinants/total progeny scored.

† Normalization procedure: Cross 1—% recombination $\times \dfrac{1\cdot35\ (\text{cross 3})}{2\cdot1\ (\text{cross 1})}\ ac_2\text{-}ac_1$

Cross 2—% recombination $\times \dfrac{21\cdot3\ (\text{cross 3})}{11\cdot3\ (\text{cross 2})}\ ac_2\text{-}sm_2$

Cross 4—% recombination $\times \dfrac{18\cdot6\ (\text{cross 3})}{8\cdot3\ (\text{cross 4})}\ ac_1\text{-}sm_2$

‡ Only part of the experiment scored for *csd*.

§ Per cent recombination $\times 2$ because only one of the two recombinant types was detected.

‖ Unequal numbers of reciprocal recombinants.

¶ Low value because of long map interval.

♯ Significance of this symbol not defined in the original paper.

distance between ac_2 and sm_2 to be equal to 20 and scaling all distances, including the distance between ac_2 and ap, accordingly (i.e. normalization

factor $= \dfrac{20}{\%\ \text{type III for}\ sm_2 - \%\ \text{type III for}\ ac_2}$). In crosses 4 and 5, ac_1

Table 5.6. Analysis of recombination data (from Table 4 of Sager & Ramanis 1970) using the R_D method. The percentages of type-III segregations for all crosses have been normalized by the two methods used by Sager and Ramanis.

Method 1 establishes a standard interval, between the marker closest to *ap* (*a*) and the one furthest from *ap* (*b*), which is equal to the recombination frequency determined by the R_B method. The distance of each marker from *ap* is then determined from a proportionality factor (*P*). $P = \%$ type-III segregation for marker $\times \dfrac{\text{standard interval } a\text{-}b}{\text{observed interval } a\text{-}b}$ where the observed interval $a\text{-}b =$ (type-III segregations for *b*) – (type-III segregations for *a*). We have used intervals and values for *P* as follows:

Cross	Interval	Standard	Observed	P
1	ac_2-sm_2	20	5·0	4·0
2	ac_2-sm_2	20	6·1	3·3
3	ac_2-ery	12	2·7	4·4
4	ac_1-nea	13·7	13·9	0·99
5	ac_1-sm_2	21·4	12·3	1·74

Method 2 assumes the ac_2-ap distance is 15·6 (see text). In crosses 1, 2 and 3, all distances have been normalized by adding the difference between 15·6 and the observed percentage of type-III segregations for ac_2 to all intervals (cross 1 = 11·7, cross 2 = 10·95, cross 3 = 5·4). In crosses 4 and 5, the distance from ap-ac_2 is assumed to be equal to 15·6 and the ac_2-ac_1 distance = 2·7 from the R_B method. Hence, ap-ac_2 = 18·3. Normalization is achieved by adding 8 in crosses 4 and 5, since there are 10% type-III segregations for ac_1.

Cross	Gene	Observed type-III segregations (%)	Relative distance of marker from *ap* following normalization		
			Method 1	Method 2	Sager
1	ac_2	3·9	15·6	15·6	15·6
	sm_2	8·9	35·6	20·6	35·6
2	ac_2	4·6	15·6	15·6	15·6
	ery	7·9	26·5	18·9	26·3
	sm_2	10·7	35·7	21·7	35·6
3	ac_2	16·2	72·0	15·6	—
	sm_4	10·8	48·0	10·2	—
	ery	18·9	84·0	18·3	—
4	ac_1	10·0	9·9	18·0	18·0
	sm_4	15·0	14·8	23·0	23·0
	nea	23·9	23·7	31·9	31·9
5	ac_1	10·0	17·4	18·0	18·0
	nea	18·0	31·3	26·0	26·0
	sm_2	22·3	38·8	30·3	30·3
	ery	30·0	52·2	38·0	—

rather than ac_2 was included and the data were normalized by assuming that ac_2 is 15·6 map units from ap and, from recombination analysis by the R_B method, that ac_1 is 2·7 map units from ac_2. This gives a map distance from ap to ac_1 of approximately 18 map units. For each marker in crosses 4 and 5 normalization is then achieved by adding 8% to the % type III segregations, since ac_1 prior to normalization yielded about 10% type III segregations. Since the first method involves multiplication by a proportionality factor whereas the second involves addition of a constant, as the % type III segregations increase the apparent distances from ap will increase much more rapidly in the first case than in the second. This point is illustrated in Table 5.6, where data from each cross have been normalized by both methods.

For all of these reasons we have serious doubts about the meaning of the apparent agreement between maps obtained by the R_B and R_D methods. Not only is the map order occasionally inconsistent (as in the case of ery and sm_2) but the apparent agreement in map distance appears more fortuitous than consequential. Our doubts also extend to the additional claim that the chloroplast genetic map is circular (Fig. 5.15; Sager & Ramanis 1971b; Sager 1972), based largely on experimental data as yet unpublished. Segregation rates of markers in liquid culture indicate the order ac_1-tm_1-spc-sm_2-ery, and, although ac_1-tm_1 and spc-sm_2 showed similar rates of segregation, they did not appear closely linked on the basis of cosegregation and recombination data. These data are consistent with both a circular map and a two-armed linear map, but Sager and Ramanis (1971b) state that the behaviour of ery with respect to the aforementioned markers is consistent only with a circular map. Since the map order with respect to ery determined by the R_D method is ambiguous, and since both the R_B and R_C methods are suspect for reasons outlined earlier, these critical pieces of evidence for circularity cannot be evaluated until the actual data and a precise description of methods are published.

The methods used by Sager have unquestionably revealed a highly significant property of chloroplast genes, i.e. their polarity of segregation, and have indicated that such chloroplast genes can be mapped. Nevertheless, the computation of recombination frequencies in pedigrees is plagued by the problem of how to correct for heterozygotes. If one is to map such genes accurately, using recombination data, we feel that all markers should be allowed to segregate to homozygosity before one scores the segregants and computes recombination frequencies by the R_A method. We have begun to do this among the progeny of biparental zygote clones scored with respect to genotype after many cell generations (Table 5.7; Boynton et al. 1976). However, this method is much more susceptible to bias due to marker selection than is Sager's octospore daughter method, since intercellular competition between different genotypes in a zygote clone could distort the true recombination frequencies. To use this method properly one would need to rule out such potential sources of error by crossing clones bearing each

Table 5.7. Recombination analysis, using the R_A method, of data from the progeny of biparental zygote clones. Data from the reciprocal crosses, between the genotypes *er-u-37+ sr-u-2-60 spr-u-1-6-2* and *er-u-37 sr-u-2-60+ spr-u-1-6-2+*, have been pooled and include both spontaneous and UV-induced biparental zygotes (see Gillham *et al.* 1974 for a description of UV induction method used, and Fig. 5.9 for analytical protocol). Each biparental zygote was scored for input of maternal and paternal alleles for each marker. As shown in Fig. 5.12, for each zygote there is a strong correlation between inputs of different paternal or maternal markers: biparental zygotes have therefore been grouped according to the relative allelic input ratios (see text for discussion). The few progeny clones still heterozygous, recognizable as mixtures of parental genotypes (see Fig. 5.9), are excluded from all calculations.

Paternal allelic ratio*	Total sampled		% Recombination†		
	Zygotes	Progeny	*er-spr*	*er-sr*	*spr-sr*
<0·01 to 0·20	80	5,072	1·9	3·2	2·4
0·21 to 0·40	36	2,281	4·6	9·6	7·4
0·41 to 0·60	37	2,366	4·2	6·1	3·6
0·61 to 0·80	14	886	4·3	3·7	1·7
0·81 to >0·99	16	1,001	2·2	2·4	1·1
overall	183	11,606	2·7	5·0	2·3

* From zygote progeny clones of the reciprocal crosses, data are pooled according to the frequency of their *er+*, *spr* and *sr* or *er*, *spr+*, *sr+* alleles respectively among the total progeny.

† Data adjusted, for inequalities in the numbers of progeny between the reciprocal crosses, by weighting their respective percentages of recombination in terms of relative numbers of progeny in each class.

set of markers in all possible combinations. This sort of approach, which involves identification of genotypes after many cell generations rather than in pedigrees, has been used successfully by Dujon *et al.* (1974) to map mitochondrial genes in *Saccharomyces cerevisiae*.

7 MODELS FOR INHERITANCE OF CHLOROPLAST GENES

Two distinct models have been proposed to explain the physical mechanism of uniparental inheritance in *Chlamydomonas*. In that of Sager and Ramanis (1973) the normal uniparental mode of gene transmission is the result of a modification-restriction system, analogous to those found in bacteria (Meselson *et al.* 1972). In this model each vegetative cell contains two copies of the chloroplast genome, which are membrane bound and undergo oriented segregation after replication (Fig. 5.18). During the formation of the gametes the number of copies is reduced to one per chloroplast. The maternal gamete produces a 'modification' enzyme capable of protecting the mt^+ chloroplast DNA, while the paternal gamete produces a 'restriction' enzyme, capable of degrading any unmodified DNA. Sager proposes that before fusion both of these enzymes are present in the gametes in an inactive form and that they are later activated by two 'regulator' molecules synthesized by the maternal gamete and released on gamete fusion (Fig. 5.18). The paternal chloroplast DNA is usually completely destroyed by the restriction enzyme, while the maternal DNA is protected by the modification enzyme, so that nearly all of the zygotes carry only maternal chloroplast genes.

Occasionally, however, the restriction enzyme fails to destroy the mt^- DNA, either because the enzyme is not activated or because the mt^- DNA is modified before it can act. Such survival of the mt^- DNA can be expected to result in a biparental zygote. Paternal zygotes could arise only by the accidental destruction of the mt^+ chloroplast DNA without concomitant destruction of the mt^- DNA. As one might expect from this hypothesis, spontaneous paternal zygotes are rarely found. The model also predicts a fourth class of zygotes, in which the modification enzyme fails to protect the mt^+ DNA in the presence of the restriction enzyme, so that chloroplast DNA from both parents is destroyed; this would presumably be lethal.

Each of the assumptions inherent in the modification-restriction model has been examined experimentally, and can be discussed separately. The presence of two copies of the genome in the vegetative cells has not yet been demonstrated (see pp. 77, 94). Evidence for the occurrence of a modification enzyme is at present circumstantial. The observed reduction of the buoyant density of the mt^+ β DNA prior to gamete fusion is consistent with its becoming methylated, which is what happens in many bacteria when the DNA is modified (Meselson *et al.* 1972). However, the magnitude of the observed shift in buoyant density (Sager & Ramanis 1973) suggests a level of methylation as high as 5%, far higher than is detected in bacteria. More exceptional zygotes are obtained after treating the mt^+ parent with DL-ethionine, a competitive inhibitor of methionine (the usual methyl donor), which also suggests that methylation may here be playing a special role.

The presence of some factor in the mt^- parent which prevents transmission of its chloroplast DNA is indicated both by the behaviour of the *mat-1* mutant and by the increase in frequency of biparental zygotes found after treating mt^- gametes before mating with inhibitors of protein synthesis. In both cases the number of surviving paternal genomes tends to be increased, suggesting that such treatments affect the restriction enzyme in the mt^- gamete.

Exposing the mt^+ parent to low doses of UV prior to mating yields an increased number of biparental zygotes, whereas higher UV doses preferentially increase the proportion of paternal zygotes. Sager and Ramanis (1973) explain this by postulating two distinct UV effects. The first effect prevents the normal transcription of mRNA for the regulator of the restriction enzyme, while the second results in inactivation of the maternal chloroplast genome. If the latter effect is achieved only after a number of UV 'hits', then this could also explain preferential photoreactivation of paternal to biparental zygotes. Repair of any one of the UV-damaged sites in the maternal chloroplast genome would convert a paternal to a biparental zygote by reducing the total number of lesions to this genome ($\geq$ N) to below the critical limit (N−1) and thereby allowing expression of the genome. Conversely a biparental zygote would be changed to a maternal zygote only by repair of a specific site in the nuclear genome necessary for the formation of the regulator. Since

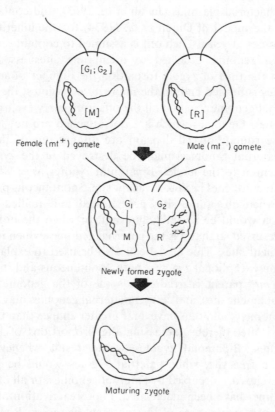

Fig. 5.18. Proposed mechanism for maternal inheritance of chloroplast DNA in *Chlamydomonas*. The female (*mt*+) gamete contains inactive modification enzyme (M) in its chloroplast, and two regulatory substances, G_1 and G_2, in the cell sap. The male (*mt*−) gamete contains inactive restriction enzyme (R) in its chloroplast. After zygote formation and before fusion of the chloroplasts, the modification enzyme is activated by G_1 to modify chloroplast DNA in the female chloroplast, and the restriction enzyme is activated by G_2 to degrade chloroplast DNA in the male chloroplast. The two chloroplasts then fuse, and only the chloroplast DNA from the female parent is available for replication. (Sager & Ramanis 1973.)

such a repair event can be assumed to occur at random, conversion of a paternal to a biparental zygote is more likely than conversion of a biparental to a maternal zygote.

The failure of mt^- gametes to respond to UV can be explained by assuming that on completion of gametogenesis the restriction enzyme in the mt^- parent is present in an inactive form and that its activation is effected by the mt^+ gamete. Treating the mt^- cells with UV before gametogenesis would be ineffective, since the UV damage could probably be repaired while the zoospores are differentiating into gametes.

In contrast to the model proposed above, Gillham et al. (1974) have suggested a theory based on the principle of competitive exclusion, similar to that found in bacterial plasmids (Jacob et al. 1963), and containing features also found in the model of Dujon et al. (1974) for the inheritance of mito-chondrial genomes in yeast. Each cell is assumed to contain several copies of the chloroplast genome, attached to membrane sites essential for their replication. At the time of zygote formation the number of such membrane attachment sites is limited and, in the majority of zygotes, they are all occu-pied by maternal genomes, so that all the offspring carry exclusively maternal chloroplast genes. Genomes which are not attached are not replicated, and with successive rounds of cell division are diluted out of the chloroplast. Unattached paternal genomes might be destroyed in the zygote, although this is not essential to the model. Biparental zygotes arise when a paternal genome finds and attaches itself to a vacant site. Spontaneous paternal zygotes appear only when all of the sites are vacated and refilled with paternal genomes, which would be an unlikely event, or when the normal exclusion preference is reversed so that it is the paternal genomes which initially occupy all the attachment sites. This model can also be used to explain most of the observations on exceptional zygotes, both spontaneous and induced.

When the mt^+ parent is irradiated, some of the genomes are detached from their attachment sites, and certain maternal genomes may be inactivated. A paternal genome would thus have a far greater chance than usual of finding an unoccupied site, thereby increasing the proportions of biparental and paternal zygotes. Biparental zygotes would result whenever a paternal genome finds a free site, while paternal zygotes would be produced only when all of the sites are occupied by paternal genomes, or all of the remaining maternal genomes have been inactivated. Photoreactivation of the cells after irradiation would repair some of the maternal genomes. Repair of any one maternal genome would be sufficient to convert a paternal to a biparental zygote, but conversion from a biparental to a maternal zygote would require the repair of all of the maternal genomes. Thus, although both models explain the differential conversion rates, they do so in somewhat different ways. The absence of any evident effect of UV on paternal gametes is readily explained by the 'exclusion' model; since the maternal genomes have not been harmed, there is no increase in the number of attachment sites available for the paternal genomes. Thus the proportion of exceptional zygotes is no higher than in a cross involving unirradiated gametes.

The two models differ essentially in the prediction that they make regarding the ratio of maternal to paternal alleles among cells derived from a biparental zygote undergoing type-II segregation. As has been stressed previously, although Sager consistently finds a 1:1 segregation, Gillham observes significant deviations from this ratio for which the 'multiple-copy' theory provides a reasonable explanation. If a biparental zygote contains an excess of maternal genomes, the majority of the offspring from a type-II segregation would be homozygous for the maternal, rather than for the paternal, allele (Fig. 5.10c). This is the simplest interpretation of the results of Gillham *et al.* (1974) using zygote clones, and it agrees well with earlier results published by Gillham (1969) using pedigrees, but it also fails to explain Sager and Ramanis's observation that given pairs of octospore daughters often show various combinations of type-I, type-II and type-III segregations for closely linked chloroplast markers. The occasionally observed reversal of the allelic ratio can be explained by a rare reversal of the direction of exclusion.

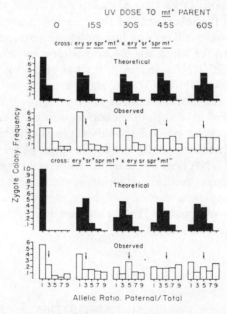

Fig. 5.19. Frequencies of theoretical and observed paternal allelic ratios from reciprocal crosses as shown in Fig. 5.11a and b. The observed paternal allelic ratios were calculated from the mean paternal allelic ratio for all three markers in each cross. Zygote clones were grouped into arbitrary classes based on the mean fraction (per clone) of cells carrying paternal markers. (Gillham *et al.* 1974.)

Treating the maternal parent with UV would be expected to reduce effectively the number of functional copies of the maternal chloroplast genome, and thereby raise the paternal : maternal ratio of alleles since the number of paternal copies would not be altered. Gillham *et al.* (1974) have compared the results of such a shift, obtained experimentally, with expected values derived mathematically from their model (Fig. 5.19). Although observed and predicted values differ, there is in fact an increase in the average paternal : maternal ratio with increasing UV dose. Sager and Ramanis (1975) observed no effect of UV on their 1 : 1 allelic ratios. Thus neither model satisfactorily explains all of the reported observations, and certain major differences remain to be resolved.

8   CONCLUSIONS

This chapter brings together observations on the physical transmission of chloroplast DNA in the sexual cycle and the behaviour of chloroplast genes in crosses, in an attempt to explain the inheritance of the chloroplast genome in *Chlamydomonas reinhardtii*. Although much has been learned since the discovery of non-Mendelian genes (Sager 1954) and the demonstration that the chloroplast contains DNA (Ris & Plaut 1962), many of the fundamental questions regarding the inheritance of the chloroplast genome remain to be answered. Also, we have still only an incomplete understanding of the specific chloroplast functions for which the chloroplast genome codes. Although the chloroplast ribosomal RNAs and certain ribosomal proteins appear to be coded by chloroplast DNA (Ch. 6), enough additional coding capacity remains to specify a significant number of other chloroplast proteins. Isolation and characterization of various acetate-requiring mutants attributable to changes in the chloroplast genome should enable us to determine what other photosynthetic functions are specified by chloroplast genes. It is already clear that the chloroplast DNA does not carry all the necessary genetic information for plastid structure and function, since many nuclear genes are known also to affect the photosynthetic capacity of this organelle (Ch. 2). There is evidence that even the chloroplast DNA polymerase is under the control of a nuclear gene (Surzycki *et al.* 1970).

ACKNOWLEDGEMENTS

This work was supported by Grants NSF GB22769 and NIH GM19427 to J.E.B. and N.W.G. as well as NIH RCDA awards GM70453 to J.E.B. and GM70437 to N.W.G. and by a predoctoral fellowship to K.V.S. on NIH training grant GM02007. We would like to thank Dr Elizabeth H. Harris for critically reading and editing this manuscript.

9 REFERENCES

ALEXANDER N.J., GILLHAM N.W. & BOYNTON J.E. (1974) The mitochondrial genome of *Chlamydomonas*: Induction of minute colony mutations by acriflavin and their inheritance. *Molec. gen. Genet.* **130**, 275–90.

BASTIA D., CHIANG K.-S. & SWIFT H. (1971) Studies on the ribosomal RNA cistrons of chloroplast and nucleus in *Chlamydomonas reinhardtii. Abstracts 11th Ann. Meeting Amer. Soc. Cell. Biol.* p. 25.

BASTIA D., CHIANG K.-S., SWIFT H. & SIERSMA P. (1971) Heterogeneity, complexity and repetition of chloroplast DNA of *Chlamydomonas reinhardtii. Proc. Nat. Acad. Sci. U.S.A.* **68**, 1157–61.

BAUR E. (1909) Das Wesen und die Erblichkeitsverhältnisse der "Varietates albomarginatae hort" von *Pelargonium zonale. Z. Vererbungs.* **1**, 330–51.

BEHN W. & ARNOLD C.G. (1972) Zur Lokalisation eines nichtmendelnden Gens von *Chlamydomonas reinhardii. Molec. gen. Genet.* **114**, 266–72.

BORST P. (1972) Mitochondrial nucleic acids. *Ann. Rev. Biochem.* **41**, 334–76.

BORST P. & KROON A.M. (1969) Mitochondrial DNA: Physiochemical properties, replication and genetic function. *Int. Rev. Cytol.* **26**, 107–89.

BOSCHETTI A. & BOGDANOV S. (1973) Different effects of streptomycin on the ribosomes from sensitive and resistant mutants of *Chlamydomonas reinhardi. Eur. J. Biochem.* **35**, 482–8.

BOYNTON J.E., ADAMS G.M.W., CONDE M.F., GILLHAM N.W., HARRIS E.H. & TINGLE C.L. (1976). In preparation.

BOYNTON J.E., BURTON W.G., GILLHAM N.W. & HARRIS E.H. (1973) Can a non-Mendelian mutation affect both chloroplast and mitochondrial ribosomes? *Proc. Nat. Acad. Sci. U.S.A.* **70**, 3463–7.

BRITTEN R.J. & KOHNE D.E. (1968) Repeated sequences in DNA. *Science*, **161**, 529–40.

BURTON W.G. (1972) Dihydrospectinomycin binding to chloroplast ribosomes from antibiotic-sensitive and resistant strains of *Chlamydomonas reinhardtii. Biochim. Biophys. Acta* **272**, 305–11.

CAVALIER-SMITH T. (1970) Electron microscopic evidence for chloroplast fusion in zygotes of *Chlamydomonas reinhardi. Nature, Lond.* **228**, 333–5.

CERDÁ-OLMEDO E., HANAWALT P.C. & GUEROLA N. (1968) Mutagenesis of the replication point by nitrosoguanidine: Map and pattern of replication of the *Escherichia coli* chromosome. *J. Mol. Biol.* **33**, 705–19.

CHIANG K.-S. (1968) Physical conservation of parental cytoplasmic DNA through meiosis in *Chlamydomonas reinhardi. Proc. Nat. Acad. Sci. U.S.A.* **60**, 194–200.

CHIANG K.-S. (1971) Replication, transmission and recombination of cytoplasmic DNAs in *Chlamydomonas reinhardi.* In *Autonomy and Biogenesis of Mitochondria and Chloroplasts.* Eds. Boardman N.K., Linnane A.W. & Smillie R.M., pp. 235–49. North-Holland, Amsterdam.

CHIANG K.-S. & SUEOKA N. (1967a) Replication of chloroplast DNA in *Chlamydomonas reinhardi* during vegetative cell cycle: its mode and regulation. *Proc. Nat. Acad. Sci. U.S.A.* **57**, 1506–13.

CHIANG K.-S. & SUEOKA N. (1967b) Replication of chromosomal and cytoplasmic DNA during mitosis and meiosis in the eucaryote *Chlamydomonas reinhardi. J. Cell. Physiol.* **70** (Suppl. 1), 89–112.

CHU-DER O.M.Y. & CHIANG K.-S. (1974) The interaction between Mendelian and non-Mendelian genes in *Chlamydomonas reinhardtii.* I. The regulation of the transmission of non-Mendelian genes by a Mendelian gene. *Proc. Nat. Acad. Sci. U.S.A.* **71**, 153–7.

CHUN E.H.L., VAUGHAN M.H. & RICH A. (1963) The isolation and characterization of DNA associated with chloroplast preparations. *J. Mol. Biol.* **7**, 130–41.

CONDE M.F., BOYNTON J.E., GILLHAM N.W., HARRIS E.H., TINGLE C.L. & WANG W.L. (1975). Chloroplast genes in *Chlamydomonas* affecting organelle ribosomes. *Molec. gen. Genet.* **140**, 183–220.

CORRENS C. (1909) Vererbungsversuche mit blass (gelb) grünen und bluntblattrigen Sippen bei *Mirabilis jalapa, Urtica pilulifera*, und *Lunaria annua. Z. Vererbungs.* **1**, 291–329.

DAVIDSON E.H. & BRITTEN R.J. (1973) Organization, transcription and regulation in the animal genome. *Quarterly Review of Biology*, **48**, 565–613.

DUJON B., SLONIMSKI P.P. & WEILL L. (1974) Mitochondrial genetics. IX: A model for recombination and segregation of mitochondrial genomes in *Saccharomyces cerevisiae. XIII International Congress of Genetics: Symposium on Non-chromosomal Inheritance*, **78**, 415–37.

EBERSOLD W.T. (1967) *Chlamydomonas reinhardi*: Heterozygous diploid strains. *Science*, **157**, 447–9.

GILLHAM N.W. (1965) Linkage and recombination between non-chromosomal mutations in *Chlamydomonas reinhardi. Proc. Nat. Acad. Sci. U.S.A.* **54**, 1560–7.

GILLHAM N.W. (1969) Uniparental inheritance in *Chlamydomonas reinhardi. American Naturalist*, **103**, 355–88.

GILLHAM N.W. (1974) Genetic analysis of the chloroplast and mitochondrial genomes. *Ann. Rev. Genet.* **8**, 347–91.

GILLHAM N.W., BOYNTON J.E. & BURKHOLDER B. (1970) Mutations altering chloroplast ribosome phenotype in *Chlamydomonas*. I. Non-Mendelian mutations. *Proc. Nat. Acad. Sci. U.S.A.* **67**, 1026–33.

GILLHAM N.W., BOYNTON J.E. & LEE R.W. (1974) Segregation and recombination of non-Mendelian genes in *Chlamydomonas. Symposium on Non-Chromosomal Inheritance: XIII International Congress of Genetics*, **78**, 439–57.

GILLHAM N.W. & FIFER W. (1968) Recombination of non-chromosomal mutations: a three-point cross in the green alga *Chlamydomonas reinhardi. Science*, **162**, 683–4.

GILLHAM N.W. & LEVINE R.P. (1962) Studies on the origin of streptomycin-resistant mutants in *Chlamydomonas reinhardi. Genetics*, **47**, 1463–74.

GRIMES G.W., MAHLER H.R. & PERLMAN P.S. (1974) Nuclear gene dosage effects on mitochondrial mass and DNA. *J. Cell. Biol.* **61**, 565–74.

HINCKLEY N., CHIANG K.-S. & GOODENOUGH U. (1974) Variations in chloroplast and cytoplasmic ribosome metabolism in gametogenesis in *Chlamydomonas reinhardtii. J. Cell. Biol.* **63**, 138a.

JACOB F., BRENNER S. & CUZIN F. (1963) On the regulation of DNA replication in bacteria. *Cold Spring Harbor Symp. Quant. Biol.* XXVIII, 329–48.

JINKS J.L. (1964) *Extrachromosomal Inheritance*, 177 pp. Prentice Hall, New Jersey.

KIRK J.T.O. (1971) Will the real chloroplast DNA please stand up? In *Autonomy and Biogenesis of Mitochondria and Chloroplasts*. Eds. Boardman N.K., Linnane A.W. & Smillie R.M., pp. 267–76. North-Holland, Amsterdam.

KIRK J.T.O. & TILNEY-BASSETT R.A.E. (1967) *The Plastids*. Freeman, San Francisco. 608 pp.

LEE R.W. & JONES R.F. (1973) Induction of Mendelian and non-Mendelian streptomycin resistant mutants during the synchronous cell cycle of *Chlamydomonas reinhardtii. Molec. gen. Genet.* **121**, 99–108.

LEE R.W., GILLHAM N.W., VAN WINKLE K.P. & BOYNTON J.E. (1973) Preferential recovery of uniparental streptomycin resistant mutants from diploid *Chlamydomonas reinhardtii. Molec. gen. Genet.* **121**, 109–16.

LEFF J., MANDEL M., EPSTEIN H.T. & SCHIFF J.A. (1963) DNA satellites from cells of green and aplastidic algae. *Biochem. Biophys. Res. Commun.* **13**, 126–30.

LEVINE R.P. & EBERSOLD W.T. (1960) The genetics and cytology of *Chlamydomonas*. *Ann. Rev. Microbiol.* **14**, 197–216.

LUCK D.J.L. & REICH E. (1964) DNA in mitochondria of *Neurospora crassa*. *Proc. Nat. Acad. Sci. U.S.A.* **52**, 931–8.

MAHLER H. (1973) Biogenetic autonomy of mitochondria. *Critical Reviews in Biochemistry* **1**, 381–460.

MANNING J.E., WOLSTENHOLME D.R., RYAN R.S., HUNTER J.A. & RICHARDS O.C. (1971) Circular chloroplast DNA from *Euglena gracilis*. *Proc. Nat. Acad. Sci. U.S.A.* **68**, 1169–73.

MESELSON M. & STAHL F.W. (1958) The replication of DNA in *Escherichia coli*. *Proc. Nat. Acad. Sci. U.S.A.* **44**, 671–82.

MESELSON M., YUAN R. & HEYWOOD J. (1972) Restriction and modification of DNA. *Ann. Rev. Biochem.* **41**, 447–66.

METS L.J. & BOGORAD L. (1971) Mendelian and uniparental alterations in erythromycin binding by plastid ribosomes. *Science*, **174**, 707–9.

RHOADES M.M. (1955) Interaction of genic and non-genic hereditary units and the physiology of non-genic inheritance. In *Encyclopedia of Plant Physiology*, ed. Ruhland W. Vol. 1, pp. 19–57. Springer, Berlin.

RIS H. & PLAUT W. (1962) Ultrastructure of DNA-containing areas in the chloroplast of *Chlamydomonas*. *J. Cell. Biol.* **13**, 383–91.

RYAN R.S., CHIANG K.-S. & SWIFT H. (1974) Circular DNA from mitochondria of *Chlamydomonas reinhardtii*. *J. Cell. Biol.* **63**, 293a.

RYAN R.S., GRANT D., CHIANG K.-S. & SWIFT H. (1973) Isolation of mitochondria and characterization of the mitochondrial DNA of *Chlamydomonas reinhardtii*. *J. Cell. Biol.* **59**, 297a.

SAGER R. (1954) Mendelian and non-Mendelian inheritance of streptomycin resistance in *Chlamydomonas*. *Proc. Nat. Acad. Sci. U.S.A.* **40**, 356–62.

SAGER R. (1962) Streptomycin as a mutagen for nonchromosomal genes. *Proc. Nat. Acad Sci. U.S.A.* **48**, 2018–26.

SAGER R. (1972) *Cytoplasmic Genes and Organelles*, 405 pp. Academic Press, New York.

SAGER R. & ISHIDA M.R. (1963) Chloroplast DNA in *Chlamydomonas*. *Proc. Nat. Acad. Sci. U.S.A.* **50**, 725–30.

SAGER R. & LANE D. (1972) Molecular basis of maternal inheritance. *Proc. Nat. Acad. Sci. U.S.A.* **69**, 2410–3.

SAGER R. & RAMANIS Z. (1963) The particulate nature of nonchromosomal genes in *Chlamydomonas*. *Proc. Nat. Acad. Sci. U.S.A.* **50**, 260–8.

SAGER R. & RAMANIS Z. (1965) Recombination of nonchromosomal genes in *Chlamydomonas*. *Proc. Nat. Acad. Sci. U.S.A.* **53**, 1053–61.

SAGER R. & RAMANIS Z. (1967) Biparental inheritance of non-chromosomal genes induced by ultraviolet irradiation. *Proc. Nat. Acad. Sci. U.S.A.* **58**, 931–7.

SAGER R. & RAMANIS Z. (1968) The pattern of segregation of cytoplasmic genes in *Chlamydomonas*. *Proc. Nat. Acad. Sci. U.S.A.* **61**, 324–31.

SAGER R. & RAMANIS Z. (1970) A genetic map of non-Mendelian genes in *Chlamydomonas*. *Proc. Nat. Acad. Sci. U.S.A.* **65**, 593–600.

SAGER R. & RAMANIS Z. (1971a) Methods of genetic analysis of chloroplast DNA in *Chlamydomonas*. In *Autonomy and Biogenesis of Mitochondria and Chloroplasts*, pp. 250–9. Eds. Boardman N.K., Linnane A.W. & Smillie R.M. North-Holland, Amsterdam.

SAGER R. & RAMANIS Z. (1971b) Formal genetic analysis of organelle genetic systems. *Stadler Symposia*. Vol. 1 & 2, pp. 65–77.

SAGER R. & RAMANIS Z. (1971c) Persistent cytoplasmic heterozygotes in *Chlamydomonas*. *Genetics*, **68**, s56.

SAGER R. & RAMANIS Z. (1972) In preparation.

F

SAGER R. & RAMANIS Z. (1973) The mechanism of maternal inheritance in *Chlamydomonas*: biochemical and genetic studies. *Theoret. Applied Genet.* **43,** 101–8.

SAGER R. & RAMANIS Z. (1974) Mutations that alter the transmission of chloroplast genes in *Chlamydomonas*. *Proc. Nat. Acad. Sci. U.S.A.* **71,** 4698–702.

SAGER R. & RAMANIS Z. (1975) Effects of UV irradiation on segregation and recombination of chloroplast genes in *Chlamydomonas*. *Genetics* **80,** s72.

SCHIMMER O. & ARNOLD C.G. (1969) Untersuchungen zur Lokalisation eines äusserkaryotischen Gens bei *Chlamydomonas reinhardi*. *Arch. Mikrobiol.* **66,** 199–202.

SCHIMMER O. & ARNOLD C.G. (1970a) Untersuchungen über Reversions- und Segregations verhalten eines äusserkaryotischen Gens von *Chlamydomonas* zur Bestimmung des Erbträgers. *Molec. gen. Genet.* **107,** 281–90.

SCHIMMER O. & ARNOLD C.G. (1970b) Über die Zahl der Kopien eines äusserkaryotischen Gens bei *Chlamydomonas reinhardii*. *Molec. gen. Genet.* **107,** 366–71.

SCHIMMER O. & ARNOLD C.G. (1970c) Hin -und Rücksegregation eines äusserkaryotischen Gens bei *Chlamydomonas reinhardii*. *Molec. gen. Genet.* **108,** 33–40.

SCHLANGER G. & SAGER R. (1974) Correlation of chloroplast DNA and cytoplasmic inheritance in *Chlamydomonas* zygotes. *J. Cell. Biol.* **63,** 301a.

SIERSMA P.W. & CHIANG K.-S. (1971) Conservation and degradation of cytoplasmic and chloroplast ribosomes in *Chlamydomonas reinhardtii*. *J. Mol. Biol.* **58,** 167–85.

STUTZ E. (1970) The kinetic complexity of *Euglena gracilis* chloroplast DNA. *FEBS Lett.* **8,** 25–8.

SURZYCKI S.J. & GILLHAM N.W. (1971) Organelle mutations and their expression in *Chlamydomonas reinhardi*. *Proc. Nat. Acad. Sci. U.S.A.* **68,** 1301–6.

SURZYCKI S.J., GOODENOUGH U.W., LEVINE R.P. & ARMSTRONG J.J. (1970) Nuclear and chloroplast control of chloroplast structure and function in *Chlamydomonas reinhardi*. *Symp. Soc. Expt. Biol.* **XXIV,** 13–37.

SWINTON D., EVES E. & CHIANG K.-S. (1974) Cyclic variation of the pattern of thymidine incorporation through the life cycle of *Chlamydomonas*. *J. Cell Biol.* **63,** 340a.

WELLS R. & SAGER R. (1971) Denaturation and the renaturation kinetics of chloroplast DNA from *Chlamydomonas reinhardi*. *J. Mol. Biol.* **58,** 611–22.

WETMUR J.G. & DAVIDSON N. (1968) Kinetics of renaturation of DNA. *J. Mol. Biol.* **31,** 349–70.

CHAPTER 6

GENETICS OF CHLOROPLAST RIBOSOME BIOGENESIS IN *CHLAMYDOMONAS REINHARDTII*

ELIZABETH W. HARRIS†, JOHN E. BOYNTON† AND
NICHOLAS W. GILLHAM*

Departments of Botany† and Zoology*, Duke University,
Durham, North Carolina 27706, U.S.A.

1 Introduction 119

2 The chloroplast protein-synthesis-
ing system of *Chlamydomonas*
121

3 Mutants with altered sensitivity to
antibiotics which inhibit chloroplast protein synthesis 125

4 Mutants blocked in ribosome
assembly 133

5 Interrelationship between nuclear
and chloroplast genomes in the
biogenesis of organelle ribosomes
136

6 References 140

1 INTRODUCTION

Chloroplast ribosome biogenesis in *Chlamydomonas reinhardtii* appears to require participation of both nuclear and chloroplast genes. Mutations which directly affect chloroplast ribosome structure or function are known in both genomes and can be distinguished by their Mendelian (nuclear) and uni-parental (chloroplast) patterns of inheritance (see Chapter 5). Mutants with altered sensitivity to antibiotics which inhibit chloroplast protein synthesis (i.e. protein synthesis on chloroplast ribosomes) are known in both genomes (Gillham 1965, 1969; Mets & Bogorad 1971; Sager 1972; Surzycki & Gillham 1971). Among these are certain mutant strains specifically resistant to such antibacterial compounds as streptomycin, spectinomycin, neamine, and erythromycin, and others dependent on streptomycin or neamine (Gillham 1965, 1969; Sager 1972; Sager & Tsubo 1962). A number of other mutants, deficient in chloroplast ribosomes, are unable to grow photosynthetically but can grow when supplied with acetate as a carbon source. Because its chloroplast ribosome functions are dispensable, *C. reinhardtii* has unique experimental potential for the identification of genes involved in ribosome formation, mutation of which might be lethal in other organisms (Harris *et al.* 1974). Five mutants of this type have been characterized so far; all are deficient in chloroplast ribosome monomers; and all show Mendelian inheritance (Boynton *et al.* 1970; Goodenough & Levine 1970; Harris *et al.* 1974). It may also be possible to find mutants blocked in chloroplast protein synthesis in which the responsible gene is uniparentally inherited. Until recently, the search for such mutants has been hindered by lack of a positive selection technique for acetate-requiring mutants and by the high background of Mendelian acetate-requiring mutants obtained after mutagenesis (Gillham & Levine 1962a). These difficulties may now be overcome, as discussed by Harris *et al.* (1974), by using arsenate as a selective inhibitor of photosynthesizing cells (Hudock & Togasaki 1974), and by mutagenizing diploid cultures to eliminate expression of recessive Mendelian mutations while allowing vegetative segregation of mutations in chloroplast DNA (Lee *et al.* 1973).

Cells blocked in chloroplast ribosome formation, like cells in which chloroplast protein synthesis is inhibited by treatment with antibiotics, show a syndrome of chloroplast defects which include deficiencies in ribulose diphosphate carboxylase (RuDPCase) activity and in photosystem II electron transport (Hill reaction activity), as well as characteristic alterations in lamellar organization (cf. Figs. 6.4A, B & 6.9A, B; Boynton *et al.* 1972; Boynton *et al.* 1973; Goodenough *et al.* 1971; Hoober *et al.* 1969; Surzycki *et al.* 1970). The photosystem II deficiency evidently involves the loss of cytochrome 553 and 563 activity (Armstrong *et al.* 1971; Levine & Armstrong 1972). In contrast, other components involved in photosynthesis, such as ferredoxin, plastocyanin, and carbon-cycle enzymes (apart from RuDPCase), are not affected (Table 6.1) and therefore are presumed to be synthesized on cytoplasmic ribosomes.

Other types of mutants blocked in chloroplast protein synthesis would be expected also to exhibit this syndrome. For example, mutants with defective chloroplast ribosomal RNA should be acetate-requiring (that is, unable to grow photosynthetically with CO_2 as sole carbon source) and would presumably show uniparental inheritance, since evidence from nucleic acid hybridization studies (Bastia *et al.* 1971) and from work with transcriptional inhibitors (Surzycki & Rochaix 1971) indicates that the cistrons for chloroplast rRNA are located in chloroplast DNA. One should also be able to obtain mutants affecting auxiliary components of chloroplast protein synthesis, such as transfer RNAs, initiation factors, aminoacyl-tRNA synthetases, etc. These mutants should form structurally normal ribosomes but would show the syndrome of deficiency in chloroplast protein synthesis.

Table 6.1. Chloroplast components affected or not affected in the chloroplast ribosome-deficient mutant *ac-20* of *Chlamydomonas reinhardtii* (Surzycki *et al.* 1970, modified in accordance with Levine & Armstrong 1972).

Affected	More or less unaffected
	1. Phosphoribulokinase
	2. Phosphoriboisomerase
	3. 3-PGA kinase
1. RuDP carboxylase	4. G-3-P dehydrogenase (NAD)
	5. G-3-P dehydrogenase (NADP)
	6. Triosephosphate isomerase
	7. FDP aldolase
	8. Total quinone
	9. Plastocyanin
2. Cytochromes 553 and 563	10. P-700
3. Cytochrome Q	11. Ferredoxin
	12. Ferredoxin-NADP reductase
	13. Chlorophyll (reduced at most by half)
	14. Carotenoid (reduced by half)
4. Chloroplast membrane	15. Membrane formation (reduced by half)
organization	16. Eyespot formation
	17. Starch synthesis
5. Pyrenoid formation	

2   THE CHLOROPLAST PROTEIN-SYNTHESIZING SYSTEM OF *CHLAMYDOMONAS*

Wild-type cells of *Chlamydomonas reinhardtii* contain two major classes of ribosome monomers: 83S* cytoplasmic ribosomes (Sager & Hamilton 1967) containing 25S and 18S rRNA (Bourque *et al.* 1969, 1971; Goodenough & Levine 1970; Hoober & Blobel 1969), and 70S chloroplast ribosomes containing 23S and 16S rRNA (Figs. 6.1A, B; Bourque *et al.* 1969, 1971; Hoober & Blobel 1969; Goodenough & Levine 1970). Mitochondrial ribosomes form a minority class, detectable in electron micrographs but not yet characterized physically or biochemically. Minor components sedimenting at 41S and 54S, which are consistently seen in sucrose gradient profiles of wild-type cells, can respectively be attributed to the small and large subunits of chloroplast

* We have referred to ribosome peaks in accordance with the generic classification of Bourque *et al.* (1971), who defined five principal classes of ribosomes, respectively sedimenting on sucrose gradients at 83, 70, 66, 54, and 41S, as determined by linear extrapolation using the 83s cytoplasmic monomer peak as standard. By analytical ultracentrifugation, Siersma and Chiang (1971) found major peaks with $S_{20,w}$ values of 79 and 70S, and minor particle peaks at 67, 57, 51, and 30S. Hoober and Blobel (1969) refer to 80S (cytoplasmic) and 68S (chloroplast) monomers, with chloroplast ribosomal subunits sedimenting at 33S and 28S. Mets and Bogorad (1972) assign S_{20}, w values of 52S and 37S to the chloroplast subunits.

ribosomes (Boynton *et al.* 1972). These subunits (as well as the small and
large subunits of cytoplasmic ribosomes) can be prepared quantitatively by
dissociation of monomer ribosomes in high salt concentrations (Chua *et al.*
1973), and can be reassociated in low concentrations into active monosomes
capable of amino-acid incorporation *in vitro* (Chua *et al.* 1973; Schlanger &
Sager 1974).

Individual ribosomal proteins of the subunits of both cytoplasmic and
chloroplast ribosomes of wild-type *Chlamydomonas* have been separated and

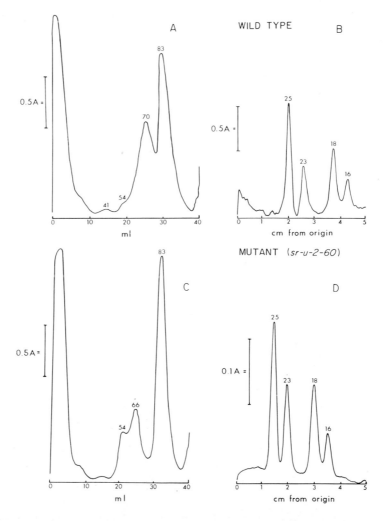

Fig. 6.1. Ribosome gradients (A_{254}) (A, C) and ribosomal RNA separated on aga-
rose-acrylamide gels (A_{260}) (B, D) from wild-type and streptomycin-resistant
mutant cells of *Chlamydomonas reinhardtii*. (Gillham *et al.* 1970.)

identified by two-dimensional gel electrophoresis (Hanson *et al.* 1974). The small chloroplast subunit is estimated to comprise 22 proteins (Fig. 6.2a) while the large chloroplast subunit comprises about 26 (Fig. 6.2b). Ohta, Inouye and Sager (1975) have identified 19 distinct components by carboxy-methylcellulose column chromatography of the proteins in the small chloroplast ribosomal subunit. In comparison, the small (30S) and large (50S) subunits of *Escherichia coli* ribosomes contain 21 and 34 proteins, respectively (Kaltschmidt & Wittmann 1970).

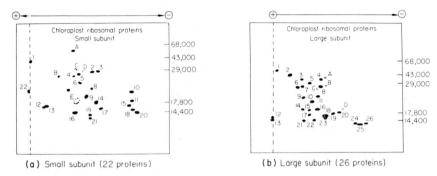

(a) Small subunit (22 proteins)   (b) Large subunit (26 proteins)

Fig. 6.2. Two-dimensional electrophoretograms of proteins from chloroplast ribosomal subunits of *Chlamydomonas reinhardtii*. (A–E: contaminants?) (Hanson *et al.* 1974.)

Transcriptional mapping studies using actinomycin D, rifampicin, and synchronous initiation of RNA synthesis suggest that each chloroplast DNA molecule of *Chlamydomonas* contains two to three copies of each rRNA cistron. These appear to be arranged in tandem repeats of a transcriptional unit coding for 16S and 23S rRNA, with the 16S rRNA gene located proximal to the promoter in each unit. Hybridization data also indicate three copies of each chloroplast rRNA cistron per unique copy of the chloroplast genome (Bastia *et al.* 1971). These estimates may be compared with the range of 5–10 rRNA cistrons per bacterial chromosome (Davies & Nomura 1972).

In general, chloroplast and mitochondrial protein synthesis in eukaryotic organisms resembles that in prokaryotic cells. Chloroplast ribosomes and mitochondrial ribosomes of many organisms are similar to bacterial ribosomes in their overall size and sedimentation velocity, and in the size and organization of their RNA and protein constituents (Boardman *et al.* 1966; Hoober & Blobel 1969; Küntzel 1969a, b; Loening & Ingle 1967; Rifkin *et al.* 1967). Mitochondrial ribosomes of vertebrates, however, are smaller than those of algae and fungi, apparently because of differences in their ribosomal RNAs (Mahler 1973). Organelle and bacterial ribosomes are also similar in functional properties such as (1) magnesium optima *in vitro* (Boardman *et al.* 1966; Chua *et al.* 1973; Grivell *et al.* 1971), (2) mechanism of chain initiation with natural messenger RNA (Bachmayer 1970; Burkhard *et al.* 1969; Epler *et al.* 1970; Galper & Darnell 1969; Halbreich & Rabinowitz 1971; Schwartz *et al.* 1967; Smith & Marcker 1968), and (3) sensitivity to certain antibiotic substances (Ellis 1970; Eisenstadt & Brawerman 1964; Hoober & Blobel 1969; Kroon 1969; Lamb *et al.* 1968; Nomura 1970; Pestka 1971; Sebald *et al.* 1969; Vazquez *et al.* 1969; Weisblum & Davies 1968). Hybrid ribosomes, reconstituted from one chloroplast ribosomal subunit (from *Euglena* or spinach) and the reciprocal subunit from *E. coli*, have been found to be active in poly-U-directed phenylalanine incorporation *in vitro* (Grivell & Walg 1972; Lee & Evans 1971). The comparable hybrids between bacterial and eukaryotic cytoplasmic ribosomal subunits are inactive.

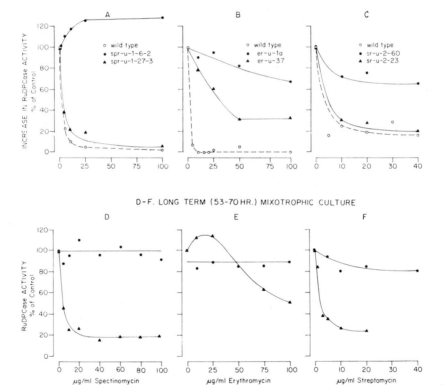

Fig. 6.3. Effects of antibiotics which inhibit chloroplast protein synthesis on activity of ribulose diphosphate carboxylase (RuDPCase) in wild-type and antibiotic-resistant mutants of *Chlamydomonas reinhardtii*. (A—C. 100% value represents a 2- to 3-fold increase in activity per cell of each genotype attributable to synthesis of new enzyme in absence of antibiotic. D—F. 100% value represents activity per cell of untreated cultures of each genotype.) (Boynton *et al.* 1973; Conde *et al.* 1975.)

3 MUTANTS WITH ALTERED SENSITIVITY TO ANTIBIOTICS WHICH INHIBIT CHLOROPLAST PROTEIN SYNTHESIS

One can readily isolate mutants of *Chlamydomonas* which are resistant to specific antibiotics known to inhibit protein synthesis in organelles and in prokaryotic cells. Several different methods have been used to demonstrate that in a number of these mutants antibiotic resistance is attributable to direct effects of the agent on chloroplast ribosomes. Resistance of chloroplast protein synthesis *in vivo* can be assessed by the ability of cells to synthesize RuDPCase (Fig. 6.3), and to maintain their capacity for photosynthetic electron transport and their normal chloroplast lamellar structure (Figs. 6.4A, B) in the presence of an antibiotic (Boynton *et al.* 1973; Conde *et al.* 1975). Some mutations to antibiotic resistance alter chloroplast ribosomal profiles, as displayed on sucrose gradients, by shifting the sedimentation value of the ribosome monomer or by causing accumulation of one of the ribosomal subunits, presumably by interfering with assembly of the other subunit (Fig. 6.1C; Gillham *et al.* 1970). Isolated mutant ribosomes may also fail to bind radioactively labelled antibiotic (Fig. 6.5; Boynton *et al.* 1973; Burton 1972; Mets & Bogorad 1971) and may be insensitive *in vitro* to inhibitors of protein synthesis (polynucleotide-directed amino-acid incorporation) (Fig. 6.6; Conde *et al.* 1975; Schlanger & Sager 1974; Schlanger *et al.* 1972). In some cases the basis of this resistance has been shown to reside in one of the ribosomal subunits by reciprocal reconstitution experiments with wild-type subunits (Fig. 6.7; Schlanger & Sager 1974). In a few instances the mutant effect has been localised at the level of a specific ribosomal protein by alterations of the banding patterns obtained by electrophoresis or column chromatography (Davidson *et al.* 1974; Mets & Bogorad 1972; Ohta *et al.* 1975). Table 6.2 summarizes the results of these analyses of both Mendelian and uniparental antibiotic-resistant mutants of *Chlamydomonas*.

The uniparental spectinomycin-resistant mutant *spr-u-1-27-3* has been studied in all of these respects. Although *in vivo* its chloroplast protein synthesis appears to be as sensitive to spectinomycin as that of wild-type cells (Figs. 6.3A, D, 6.4B; Boynton *et al.* 1973; Conde *et al.* 1975), its behaviour *in vitro* clearly indicates a direct effect of the mutation on chloroplast ribosome structure and function. Chloroplast ribosome monomers of *spr-u-1-27-3* sediment at approximately 66S rather than 70S (Gillham *et al.* 1970), and fail to bind labelled dihydrospectinomycin (Fig. 6.5; Boynton *et al.* 1973; Burton 1972). Chloroplast ribosomes of *spr-u-1-27-3 in vitro* are somewhat more resistant than those of wild-type cells to inhibition by spectinomycin of poly-U-directed phenylalanine incorporation (Fig. 6.6A; Conde *et al.* 1975). Polyacrylamide-gel electrophoresis of proteins of the small subunit of chloroplast ribosomes from *spr-u-1-27-3* reveals that a protein band found in wild-type cells is missing in this mutant (Boynton *et al.* 1973).

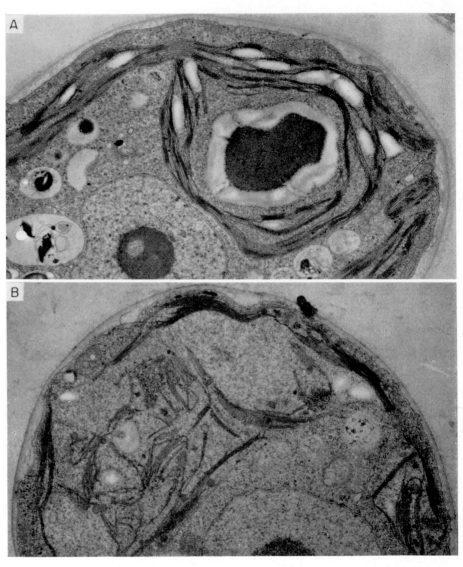

Fig. 6.4. Median sections through cells of *Chlamydomonas reinhardtii* mutants grown mixotrophically to late log phase in presence of 90 μg/ml spectinomycin (× 9,000). A. *spr-u-1-6-2*. (Pyrenoid present; chloroplast lamellar organization normal.) B. *spr-u-1-27-3*. (Pyrenoid absent; chloroplast lamellar organization abnormal; other cell organelles appear unaffected.) (Conde *et al.* 1975.)

The subunit site of spectinomycin resistance in chloroplast ribosomes of *spr-u-1-6-2* has not yet been specifically established. Although this mutant appears to be allelic with *spr-u-1-27-3* (Conde *et al.* 1975), it differs greatly from *spr-u-1-27-3* in that spectinomycin has very little inhibitory effect on its chloroplast protein synthesis *in vivo* (Figs. 6.3A, D; 6.4A) or *in vitro* (Fig. 6.6A; Conde *et al.* 1975). Like *spr-u-1-27-3*, this mutant displays an altered ribosomal profile in sucrose gradients.

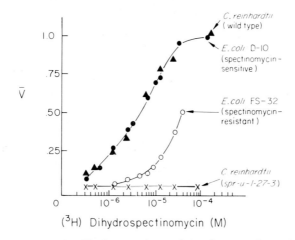

Fig. 6.5. Binding of ³H-dihydro-spectinomycin to ribosomes from algal chloroplasts and from bacteria. ($\bar{V}$ = mol antibiotic bound per mol ribosome) (Boynton *et al.* 1973).

Another mutant, *spc*, has been found by Schlanger and Sager (1974) to have small chloroplast-ribosomal subunits resistant to spectinomycin in poly-U-directed phenylalanine incorporation (Fig. 6.7). Since this mutant also appears to be allelic with *spr-u-1-27-3* (Conde *et al.* 1975) only one chloroplast gene locus for spectinomycin resistance has been identified.

Chloroplast ribosomes from a uniparental neamine-resistant mutant, *nea* (*nr-u-2-1*), are resistant to neamine inhibition of phenylalanine incorporation *in vitro*, and this resistance is a property of the small chloroplast-ribosomal subunit (Fig. 6.7; Schlanger & Sager 1974).

At least five independently isolated uniparental mutants resistant to streptomycin have demonstrably altered chloroplast ribosomes. Some of these are distinguishable phenotypically. For example, *sr-u-2-23* cannot be distinguished from wild type with respect to streptomycin inhibition of RuDPCase synthesis *in vivo*, whereas *sr-u-2-60* is clearly resistant (Fig. 6.3C, F; Conde *et al.* 1975). Chloroplast ribosomes of *sr-u-2-60* are highly resistant to streptomycin-induced misreading of isoleucine for phenylalanine with poly-U as message, whereas chloroplast ribosomes of *sr-u-2-23* misread in the presence of streptomycin, although less extensively than those of wild-

Table 6.2. *Chlamydomonas reinhardtii* mutants with altered antibiotic sensitivity resulting from a direct effect on chloroplast ribosomes.

Mutant	Phenotype	Mode of inheritance	Evidence for effect on chloroplast ribosomes
spr-u-1-27-3	spectinomycin-resistant	UP (Boynton et al. 1973)	G (Gillham et al. 1970) B (Boynton et al. 1973; Burton 1972) *RM (Conde et al. 1975) PS (Boynton et al. 1973)
spr-u-1-6-2	spectinomycin-resistant	UP (Conde et al. 1975)	S, G, RM (Conde et al. 1975)
spc	spectinomycin-resistant	UP (Sager 1972)	RS (Schlanger & Sager 1974)
nea (nr-u-2-1)	neamine-resistant	UP (Gillham 1965; Sager 1972)	RS (Schlanger & Sager 1974)
sr-u-2-60	streptomycin-resistant	UP (Gillham & Levine 1962b)	S, RM (Conde et al. 1975) G (Gillham et al. 1970)
sr-u-2-23	streptomycin-resistant	UP (Conde et al. 1975)	*RM (Conde et al. 1975)
sm-2	streptomycin-resistant	UP (Sager 1954)	RS (Schlanger & Sager 1974) PS (Ohta et al. 1975)
sr-u-2-281	streptomycin-resistant	UP (Gillham & Levine 1962b)	G (Gillham et al. 1970)
sr_{35}	streptomycin-resistant	UP (Boschetti & Bogdanov 1973)	S (Boschetti & Bogdanov 1973)
sd-u-3-18	streptomycin-dependent	UP (Gillham 1965)	G (Gillham et al. 1970)
ery-M1	erythromycin-resistant	M (Mets & Bogorad 1971)	B (Mets & Bogorad 1971) PL (Davidson et al. 1974)
ery-M2	erythromycin-resistant	M (Mets & Bogorad 1971)	B (Mets & Bogorad 1971) PL (Mets & Bogorad 1972)

er-u-1a	erythromycin-resistant	UP (Mets & Bogorad 1971)	B (Mets & Bogorad 1971) PL (Mets & Bogorad 1972) S, RM (Conde *et al.* 1975)
er-u-37	erythromycin-resistant	UP (Conde *et al.* 1975)	S, RM (Conde *et al.* 1975)
car	carbomycin-resistant	UP (Schlanger *et al.* 1972; Sager 1972)	RL (Schlanger *et al.* 1972; Schlanger & Sager 1974)
cle	cleocin-resistant	UP (Sager 1972)	RL (Schlanger & Sager 1974)

B = chloroplast ribosomes fail to bind labelled antibiotic *in vitro*
G = altered chloroplast-ribosomal profile on sucrose gradients
M = Mendelian inheritance
PL = altered ribosomal protein in large subunit
PS = altered ribosomal protein in small subunit
RL = resistance localized to large subunit
RM = ribosome monomers resistant in amino-acid incorporation *in vitro*
RS = resistance localized to small subunit
S = chloroplast protein synthesis resistant *in vivo*
UP = uniparental inheritance

* Ribosomes only slightly more resistant than those of wild-type cells.

type cells (Fig. 6.6C; Conde *et al.* 1975). Sucrose-gradient analyses of *sr-u-2-60* and *sr-u-2-281* show that the 70S ribosomes normally present are entirely replaced by 66S ribosomes (which still contain both 23S and 16S rRNA), and that large amounts of 54S particles are also accumulated (Fig. 6.1C, D; Gillham *et al.* 1970). Ribosome profiles of *sr-u-2-23* are not noticeably different from those of wild type (Conde *et al.* 1975). Since *sr-u-2-60* shows about 3% recombination with *sr-u-2-23* and about 14% recombination with Sager's *sm-2* mutant (Conde *et al.* 1975), there appear to be at least three discrete loci for streptomycin resistance in the chloroplast genome.

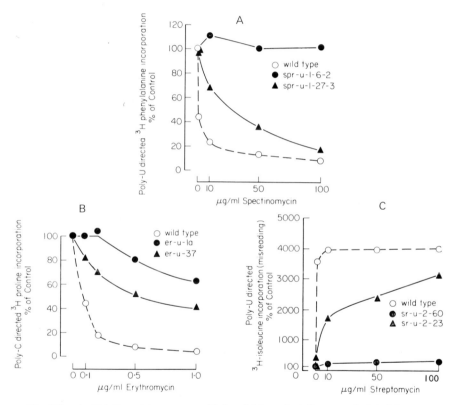

Fig. 6.6. Antibiotic resistance or sensitivity of chloroplast ribosomes isolated from wild type and antibiotic-resistant mutants of *Chlamydomonas reinhardtii*. Amino-acid incorporation *in vitro* as described by Schlanger and Sager (1974), with ³H-isoleucine substituted for ³H-phenylalanine to measure streptomycin-induced misreading, and with poly-C and ³H-proline substituted for poly-U and ³H-phenylalanine to measure erythromycin resistance. (Conde *et al.* 1975.)

In the case of *sm-2*, streptomycin resistance of phenylalanine incorporation *in vitro* has been traced to the small chloroplast-ribosomal subunit (Fig. 6.7; Schlanger & Sager 1974), and it is presumably the result of an

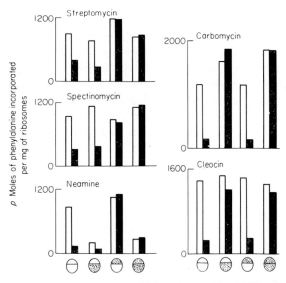

Fig. 6.7. Antibiotic resistance or sensitivity of recombined 30S and 50S chloroplast ribosomal subunits from wild-type and resistant mutants of *Chlamydomonas reinhardtii*. (Subunits from wild-type (white) and mutant corresponding to antibiotic indicated (dotted) were reassociated in the 4 combinations shown. Phenylalanine-incorporating activity of reassociated ribosomes was examined in the absence (white bars) and presence (black bars) of the antibiotic corresponding to the mutant ribosomes.) (Schlanger & Sager 1974.)

alteration of a specific protein, as revealed by carboxymethylcellulose chromatography of dissociated small-subunit proteins (Ohta *et al.* 1975).

In another mutant, sr_{35}, RuDPCase synthesis has been found to be resistant to streptomycin (Boschetti & Bogdanov 1973). Although the ribosome-sedimentation profile of this mutant resembles that of wild-type cells, it differs from wild type in the extent to which streptomycin pretreatment *in vivo* can induce dimerization of the 70S ribosomes (Boschetti & Bogdanov 1973).

The streptomycin-dependent mutant *sd-u-3-18* has been shown to contain fewer 70S ribosomes than wild type and to accumulate large ribosomal subunits (Gillham *et al.* 1970). Under starvation conditions the proportion of 70S ribosomes in this mutant is reduced even further, and additional large subunits are accumulated, suggesting that streptomycin dependence results from a specific defect in the small subunit or its assembly that can be corrected by streptomycin (Gillham *et al.* 1970). Sensitive revertants of this mutant, isolated by Schimmer and Arnold (1970), contain normal levels of 70S ribosomes and do not accumulate large subunits (Gillham *et al.* 1970).

Both Mendelian and uniparentally inherited mutations inducing resistance to erythromycin affect chloroplast ribosomes. Neither of the nonallelic

1971). A second ribosome-assembly mutant, *cr-1*, arose spontaneously in a stock of *ac-20*, but differs from the latter in that it accumulates 54S ribosomal particles (Fig. 6.8) containing 23S rRNA, which represent the large chloroplast ribosomal subunits (Boynton *et al.* 1970, 1972). Three other Mendelian mutants, induced by nitrosoguanidine, have more recently been characterized (Harris *et al.* 1974). The ribosome profile of *cr-2* is similar to that of *cr-1*; *cr-3* also accumulates 54S particles, but makes some 70S ribosome monomers as well; *cr-4* is similar in ribosome phenotype to *ac-20*, but with slightly more 70S ribosomes (Fig. 6.8).

In these mutants the content of 70S ribosomes is correlated with RuDP-Case and Hill-reaction activities, but not with chlorophyll content (Harris *et al.* 1974). This further supports previous conclusions that chloroplast ribosomal function *in vivo* is specifically required for RuDPCase synthesis and for the formation of some component(s) of photosystem II, but is not necessary for chlorophyll accumulation (Table 6.1; Boynton *et al.* 1972, 1973; Goodenough & Levine 1970). The severity of defects in the organization of chloroplast lamellae also appears to be related to the content of chloroplast ribosomes (Harris *et al.* 1974); the defects are similar to those resulting from antibiotic inhibition of chloroplast protein synthesis (cf. Figs. 6.4A, B & 6.9A, B).

These five Mendelian ribosome-deficient mutants have been analysed genetically (Harris *et al.* 1974) by crossing them in all possible combinations to generate double mutants and vegetative diploids (Ebersold 1967). Recovery of double-mutant and wild-type recombinants at an appreciable frequency in every cross indicates that the mutants represent five distinct loci, and phenotypic analysis of diploids indicates that these mutants can complement one another. By comparing the photosynthetic capacity and chloroplast ribosome profiles of the individual mutants, of their double mutants, and of their diploids (with each other and with wild type) a working model can be formulated for the assembly of chloroplast ribosomes, analogous to the formal schemes used to deduce enzymatic pathways from the behaviour of auxotrophic mutants. Relations of dominance and epistasis among this group of mutants can be used to rank them in a probable sequential order of function (Fig. 6.10). Thus *ac-20* and *cr-4* appear to affect the formation of components common to both the small and the large ribosomal subunits, whereas *cr-1*, *cr-2*, and *cr-3* (all of which accumulate 54S particles) are probably impaired in a specific pathway leading to the formation of the 41S subunit. Both *ac-20* and *cr-4* are epistatic to *cr-1*, *cr-2*, and *cr-3*, suggesting that these steps affecting both subunits occur prior to a branch point rather than at a terminal stage of ribosome assembly. In diploids *cr-1* is partially dominant over wild type, whereas the other four mutants are recessive. This suggests that *cr-1* blocks synthesis of a ribosomal component needed in stoichiometric amounts, i.e. a structural component; diploids containing one *cr-1* and one *cr-1*[+] (wild-type) allele thus make fewer functional ribosomes per cell than do wild-type/wild-type diploids or wild-type/mutant diploids in which the mutant allele affects only a catalytic function.

This model for ribosome assembly indicates possible biochemical roles for each of the individual gene loci in ribosome formation, and suggests experimental approaches to verify the molecular defects in each of the mutants.

A   B

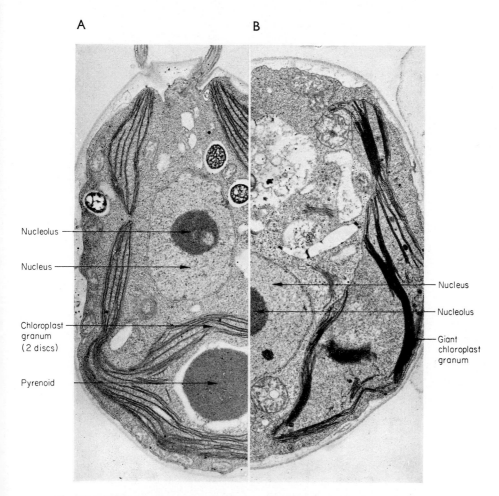

Fig. 6.9.   Median sections through mixotrophically grown cells of *Chlamydomonas reinhardtii* × 9,000. A. wild-type cell, showing the nucleus with nucleolus, cup-shaped chloroplast with a well-developed pyrenoid, and lamellar system largely organized into 2-disc grana. B. *cr-2* mutant cell. Chloroplast lacks a pyrenoid and the lamellae are stacked into giant grana. (Harris *et al.* 1974.)

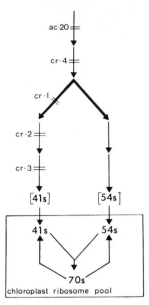

Fig. 6.10. Working model for chloroplast ribosome assembly in *Chlamydomonas reinhardtii*, based on genetic and phenotypic analysis of five non-allelic mutants deficient in chloroplast ribosomes. (Mutants *ac-20* and *cr-4* appear to be blocked early in ribosome assembly. Mutants *cr-1*, *cr-2*, and *cr-3* all accumulate 54S subunits, and are therefore thought to be blocked in 41S subunit formation.) (Harris *et al.* 1974.)

5 INTERRELATIONSHIP BETWEEN NUCLEAR AND CHLOROPLAST GENOMES IN THE BIOGENESIS OF ORGANELLE RIBOSOMES

Clearly both Mendelian and uniparental genes directly affect the structure, assembly, and function of chloroplast ribosomes in *Chlamydomonas*. At least two unlinked Mendelian loci specify protein alterations conferring antibiotic resistance, and mutations at five Mendelian loci which are not closely linked block the assembly of ribosomes. Thus no single nuclear operon contains information for all of the chloroplast ribosome determinants. The chloroplast genome of *Chlamydomonas* specifies chloroplast rRNA and a number of different proteins of both the large and small chloroplast ribosomal subunits. Sager and co-workers have found linkage between *nea*, *sm-2*, *ery-1*, *spc*, *car*, *cle*, and other chloroplast mutants. The mutants *er-u-37*, *er-u-1a*, *spr-u-1-27-3*, *spr-u-1-6-2*, *sr-u-2-60*, *sr-u-2-23*, and *nr-u-2-1* (which is identical with Sager's *nea*) also map in a single linkage group (see Ch. 5). Thus all the known chloroplast genes which affect chloroplast ribosomes appear to fall in a single linkage group, although some of the discrepancies in map order and distances have not yet been reconciled.

The relationship of the chloroplast rRNA cistrons to the chloroplast ribosomal-protein genes in *Chlamydomonas* has not yet been explored, nor have rRNA mutants been identified. Nomura and Engbaek (1972) have postulated that in *E. coli* at least four, and probably many other, ribosomal-protein genes are tightly linked and transcribed as a single unit or operon but are not closely linked to the genes for ribosomal RNAs. Clearly not all *Chlamydomonas* genes specifying chloroplast ribosomal proteins constitute a single operon. The known uniparental genes conferring antibiotic resistance on chloroplast ribosomes are linked, but it remains to be determined whether they are transcribed as a single unit and how they are related to the rRNA genes in the chloroplast genome.

The subunit localizations of specific antibiotic resistance in chloroplast ribosomes of *Chlamydomonas* mutants mirror those described in bacteria; that is, resistance to erythromycin, carbomycin, and cleocin can be ascribed to the large ribosomal subunit, whereas resistance to spectinomycin, strepto-mycin, and neamine results from protein alterations in the small subunit (Davies & Nomura 1972). Sager and co-workers have also noted similarities between their linkage map for uniparental genes of *Chlamydomonas* and the map positions of certain antibiotic-resistance genes in *Bacillus subtilis* (Schlanger & Sager 1974; Ohta *et al.* 1975). Ohta *et al.* further discuss similarities in chromatographic behaviour of the *E. coli* protein S-12, which is associated with streptomycin resistance, and the 'peak 17' protein they find to be altered in the *sm-2* mutant of *Chlamydomonas*. Moreover, the electrophoretic properties of the *spr* protein, found by Boynton *et al.* (1973) to be missing from ribosomes of the algal mutant *spr-u-1-27-3*, are not unlike those of S-5, the *spr* protein of the *E. coli* 30S subunit (Davies & Nomura 1972; Funatsu *et al.* 1971).

Antibiotic resistance in chloroplast ribosomes of certain *Chlamydomonas* mutants is accompanied by alterations extensive enough to interfere with ribosome assembly. For example, the mutants *sr-u-2-60* and *sd-u-3-18* accumulate 54S ribosomal subunits (Gillham *et al.* 1970), suggesting that assembly of the small subunits is affected by these mutations. Other mutations, at loci known to confer antibiotic resistance or in other chloroplast cistrons unrelated to antibiotic resistance, could produce non-functional ribosomes or ribosomal precursors unable to assemble, and they would then be recognizable as uniparental acetate-requiring mutants. Reconstitution studies with bacterial ribosomal proteins indicate that a precise topological relationship exists among the individual proteins of the small subunit, and that assembly of ribosomes proceeds in a sequential and ordered fashion (Nomura 1970). In analogous studies with *Chlamydomonas* chloroplast ribosomes, one should be able to combine the extensive bacterial methodology in this area with the potentially unrestricted range of mutants made possible because chloroplast protein synthesis in *Chlamydomonas* is dispensable.

Specification of certain chloroplast ribosome components in the chloroplast DNA does not necessarily imply that these components must be translated on chloroplast ribosomes. In fact, experiments with selective translational inhibitors in *Chlamydomonas* suggest otherwise. When *spr-u-1-27-3* is grown mixotrophically in the presence of spectinomycin at concentrations which totally inhibit RuDPCase synthesis, normal amounts of chloroplast ribosome particles are still found by sucrose-gradient analysis (Conde *et al.* 1975). Similar conclusions have been reached (Goodenough 1971) with a Mendelian spectinomycin-resistant mutant, *spA-2*, in which chloroplast protein synthesis *in vivo* is sensitive to spectinomycin. Using wild-type cells and the *arg-1* mutant, Honeycutt and Margulies (1973) have found that incorporation of arginine into chloroplast ribosomal protein can be inhibited by cycloheximide, which specifically inhibits the activity of eukaryotic cytoplasmic ribosomes, but not by chloramphenicol or spectinomycin, which affect only organelle ribosomes. All these studies indicate that chloroplast ribosomes continue to be made in the absence of chloroplast protein synthesis, and that therefore not only the chloroplast ribosomal proteins, but also the chloroplast DNA and RNA polymerases, must be synthesized on cytoplasmic ribosomes.

In spite of the fact that mutants deficient in chloroplast ribosomes are able to grow in the presence of acetate, suggesting that chloroplast protein synthesis is largely dispensable, Blamire, Flechtner and Sager (1974) have hypothesized that chloroplast protein synthesis is essential for replication of nuclear DNA. This hypothesis is based on the observation that incorporation of adenine into nuclear DNA is inhibited by antibiotics which block protein synthesis on organelle ribosomes. This inhibition is not observed, however, when certain antibiotic-resistant strains known to have altered chloroplast ribosomes are treated with their respective antibiotics. Blamire *et al.* (1974) conclude that a product of chloroplast protein synthesis regulates nuclear DNA replication, and that chloroplast ribosomes are the primary cellular target of antibiotics such as streptomycin, neamine, and spectinomycin.

In *Chlamydomonas*, these antibiotics probably also inhibit mitochondrial protein synthesis as they do in other organisms (Kroon 1969; Lamb *et al.* 1968; Sebald *et al.* 1969). Since mutants such as *spr-u-1-27-3* and *sr-u-2-23* can grow, in the presence of acetate, at concentrations of antibiotic which appear to inhibit totally their chloroplast protein synthesis *in vivo* and which are lethal to wild-type cells, the primary site of antibiotic resistance in both mutants may be the mitochondria rather than the chloroplast (Boynton *et al.* 1973; Conde *et al.* 1975). However, chloroplast ribosomes of *spr-u-1-27-3* and *sr-u-2-23* are slightly more resistant than those of wild-type cells *in vitro* (Table 6.2, Fig. 6.6C). In addition, *spr-u-1-27-3* has been found to have a protein alteration in the small subunit of the chloroplast ribosomes (Table 6.2, Boynton *et al.* 1973). Thus in each of these mutants a single mutational event seems to have altered both chloroplast and mitochondrial ribosomes, although the alteration in chloroplast ribosomes does not result in resistance of that organelle *in vivo*. Since both of these mutations are linked to other chloroplast genes for antibiotic resistance, it would appear that the chloroplast DNA codes for several structural components shared by both chloroplast and mitochondrial ribosomes (Boynton *et al.* 1973; Conde *et al.* 1975). The high incidence of one-step mutations which confer antibiotic resistance under phototrophic, mixotrophic, and heterotrophic growth conditions suggests that *spr-u-1-27-3* and *sr-u-2-23* are not unique, but rather that this phenomenon is general among those uniparental mutations which confer resistance to antibiotics inhibitory to both chloroplast and mitochondrial protein synthesis (Boynton *et al.* 1973; Conde *et al.* 1975; Surzycki & Gillham 1971).

This hypothesis is not unreasonable in view of the known differences between chloroplast and mitochondrial ribosomes (Borst & Grivell 1971; Mahler 1973; Sager 1972). Thus it is quite possible that a gene for a single protein, normally functional as a component of both chloroplast and mitochondrial ribosomes, could mutate to confer resistance in one case but not in the other. Furthermore, as discussed above, it appears that chloroplast ribosomal proteins are themselves synthesized on cytoplasmic ribosomes. (Evidence from studies with the fungus *Neurospora* (Küntzel 1969b; Lizardi & Luck 1972) indicates that the same is true for the proteins of mitochondrial ribosomes in that organism.) Thus a message transcribed from chloroplast DNA could be translated on cytoplasmic ribosomes, and the newly synthesized proteins could be returned to both chloroplast and mitochondria for assembly, along with other proteins (perhaps different for the two organelles) whose structure was specified by nuclear DNA.

The data of Blamire *et al.* (1974), showing that nuclear DNA synthesis in wild-type cells is inhibited by organelle-specific antibiotics such as streptomycin and is resistant in drug-resistant mutant strains which have resistant chloroplast ribosomes, are entirely consistent with the foregoing hypothesis. Furthermore, nuclear DNA synthesis must be resistant in cells of *spr-u-1-27-3* and *sr-u-2-23*, which have sensitive chloroplast protein synthesis *in vivo*, or else cells of these mutants could not grow on acetate-containing medium in the presence of these antibiotics.

A further implication of the hypothesis that chloroplast and mitochondrial ribosomes share certain genetic determinants located in the chloroplast DNA is that, if mitochondrial function is essential for cell survival under all growth conditions, the loss or deletion of a considerable part of the chloroplast DNA would be lethal, even though photosynthesis and chloroplast protein synthesis are dispensable. Thus, the interactions of nuclear and organelle genomes may be even more subtle than was previously imagined.

ACKNOWLEDGEMENTS

We wish to thank the following persons who have made significant contributions to this manuscript: Dr Lawrence Bogorad, Mr Jeffrey Davidson, and Ms Maureen Hanson, for providing us with stocks of er-u-1a and preprints of their papers; Dr Gladys Schlanger and Dr Ruth Sager, for stocks of the mutants er-11, sp-23 and sm-2, for preprints of their papers and for additional information regarding the poly-U assays.

This work was supported by Grants NSF-GB22769, NSF-GB43559 and NIH-GM19427 to J.E.B. and N.W.G. as well as NIH RCDA awards GM70453 to J.E.B. and GM70437 to N.W.G.

6 REFERENCES

ARMSTRONG J.J., SURZYCKI S.J., MOLL B. & LEVINE R.P. (1971) Genetic transcription and translation specifying chloroplast components in *Chlamydomonas reinhardi*. *Biochemistry*, **10**, 692–701.

BACHMAYER H. (1970) Initiation of protein synthesis in intact cells and in isolated chloroplasts of *Acetabularia mediterranea*. *Biochim. Biophys. Acta*, **209**, 584–6.

BASTIA D., CHIANG K.-S. & SWIFT H. (1971) Studies on the ribosomal RNA cistrons of chloroplast and nucleus in *Chlamydomonas reinhardtii*. Abstracts 11th Ann. Meeting Amer. Soc. Cell Biol. p. 25.

BIRGE E.A. & KURLAND C.G. (1969) Altered ribosomal protein in streptomycin-dependent *Escherichia coli*. *Science*, **166**, 1282–4.

BISWAS D.K. & GORINI L. (1972) The attachment site of streptomycin to the 30 S ribosomal subunit. *Proc. Nat. Acad. Sci. U.S.A.* **69**, 2141–4.

BLAMIRE J., FLECHTNER V.R. & SAGER R. (1974) Regulation of nuclear DNA replication by the chloroplast in *Chlamydomonas*. *Proc. Nat. Acad. Sci. U.S.A.* **71**, 2867–71.

BOARDMAN N.K., FRANCKI R.I.B. & WILDMAN S.G. (1966) Protein synthesis by cell-free extracts of tobacco leaves. III. Comparison of the physical properties and protein synthesizing activities of 70S chloroplast and 80S cytoplasmic ribosomes. *J. Mol. Biol.* **17**, 470–89.

BORST P. & GRIVELL L.A. (1971) Mitochondrial ribosomes. *FEBS Lett.* **13**, 73–88.

BOSCHETTI A. & BOGDANOV S. (1973) Different effects of streptomycin on the ribosomes from sensitive and resistant mutants of *Chlamydomonas reinhardi*. *Eur. J. Biochem.* **35**, 482–8.

BOURQUE D.P., BOYNTON J.E. & GILLHAM N.W. (1969) Studies on the synthesis of different species of ribosomal RNA in mutants of *Chlamydomonas reinhardi*. *J. Cell. Biol.* **43**, 15a.

BOURQUE D.P., BOYNTON J.E. & GILLHAM N.W. (1971) Studies on the structure and cellular location of various ribosome and ribosomal RNA species in the green alga *Chlamydomonas reinhardi*. *J. Cell. Sci.* **8**, 153–83.

BOYNTON J.E., BURTON W., GILLHAM N.W. & HARRIS E.H. (1973) Can a non-Mendelian mutation affect both chloroplast and mitochondrial ribosomes? *Proc. Nat. Acad. Sci. U.S.A.* **70**, 3463–7.

BOYNTON J.E., GILLHAM N.W. & BURKHOLDER B. (1970) Mutations altering chloroplast ribosome phenotype in *Chlamydomonas*. II. A new Mendelian mutation. *Proc. Nat. Acad. Sci. U.S.A.* **67**, 1505–12.

BOYNTON J.E., GILLHAM N.W. & CHABOT J.F. (1972) Chloroplast ribosome deficient mutants in the green alga *Chlamydomonas reinhardi* and the question of chloroplast ribosome function. *J. Cell. Sci.* **10,** 267–305.

BRECKENRIDGE L. & GORINI L. (1970) Genetic analysis of streptomycin resistance in *E. coli. Genetics,* **65,** 9–25.

BURKARD G., ECLANCHER B. & WEIL J.H. (1969) Presence of N-formyl-methionyl-transfer RNA in bean chloroplasts. *FEBS Lett.* **4,** 285–7.

BURTON W.G. (1972) Dihydrospectinomycin binding to chloroplast ribosomes from antibiotic-sensitive and resistant strains of *Chlamydomonas reinhardtii. Biochim. Biophys. Acta,* **272,** 305–11.

CHUA N.-H. BLOBEL G. & SIEKEVITZ P. (1973) Isolation of cytoplasmic and chloroplast ribosomes and their dissociation into active subunits from *Chlamydomonas reinhardtii. J. Cell. Biol.* **57,** 798–814.

CONDE M.F., BOYNTON J.E., GILLHAM N.W., HARRIS E.H., TINGLE C.L. & WANG W.L. (1975) Chloroplast genes in *Chlamydomonas* affecting organelle ribosomes. Genetic and biochemical analysis of antibiotic-resistant mutants at several gene loci. *Molec. Gen. Genet.* **140,** 183–220.

DAVIDSON J.N., HANSON M.R. & BOGORAD L. (1974) An altered chloroplast ribosomal protein in ery-M1 mutants of *Chlamydomonas reinhardi. Molec. Gen. Genet.* **132,** 119–29.

DAVIES J. & NOMURA M. (1972) The genetics of bacterial ribosomes. *Ann. Rev. Genet.* **6,** 203–34.

EBERSOLD W.T. (1967) *Chlamydomonas reinhardi*: Heterozygous diploid strains. *Science,* **157,** 447–9.

EISENSTADT J. & BRAWERMAN G. (1964) The protein-synthesizing systems from the cytoplasm and the chloroplasts of *Euglena gracilis. J. Mol. Biol.* **10,** 392–402.

ELLIS R.J. (1970) Further similarities between chloroplast and bacterial ribosomes. *Planta,* **91,** 329–35.

EPLER J.L., SHUGART L.R. & BARNETT W.E. (1970) N-formyl methionyl transfer ribonucleic acid in mitochondria from *Neurospora. Biochemistry,* **9,** 3575–9.

FUNATSU G., NIERHAUS K. & WITTMANN H.G. (1972) Ribosomal proteins. XXXVII. Determination of allele types and amino acid exchanges in protein S12 of three streptomycin-resistant mutants of *Escherichia coli. Biochim. Biophys. Acta,* **287,** 282–91.

FUNATSU G., SCHILTZ E. & WITTMANN H.G. (1971) Ribosomal proteins XXVII. Localization of the amino acid exchanges in protein S5 from two *Escherichia coli* mutants resistant to spectinomycin. *Molec. Gen. Genet.* **114,** 106–11.

FUNATSU G. & WITTMANN H.G. (1972) Ribosomal proteins XXXIII. Location of amino acid replacements in protein S12 isolated from *Escherichia coli* mutants resistant to streptomycin. *J. Mol. Biol.* **68,** 547–50.

GALPER J.B. & DARNELL J.E. (1969) The presence of N-formyl-methionyl tRNA in HeLa cell mitochondria. *Biochem. Biophys. Res. Comm.* **34,** 205–14.

GILLHAM N.W. (1965) Induction of chromosomal and nonchromosomal mutations in *Chlamydomonas reinhardi* with N-methyl-N′-nitro-N-nitrosoguanidine. *Genetics,* **52,** 529–37.

GILLHAM N.W. (1969) Uniparental inheritance in *Chlamydomonas reinhardi. Amer. Natur.* **103,** 355–88.

GILLHAM N.W., BOYNTON J.E. & BURKHOLDER B. (1970) Mutations altering chloroplast ribosome phenotype in *Chlamydomonas.* I. Non-Mendelian mutations. *Proc. Nat. Acad. Sci. U.S.A.* **67,** 1026–33.

GILLHAM N.W. & LEVINE R.P. (1962a) Pure mutant clones induced by ultra-violet light in the green alga, *Chlamydomonas reinhardi. Nature,* **194,** 1165–6.

GILLHAM N.W. & LEVINE R.P. (1962b) Studies on the origin of streptomycin-resistant mutants in *Chlamydomonas reinhardi. Genetics,* **47,** 1463–74.

GOODENOUGH U.W. (1971) The effects of inhibitors of RNA and protein synthesis on chloroplast structure and function in wild-type *Chlamydomonas reinhardi*. *J. Cell. Biol.* **50,** 35–49.

GOODENOUGH U.W. & LEVINE R.P. (1970) Chloroplast structure and function in *ac-20*, a mutant strain of *Chlamydomonas reinhardi*. III. Chloroplast ribosomes and membrane organization. *J. Cell. Biol.* **44,** 547–62.

GOODENOUGH U.W. & LEVINE R.P. (1971) The effects of inhibitors of RNA and protein synthesis on the recovery of chloroplast ribosomes, membrane organization, and photosynthetic electron transport in the *ac-20* strain of *Chlamydomonas reinhardi*. *J. Cell. Biol.* **50,** 50–62.

GOODENOUGH U.W., TOGASAKI R.K., PASZEWSKI A. & LEVINE R.P. (1971) Inhibition of chloroplast ribosome formation by gene mutation in *Chlamydomonas reinhardi*. In *Autonomy and Biogenesis of Mitochondria and Chloroplasts*, eds. Boardman N.K., Linnane A.W. & Smillie R.M., pp. 224–34. North-Holland, Amsterdam.

GRIVELL L.A., REIJNDERS L. & BORST P. (1971) Isolation of yeast mitochondrial ribosomes highly active in protein synthes is. *Biochim. Biophys. Acta*, **247,** 91–103.

GRIVELL L.A. & WALG H.L. (1972) Subunit homology between *Escherichia coli*, mitochondrial and chloroplast ribosomes. *Biochem. Biophys. Res. Comm.* **49,** 1452–1458.

HALBREICH A. & RABINOWITZ M. (1971) Isolation of *Saccharomyces cerevisiae* mitochondrial formyltetrahydrofolic acid: methionyl tRNA transformylase and the hybridization of mitochondrial fmet-tRNA with mitochondrial DNA. *Proc. Nat. Acad. Sci. U.S.A.* **68,** 294–8.

HANSON M.R., DAVIDSON J.N., METS L.J. & BOGORAD L. (1974) Characterization of chloroplast and cytoplasmic ribosomal proteins of *Chlamydomonas reinhardi* by two-dimensional gel electrophoresis. *Molec. Gen. Genet.* **132,** 105–18.

HARRIS E.H., BOYNTON J.E. & GILLHAM N.W. (1974) Chloroplast ribosome biogenesis in *Chlamydomonas*: selection and characterization of mutants blocked in ribosome formation. *J. Cell. Biol.* **63,** 160–79.

HOOBER J.K. & BLOBEL G. (1969) Characterization of the chloroplast and cytoplasmic ribosomes of *Chlamydomonas reinhardi*. *J. Mol. Biol.* **41,** 121–38.

HOOBER J.K., SIEKEVITZ P. & PALADE G.E. (1969) Formation of chloroplast membranes in *Chlamydomonas reinhardi y-1*. Effects of inhibitors of protein synthesis. *J. Biol. Chem.* **244,** 2621–31.

HONEYCUTT R.C. & MARGULIES M.M. (1973) Protein synthesis in *Chlamydomonas reinhardi*. Evidence for synthesis of proteins of chloroplastic ribosomes on cytoplasmic ribosomes. *J. Biol. Chem.* **248,** 6145–53.

HUDOCK M.O. & TOGASAKI R.K. (1974) Photosynthetic mutants of *Chlamydomonas reinhardi*. *J. Cell. Biol.* **63,** 149a.

KALTSCHMIDT E. & WITTMANN H.-G. (1970) Ribosomal proteins, XII. Number of proteins in small and large ribosomal subunits of *Escherichia coli* as determined by two-dimensional gel electrophoresis. *Proc. Nat. Acad. Sci. U.S.A.* **67,** 1276–82.

KROON A.M. (1969) On the effects of antibiotics and intercalating dyes on mitochondrial biosynthesis. In *Inhibitors. Tools in Cell Research*, eds. Bücher Th. & Sies H. pp. 159–66. Springer-Verlag, New York.

KÜNTZEL H. (1969a) Mitochondrial and cytoplasmic ribosomes from *Neurospora crassa*: characterization of their subunits. *J. Mol. Biol.* **40,** 315–20.

KÜNTZEL H. (1969b) Proteins of mitochondrial and cytoplasmic ribosomes from *Neurospora crassa. Nature*, **222,** 142–6.

LAMB A.J., CLARK-WALKER G.D. & LINNANE A.W. (1968) The biogenesis of mitochondria. 4. The differentiation of mitochondrial and cytoplasmic protein synthesizing systems *in vitro* by antibiotics. *Biochim. Biophys. Acta*, **161,** 415–27.

LEE R.W., GILLHAM N.W., VAN WINKLE K.P. & BOYNTON J.E. (1973) Preferential recovery of uniparental streptomycin resistant mutants from diploid *Chlamydomonas reinhardtii*. *Molec. Gen. Genet.* **121,** 109–16.

LEE S.G. & EVANS N.R. (1971) Hybrid ribosome formation from *Escherichia coli* and chloroplast ribosome subunits. *Science*, **173,** 241–2.

LEVINE R.P. & ARMSTRONG J. (1972) The site of synthesis of two chloroplast cytochromes in *Chlamydomonas reinhardi*. *Plant Physiol.* **49,** 661–2.

LEVINE R.P. & PASZEWSKI A. (1970) Chloroplast structure and function in *ac-20*, a mutant strain of *Chlamydomonas reinhardi*. II. Photosynthetic electron transport. *J. Cell Biol.* **44,** 540–6.

LEVINE R.P. & TOGASAKI R.K. (1965) A mutant strain of *Chlamydomonas reinhardi* lacking ribulose diphosphate carboxylase activity. *Proc. Nat. Acad. Sci. U.S.A.* **53,** 987–90.

LIZARDI P.M. & LUCK D.J.L. (1972) The intracellular site of synthesis of mitochondrial ribosomal proteins in *Neurospora crassa*. *J. Cell Biol.* **54,** 56–74.

LOENING U.E. & INGLE J. (1967) Diversity of RNA components in green plant tissues. *Nature*, **215,** 363–7.

MAHLER H. (1973) Biogenetic autonomy of mitochondria. *Critical Rev. in Biochem.* **1,** 381–460.

METS L.J. & BOGORAD L. (1971) Mendelian and uniparental alterations in erythromycin binding by plastid ribosomes. *Science*, **174,** 707–9.

METS L. & BOGORAD L. (1972) Altered chloroplast ribosomal proteins associated with erythromycin-resistant mutants in two genetic systems of *Chlamydomonas reinhardi*. *Proc. Nat. Acad. Sci. U.S.A.* **69,** 3779–83.

NOMURA M. (1970) Bacterial ribosome. *Bact. Rev.* **34,** 228–77.

NOMURA M. & ENGBAEK F. (1972) Expression of ribosomal protein genes as analyzed by bacteriophage mu-induced mutations. *Proc. Nat. Acad. Sci. U.S.A.* **69,** 1526–30.

OHTA N., INOUYE M. & SAGER R. (1975) Identification of a chloroplast ribosomal protein altered by a chloroplast mutation in *Chlamydomonas*. *J. Biol. Chem.* **250,** 3655–9.

OZAKI M., MIZUSHIMA S. & NOMURA M. (1969) Identification and functional characterization of the protein controlled by the streptomycin-resistant locus of *Escherichia coli*. *Nature*, **222,** 333–9.

PESTKA S. (1971) Inhibitors of ribosome functions. *Ann. Rev. Microbiol.* **25,** 487–562.

RIFKIN M.R., WOOD D.D. & LUCK D.J.L. (1967) Ribosomal RNA and ribosomes from mitochondria of *Neurospora crassa*. *Proc. Nat. Acad. Sci. U.S.A.* **58,** 1025–32.

SAGER R. (1954) Mendelian and non Mendelian inheritance of streptomycin resistance in *Chlamydomonas reinhardi*. *Proc. Nat. Acad. Sci. U.S.A.* **40,** 356–63.

SAGER R. (1972) *Cytoplasmic Genes and Organelles*. Academic Press, New York.

SAGER R. & HAMILTON M.G. (1967) Cytoplasmic and chloroplast ribosomes of *Chlamydomonas*: ultracentrifugal characterization. *Science*, **157,** 709–11.

SAGER R. & RAMANIS Z. (1965) Recombination of nonchromosomal genes in *Chlamydomonas*. *Proc. Nat. Acad. Sci. U.S.A.* **53,** 1053–61.

SAGER R. & TSUBO Y. (1962) Mutagenic effects of streptomycin in *Chlamydomonas*. *Arch. Mikrobiol.* **42,** 159–75.

SCHIMMER O. & ARNOLD C.G. (1970) Untersuchungen über Reversions- und Segregationsverhalten eines ausserkaryotischen Gens von *Chlamydomonas reinhardii* zur Bestimmung des Erbträgers. *Molec. Gen. Genet.* **107,** 281–90.

SCHLANGER G. & SAGER R. (1974) Localization of five antibiotic resistances at the subunit level in chloroplast ribosomes of *Chlamydomonas*. *Proc. Nat. Acad. Sci. U.S.A.* **71,** 1715–9.

SCHLANGER G., SAGER R. & RAMANIS Z. (1972) Mutation of a cytoplasmic gene in *Chlamydomonas* alters chloroplast ribosome function. *Proc. Nat. Acad. Sci. U.S.A.* **69,** 3551–3555.

SCHWARTZ J.H., MEYER R., EISENSTADT J.M. & BRAWERMAN G. (1967) Involvement of N-formylmethionine in initiation of protein synthesis in cell-free extracts of *Euglena gracilis. J. Mol. Biol.* **25,** 571–4.

SEBALD W., SCHWAB A. & BÜCHER TH. (1969) Incorporation of amino acids into mitochondrial proteins. In *Inhibitors. Tools in Cell Research*, eds. Bücher Th. & Sies H. pp. 140–54. Springer-Verlag, New York.

SIERSMA P.W. & CHIANG K.-S. (1971) Conservation and degradation of cytoplasmic and chloroplast ribosomes in *Chlamydomonas reinhardtii. J. Mol. Biol.* **58,** 167–85.

SMITH A.E. & MARCKER K.A. (1968) N-formylmethionyl transfer RNA in mitochondria from yeast and rat liver. *J. Mol. Biol.* **38,** 241–3.

SURZYCKI S.J., GOODENOUGH U.W., LEVINE R.P. & ARMSTRONG J.J. (1970) Nuclear and chloroplast control of chloroplast structure and function in *Chlamydomonas reinhardi. Symp. Soc. Exp. Biol.* **24,** 13–27.

SURZYCKI S.J. & ROCHAIX J.D. (1971) Transcription mapping of ribosomal RNA genes of the chloroplast and nucleus of *Chlamydomonas reinhardi. J. Mol. Biol.* **62,** 89–109.

SURZYCKI S.J. & GILLHAM N.W. (1971) Organelle mutations and their expression in *Chlamydomonas reinhardi. Proc. Nat. Acad. Sci. U.S.A.* **68,** 1301–6.

TOGASAKI R.K. & LEVINE R.P. (1970) Chloroplast structure and function in *ac-20*, a mutant strain of *Chlamydomonas reinhardi*. I. CO_2 fixation and ribulose-1,5-diphosphate carboxylase synthesis. *J. Cell Biol.* **44,** 531–9.

VAZQUEZ D., STAEHELIN T., CELMA M.L., BATTANER E., FERNANDEZ-MUNOZ R. & MONRO R.E. (1969) Inhibitors as tools in elucidating ribosomal function. In *Inhibitors. Tools in Cell Research*, eds. Bücher Th. & Sies H. pp. 100–25. Springer-Verlag, New York.

WEISBLUM B. & DAVIES J. (1968) Antibiotic inhibitors of the bacterial ribosome. *Bact. Rev.* **32,** 493–528.

CHAPTER 7

GENETICS OF
CHLAMYDOMONAS MOEWUSII
AND *CHLAMYDOMONAS EUGAMETOS*

C. SHIELDS GOWANS

Division of Biological Sciences,
University of Missouri-Columbia, U.S.A.

1 Introduction 146

2 Conspecificity of *C. moewusii* and *C. eugametos* 146

3 Morphological mutants 148
  3.1 Flagella (and sterility) 148
  3.2 Contractile vacuoles 151
  3.3 Misdividing mutants 152
  3.4 Others 152

4 Auxotrophic mutants 153
  4.1 Niacin 153
  4.2 Para-aminobenzoic acid 154
  4.3 Thiamine 155
  4.4 Purines 155

5 Mutants with differential responses to supplements 155

6 Inhibitor-resistant, -sensitive, and -dependent mutants 156
  6.1 Antibiotics 156
  6.2 Analogues of metabolites 157

7 Obligate mixotrophs 159

8 Modifier and suppressor mutants 160

9 Mutation 161
  9.1 Response to irradiation 161
  9.2 Simultaneous mutations 162
  9.3 Reversion, back-mutation and chemical mutagens 162
  9.4 Chromosome aberrations 164
  9.5 Mutator gene 165
  9.6 Limited mutation spectrum 165

10 Formal genetics 167
  10.1 'False linkage' and 'negative linkage' 167
  10.2 High tetratype-tetrad frequencies 168
  10.3 Temperature and crossing-over 169
  10.4 'Non-chromosomal' genes 169

11 Conclusion 169

12 References 170

Contribution from the Missouri Agricultural Experiment Station. Journal Series Number 6942. Approved Feb. 28, 1974.

1 INTRODUCTION

Although most of the current genetic work on *Chlamydomonas* is carried out with *C. reinhardtii*, a considerable amount of genetic information has accumulated through the efforts of a few investigators on *C. moewusii* and *C. eugametos*. These two species (see below) differ from *C. reinhardtii* in several ways. Important to genetic studies are differences in (1) the life cycles, (2) heterotrophic ability, and (3) factors controlling the induction of the sexual state.

(1)   Whereas *C. reinhardtii* undergoes cytogamy by a lateral fusion of the two gametes, with karyogamy quickly following plasmogamy to form a quadriflagellated planozygote, *C. eugametos* and *C. moewusii* form tandems or vis-à-vis pairs in which, for at least four to eight hours (in light), the two cells are united apically by only a small cytoplasmic isthmus. During this time the tandems are dikaryotic and motile, the flagella of the male (or +) gamete remaining active while those of the female (or −) gamete become immotile shortly after the establishment of the cytoplasmic bridge. Gowans (1963) used this transitory difference to designate the active partner as male and the passive partner as female, whereas Lewin (1974) used the more neutral terms *plus* (*mt*⁺) and *minus* (*mt*⁻). A second difference in the life cycles is seen at the time of zygote germination. Germinating zygotes of *C. moewusii* and *C. eugametos* typically form four haploid zoospores, whereas in *C. reinhardtii* the two meiotic divisions are generally followed by a mitotic division, resulting in the formation of eight haploid zoospores. (However, this observed difference may result from different laboratory techniques for maturing and germinating the zygospores.)

(2)   *C. reinhardtii* is capable of heterotrophic growth in the dark with acetate as the carbon source. *C. moewusii* and *C. eugametos*, on the other hand, are obligate phototrophs (J. Lewin 1950; Wetherell 1958).

(3)   A low-nitrogen medium is effective in inducing the sexual state in *C. reinhardtii* cultures (Sager & Granick 1953). Depletion of available nitrogen has little or no effect in inducing sexual activity of *C. moewusii* or *C. eugametos* (Bernstein & Jahn 1955; Lewin 1956; Trainor 1959), which is apparently influenced by other factors: calcium, temperature, light, oxygen, and carbon dioxide have all been implicated (see review by Coleman 1962).

2 CONSPECIFICITY OF *C. MOEWUSII* AND *C. EUGAMETOS*

Chlamydomonas eugametos was isolated and described by Moewus (1931), who stated (personal communication) that all stocks of *C. eugametos* made available to individuals and culture collections belonged to his f. *typica*. However, Czurda (1935) and Gerloff (1940) found that their cultures of *C.*

eugametos did not agree with the description given by Moewus (see Gowans 1963 for a more detailed discussion). Later Gerloff isolated clones which he considered to be identical to the available cultures of *C. eugametos*; but because the cells did not agree with Moewus' description, he redescribed them formally under a new name, *Chlamydomonas moewusii* (Gerloff 1940). Cells of the wild-type strains of *C. moewusii* and *C. eugametos* are morphologically identical (Gowans 1963), and all conform to the description given by Gerloff.

The strains of *C. moewusii* used by Lewin (syngen I of Wiese and Hayward 1972) were isolated by Provasoli (Lewin 1952a) and are maintained, now as numbers 96 (*mt⁻* strain) and 97 (*mt⁺* strain), in the Indiana University Culture Collection of Algae (Starr 1964). The strains of *C. eugametos* utilized by Gowans and associates were originally obtained from G. M. Smith, who had received them directly from Moewus. The male (*mt⁺*) strain was lost in 1954 and replaced with the male strain (number 9) from the Indiana Collection, which had originally obtained it from Moewus via Pringsheim. In 1959 both male and female strains, along with many derivative strains under investigation in Columbia, Mo., were lost in a culture-chamber failure; they were later replaced by numbers 9 (male or *mt⁺*) and 10 (female or *mt⁻*) from the Indiana Collection.

C. moewusii and *C. eugametos* mate readily and reciprocally (Gerloff 1940; Hutner & Provasoli 1951; Bernstein & Jahn 1955; Gowans 1963). Gowans (1963) found that the hybrid zygotes germinated at a high frequency (97·5%), and that each zygote generally produced four zoospores, but that there was considerable subsequent lethality. In most cases only one of the meiotic products from each zygote survived to produce a gone, although from 10% of the zygotes two products survived. The observed lethality evidently did not disturb the segregation ratios for alternate alleles of five marker genes in the cross, and the lethality pattern could not be explained by postulating meiosis of zygotes heterozygous for any single simple chromosomal aberration. At the time Gowans reported these results, it was assumed that all of the lethality was in some other way due to hybridity. However, Lewin (1974) observed progeny lethality when the *C. moewusii* partner in the hybrid cross (M.02) was crossed with wild type of the same species, indicating that some source of product lethality seems to be inherent in the M.02 strain. The pattern of lethality in the *C. moewusii* × *C. moewusii* cross is different, and the percentage of lethal progeny is less (40%) than in the *C. moewusii* × *C. eugametos* cross (72·5%). Obviously hybrid lethality can be explained only after further studies of the viability of progeny of zygotes formed by crossing the wild types of both species.

Some physiological differences have been found in the responses of *C. moewusii* and *C. eugametos* to environmental factors (Bernstein & Jahn 1955), and in their mating reactions (Trainor 1959). In their sensitivity to analogues of growth factors no greater similarity was found between these species than between either and *C. reinhardtii* (McBride A.C. & Gowans 1970).

Comparative studies of the sexual agglutinins (gamones) in several chlamydomonads have indicated that these substances are glycoproteins (Wiese 1969). Investigations of the effects of enzymes on isoagglutination, induced in cell suspensions of one mating type by the supernatant from a suspension of the opposite mating type, showed that in both *C. eugametos* and *C. moewusii* the active contact site was protein in the female (*mt⁻*) strain and polysaccharide in the male (*mt⁺*) strain; whereas in both mating types of *C. reinhardtii* the active contact site was found to be protein (Wiese & Hayward 1972; see also Chapter 8).

An electrophoretic study of the isoenzymes of *C. eugametos* and *C. moewusii* revealed identical patterns for malate dehydrogenase and leucine aminopeptidase, but somewhat dissimilar patterns for *alpha* esterase (Dempsey & Delcarpio 1971). Immunological studies showed a striking similarity between *C. moewusii* and *C. eugametos*. Indeed, there appears to be a closer immunological relationship between wild types of the two species than there is between different mutant strains of *C. moewusii* (Brown & Walne 1967).

Buffaloe (1958) found that the haploid chromosome number in both *C. moewusii* and *C. eugametos* is 8, and that in both species the metaphase chromosomes arrange themselves similarly, in a ring. He also reported that in both species the cells exhibit endomitosis and become autopolyploid when grown in high light intensities, and undergo somatic reduction again when placed in the dark; but these observations have not been independently confirmed.

On the basis of available evidence Gowans (1963) concluded that *C. eugametos* and *C. moewusii* are conspecific, and accordingly recommended that *C. moewusii* be redesignated as *C. eugametos* var *moewusii*. However, Lewin (1974), while accepting the evidence for conspecificity, expressed the opinion that *C. moewusii* is the most formally acceptable binomial for the species. The name applied is in fact of little consequence; but from a genetic standpoint it is important to recognize that a considerable amount of evidence supports the conclusion that the cultures generally designated as *C. eugametos* and *C. moewusii* are but varieties of a single species, and that therefore the genetic information gleaned from studies of both can be properly combined and summarized. Nevertheless, in this paper the use of the two different cognomina will be preserved in order to differentiate between studies of the respective varieties.

3   MORPHOLOGICAL MUTANTS

3.1   *Flagella (and sterility)*

The flagella of *C. moewusii*, examined by light microscopy (Lewin 1952a) and by electron microscopy (Lewin & Meinhart 1953; Gibbs *et al.* 1958; Brown *et al.* 1968), appear essentially similar to those of other protists.

Essentially they comprise a cylinder of 9 fibril pairs and a central pair, all enclosed in a membrane-limited matrix.

Lewin (1960) developed a simple device to enrich for mutants with impaired motility. Mutants of *C. moewusii* have been isolated which: (1) are flagellaless, (2) have short flagella, (3) are palmelloid, or show other evidence for abnormal cell division, (4) have normal flagella but exhibit reduced fertility or absolute sterility, and (5) have flagella of normal length but with various degrees of functional impairment, ranging from poor synchrony and/or slow beating to complete paralysis (Lewin 1974). Similarly, mutants of *C. eugametos* have been reported (Gowans 1960) which are: (1) flagellaless, (2) paralysed, (3) slow-swimming, (4) palmelloid and (5) exhibit a periodic unilateral paralysis as observed by Lewin (1952a) in old cultures of wild-type *C. moewusii*. However, even completely paralysed mutants, like any *Chlamydomonas* cell which has flagella, retain their ability to 'creep' along a solid surface, as described by Lewin (1952b).

Palmelloid mutants are effectively sterile, since they normally produce no swimming cells. Flagellaless mutants are essentially sterile, too, although Lewin was able to produce a few zygotes from a cross between an apparently flagellaless mutant (M.83 mt^-) and wild type (mt^+) by centrifuging mixtures to pack cells of the two strains together. One such zygote germinated to produce a tetratype tetrad, confirming its hybrid origin (Lewin 1974). Likewise, mutants with short flagella mate poorly, although mating frequencies may be increased by mechanical agitation of gamete mixtures (Lewin 1974). Mutants which have normal flagella but which are sterile, or have reduced fertility, apparently lack essential sexual agglutination factors (Wiese 1969). Lewin (1974) was able to produce a few zygotes in crosses involving one of these mutants (M.53-2), which is not completely sterile. However, for genetic studies, the useful mutants are those with flagella of normal length but completely paralysed, since sterility problems are avoided and segregation observations are clear-cut. In his extensive analysis of the genetics of motility, Lewin (1974) concentrated primarily on such mutants.

He reported that 17 independently isolated paralysed mutants had sustained mutations at nine different gene loci in seven different linkage groups (see Fig. 7.1). Mutations occurred twice at each of three different loci, and three more were mapped each at a single locus. One paralysis gene (mutant M.02) is inseparably linked to the mating-type locus, 13 map units from the centromere in the linkage group he designated as I. Two of the paralysis genes are completely or very closely linked to their respective centromeres (no tetratype tetrads were found among 45 zygotes produced by crossing the two); and these could thus be used as genetic markers for the first division of meiosis in the calculation of gene-centromere distances for other paralysis genes.

The formation of tandems during the mating of gametes of *C. moewusii* provides a stage of transitory dikaryosis. In the light, this dikaryotic state

H

lasts from four to eight hours, and its duration can be extended for as much as 72 hours by setting the tandems in the dark (Lewin 1954). This period of time was utilized by Lewin in tests for complementarity between different paralysis genes, as a parallel test of genetic allelism, and to classify various mutants according to the time required to recover from paralysis under these conditions. For such a recovery test the mutant gene to be tested must reside in the mt^+ or male gamete. The mt^- gamete may be of a wild type (non-paralysed) or a paralysed stock capable of complementing the mutant gene to be tested. Cells of some of the paralysed stocks tested in this way proved capable of recovery under these conditions, others did not. Lewin (1954) demonstrated by differential labelling of the mated strains that it is indeed the mt^+ or male paralysed cell which recovers motility, and not the mt^- or female partner which assumes an active function after pairing.

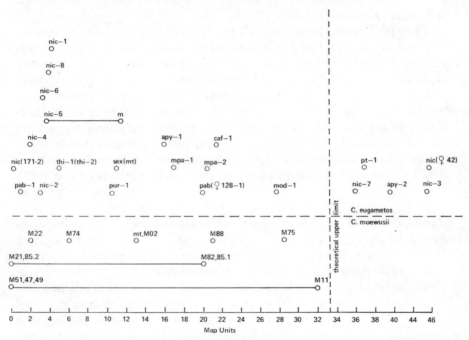

Fig. 7.1. Gene-centromere map-distances in *C. eugametos* and *C. moewusii*. Linkage groups are indicated by solid connecting lines. (Gowans 1960; Lewin 1974; McBride A.C. 1967; McBride J.C. & Gowans 1967; Nakamura & Gowans 1965; Thometz 1973; Wang 1972.)

Those strains which are 'recoverable' can be divided into three (Lewin 1954) or five (Lewin 1974) groups on the basis of the time required for the flagella of the paralysed partner to begin to beat after the establishment of the cytoplasmic bridge. The shortest recovery time was two hours, and the longest 24 to 48 hours. Recovery may be due to the diffusion of some cytoplasmic gene-controlled factor (? messenger RNA) or to an intrusion of the electron-dense microstrands observed in the cytoplasm of the connecting bridge (Brown *et al.* 1968): it apparently cannot be effected by filtrates from wild-type cells (Lewin 1954). Lewin (1954) also made tandems by crossing two complementary paralysed stocks (M.02 $mt^+ \times$ M.01 mt^-), waited for recovery of motility in the tandems, and then separated the two copulants by mechanical means. Although he was able to separate the gametes from 40% of the tandems so cut, none remained motile; either complementarity is dependent on continued physical connection of the cells, or mechanical separation results in irreparable damage and/or death of the separated cells.

Electron microscopy failed to reveal any structural differences between the flagella of paralysed mutants and those of wild-type *C. moewusii* (Gibbs *et al.* 1958), but critical fine-structure studies have not yet been made (Lewin 1974). Comparative immunological studies of three paralysed mutants of *C. moewusii* (M.02, M.75 and M.01—694, 695 and 697 respectively in the Indiana Culture Collection), one flagellaless mutant (M.14—716 in the Indiana collection), and wild-type (Brown & Walne 1967) indicated that although it is difficult or impossible to distinguish mutants from wild-type or from each other on the basis of double-diffusion reactions, the mutants are clearly distinguishable from each other, and from the wild-type, on the basis of their immunoelectrophoresis patterns. Brokaw (1960) reported that the ATPase activity of flagella isolated from the paralysed mutant M.02 was only 30 to 40% of the activity of wild-type flagella.

Wetherell and Krauss (1957), who isolated mutants of *C. eugametos* following X-ray treatment, found that some 6% of the surviving cells showed some evidence of damage to the flagella: the cells were flagellaless or paralysed, or exhibited irregular swimming patterns. No genetic studies were carried out on these mutants.

3.2 *Contractile vacuoles*

Mutants of *C. eugametos* requiring for their survival a high osmotic pressure in the medium, and having no observable contractile vacuoles, were reported by Gowans (1960). These mutants were originally isolated on a 'complete' medium, but it was found that merely the addition of 0·78 g/l NaCl to the minimal medium would permit their growth. Thus, whereas the wild type survives on minimal medium (osmotic pressure = ca. 0·4 atm.), the mutants required an osmotic pressure of ca. 1·0 atm. No genetic studies were done on these mutants, since in distilled water the cells were non-motile and soon burst, while in sodium chloride solutions, although motile, they were not sexually active.

Guillard (1960) made a more extensive study of a single vacuole-less mutant of *C. moewusii* (S-16), isolated following X-irradiation and requiring an ambient osmotic pressure of 1·5 atm. Contractile vacuoles are not seen even in wild-type cells when suspended in a medium of a similarly high osmolarity. Those substances least permeable to the cell membrane (such as glucose and NaCl) were most effective in promoting the growth of S-16, leading Guillard to conclude that the sole essential function of the contractile vacuole in *C. moewusii* is the elimination of water.

3.3 *Misdividing mutants*

Palmelloid strains are among the most common morphological mutants recovered from irradiated suspensions of *C. moewusii* or *C. eugametos* (Lewin 1953; Gowans 1960). The palmella condition is characterized by the retention of daughter cells in the wall of the mother cell, with the consequence that motile zoospores are not liberated. Such mutants are readily identifiable by their rough colonial appearance, but since they do not form sexually active cells no genetic studies are possible with them.

Other mutant strains have been found (Lewin 1953; Wetherell & Krauss 1957; Gowans 1960) in which there is a high percentage of failure to complete cell divisions. Apparently normal cells are also found in such cultures, but subculture from these once again produces a high proportion of bizarre forms. Lewin (1952b) found a gene for twinning of this sort, in which in a young liquid culture from 5 to 10% of the cells had divided incompletely. Longitudinal division, proceeding posteriorly from the papilla, was apparently frequently arrested before completion. The Siamese twins thus produced ranged from pairs attached by only a narrow bridge at the posterior end, through pairs with a much more extensive attachment, to cases where a furrowed papilla and the presence of three or four flagella were the only evidence of the double nature of the cell. Some monsters with 3–4 cell elements were also produced. The gene segregated, and showed linkage to a gene producing a lazy swimming pattern (Lewin 1953).

3.4 *Others*

Lewin (1953) isolated mutants of *C. moewusii* characterized by the presence of large refractile globules of volutin in many of the cells. These volutin bodies could be stained by neutral red, and exhibited metachromasy when stained with methylene blue. Growth (of one of these mutants) in limiting phosphate media produced cells which appeared normal, suggesting that the metachromatic material was metaphosphate. Crosses of two independent volutin mutants yielded some progeny with wild-type cells, indicating that at least two genes are involved with this morphological characteristic. Wetherell and Krauss (1957) reported similar mutants.

Mutants affecting the eyespot of *C. moewusii* and *C. eugametos*, comparable to the mutant of *C. reinhardtii* studied by Hartshorne (1953), have not been reported, although they could perhaps be sought following mutant enrichment procedures utilizing phototaxis. No reports of mutants involving the pyrenoid or the structure of the chloroplast have appeared.

4 AUXOTROPHIC MUTANTS

4.1 *Niacin*

Gowans (1960) reported studies on 18 independently induced mutants of *C. eugametos* requiring nicotinamide. None would grow on media supplemented with nicotinic acid at comparable concentrations. Although in most cases their growth responses to intermediates in the tryptophan-nicotinic acid pathway were hazy, Gowans was able to separate the 18 mutants into six groups on the basis of their different physiological responses to these intermediates. Three independent loci were identified (*nic-1, -2*, and *-3*), and their respective gene-centromere distances were calculated (Fig. 7.1). Gene-centromere distances were also calculated for two other mutants, *nic*(171–2) and *nic*(42), which were not tested for allelism. Four other mutants (not assigned a locus) proved non-allelic with *nic-1*. Strains carrying the three genes *nic-1, -2*, and *-3* fell respectively into three of the physiological groups. However, one mutant, *nic*(42), though in the same physiological group as *nic-1*, proved to be genetically non-allelic with *nic-1*. Since it had approximately the same gene-centromere distance as *nic-3*, it may have been an allele of *nic-3*, but the critical allelism test was not done.

Nakamura and Gowans (1965) reported that 12 different, independently isolated nicotinamide auxotrophs could be genetically mapped at five separate loci (*nic-4, -5, -6, -7, -8*). (It is not known whether any of these loci are the same as those identified earlier by Gowans, which were lost in 1959 when a culture chamber failed, and therefore the loci were numbered from *nic-4*). Four independent mutations were located at the *nic-4* locus, five at *nic-5*, and one each at *nic-6, -7*, and *-8*.

Attempts to determine a sequence for the action of genes at these five loci, by means of syntrophy experiments, failed. Similarly the addition to the nutrient medium of most intermediates in the tryptophan-nicotinic acid pathway failed to produce growth; the single exception was quinolinic acid. Following the discovery of a modifying gene (*mod-1*) (see section on Modifier and Suppressor mutants), Nakamura and Gowans were able to demonstrate that mutants at two of the loci (*nic-4* and *nic-6*) could utilize quinolinic acid for growth, although the concentration necessary to produce growth were very high, probably because of the relative impermeability of the cells to its anionic (dissociated) form. They concluded that these mutants were genetically blocked before the biosynthesis of this intermediate. Mutants at the other three loci, which could not utilize quinolinic acid, were therefore blocked on the biosynthetic pathway between quinolinic acid and the functional form of the vitamin.

The mutation to nicotinamide auxotrophy is the most common auxotrophic mutation in *C. eugametos* (Li *et al.* 1967). Although Lewin (1953) did not recover any niacin auxotrophs of *C. moewusii*, this can be explained by the fact that the vitamin supplement he used in his studies contained niacin in the form of nicotinic acid, at concentrations insufficient to promote the growth of *nic* mutants like those studied by Gowans.

The synthesis of niacin (like that of lysine) violates the general rule of unity in the comparative biochemistry of synthesis of metabolites, in that two quite distinct pathways are present in different types of organisms. In general, animals and fungi seem to utilize the tryptophan-nicotinic acid pathway (Henderson *et al.* 1949; Beadle *et al.* 1947; Nishizuka & Hayaishi 1963; Wilson & Henderson 1960), while higher plants (Henderson *et al.* 1959; Leete *et al.* 1955; Waller *et al.* 1966) and most bacteria (Ortega & Brown 1960) utilize a four-carbon dicarboxylic acid and glycerol as precursors of niacin synthesis.

Since *Chlamydomonas* may be considered either as a protozoan (Phytomastigophora) or as an alga in the Chlorophyta (which presumably gave rise to higher plants), and since the studies of Nakamura and Gowans did not clearly indicate which pathway was operative in *Chlamydomonas*, mutant and wild-type strains of *C. eugametos* were studied in order to establish which pathway it uses. Uhlik and Gowans (1974) studied *nic-4* strain and wild type, as well as a strain carrying the *apy-1* gene (see Inhibitor Resistant Mutants), which excretes excess nicotinic acid (Nakamura & Gowans 1964), by measuring the incorporation of ^{14}C from labelled molecules of various potential precursors into the nicotinic acid synthesized. Their results clearly indicated that the tryptophan pathway is the one used by *C. eugametos*.

4.2   *Para-aminobenzoic acid (paba)*

Lewin (1953) isolated an auxotroph of *C. moewusii* requiring an exogenous supply of paba. It showed a growth response to aniline, but not to benzoic acid. The relative efficiency of aniline to paba was 1:100, suggesting that aniline does not lie on the direct path of paba synthesis. This was corroborated by the fact that, although *para*-aminobenzoic acid could reverse sulphanilamide inhibition of wild-type growth, aniline was ineffective. The mutant gene showed linkage to the mating-type locus (Lewin 1954) at the 5% level (9PD:2NPD) (see tables in Perkins 1955 or Gowans 1965).

Gowans (1960) recovered 3 paba-less auxotrophs of *C. eugametos*. Two of the responsible gene loci were identified (*pab-1* and *pab-2*), and preliminary studies indicated that the third mutant might represent still another locus, although direct allelism tests were not done. (Unfortunately, the paba mutants of Lewin and Gowans have all been lost.)

4.3 *Thiamine*

Lewin (1953) recovered a thiamine-requiring auxotroph of *C. moewusii* which can grow on media supplemented with the pyrimidine moiety of the vitamin. The mutant of *C. eugametos* reported by Gowans (1960) (*thi-1*), on the other hand, requires the intact thiamine molecule for growth. A second thiamine auxotroph of *C. eugametos*, recovered by McBride A.C. (1967), was studied for pyrithiamine resistance (McBride J.C. & Gowans 1967). Jones (1968) isolated three other thiamine auxotrophs of *C. eugametos*, which, like *thi-1*, require intact thiamine and do not grow in media supplemented with the pyrimidine or thiazole moieties or with mixtures of the two. Since genetic tests demonstrated that *thi-1*, *thi-2*, and the three mutants recovered by Jones were all allelic (Jones 1968), *thi-2* should be redesignated as an allele of *thi-1*.

4.4 *Purine*

Gowans (1960) isolated four mutants which would grow in minimal medium supplemented with adenine, hypoxanthine, xanthine, or guanine. Genetic studies showed that at least two of the mutations concerned were non-allelic. None would grow on 4-amino imidazole 5-carboxamide with or without folic acid. (All of these mutants were lost in the culture-chamber failure of 1959, and no mutants of this type have subsequently been recovered.) Since they responded in a non-selective fashion to any purine supplied, and in view of the fact that *C. moewusii* was reported to be able to quantitatively use the nitrogen atoms from the guanine molecule (Syrett 1962), these mutants may have required merely an additional supplement to the nitrogen source.

5   MUTANTS WITH DIFFERENTIAL RESPONSES TO SUPPLEMENTS

Wetherell and Krauss (1957) recovered 47 strains, following X-irradiation, which showed differential responses to various of their selective screening media. Although 30 of these eventually reverted to normal, some had been partially characterized before their reversion, and these were included in considerations of the various kinds recovered. Among these 47 strains were some types which failed to grow on minimal medium, and others which did so only after long lag periods. Both of these types grew normally on complete medium. Some responded to a vitamin mixture only, some to vitamin mixture, casein hydrolysate, or nucleic-acid hydrolysate, some to casein hydrolysate only, some to casein hydrolysate or nucleic-acid hydrolysate, some to various organic acid or sugar supplements, and others to complete medium only. In addition some strains were recovered which were inhibited by organic supplements. Although none of the mutants responding to vitamin mixtures could be classified as requiring any single vitamin, some stimulatory responses were elicited in one or more of the strains by choline, thiamine, *para*-aminobenzoic acid, nicotinic acid, pyridoxine, and folic acid. However, genetic studies of these mutants were limited. Two of the mutants responding to organic acids or sugar proved vegetatively stable and showed 2:2 segregation in crosses with wild type. Two strains showing a response to vitamins, on the other hand, showed vegetative segregation, and in crosses they segregated to yield 1 normal:1 mutant:2 lethal. On the basis of these studies Wetherell and Krauss concluded that the mutant phenotype of many of their isolates was caused by chromosomal deletions in segments of the genome which are normally duplicated, weakening their synthetic capacity, but not producing any clear-cut auxotrophic requirement for a single supplement.

Wang (1972) recovered many strains of this species which showed a positive growth response to histidine, but all but two of these reverted to wild type. The two stable strains could grow slowly in minimal medium, but grew better in media supplemented with histidine. They were hyperinhibited (as compared to wild type) by tryptophan, this inhibition being reversible by tyrosine, phenylalanine, or shikimic acid. Uptake of tryptophan was reduced, and tryptophan uptake was facilitated by tyrosine. When these strains were streaked or plated on minimal medium, some single cells formed slow-growing colonies, while others died without dividing or after undergoing one to three mitotic divisions. Repeated attempts to isolate clear-cut histidine auxotrophs, by reisolating mitotic daughter cells of these strains, failed; all reisolated behaved like the original mutants.

6 INHIBITOR-RESISTANT, -SENSITIVE, AND -DEPENDENT MUTANTS

6.1 *Antibiotics*

Gowans (1960) isolated four mutants resistant to streptomycin at concentrations up to 50 μg/ml, but did not study them further either physiologically or genetically. A.C. McBride (unpublished) isolated a strain (*sr-1*) resistant to 50 μg/ml streptomycin, the gene for which shows a 2:2 segregation in crosses. In addition, two mutants have been isolated which are resistant to 500 μg/ml streptomycin. The first of these is male (*mt+*), and in crosses transfers streptomycin resistance to all four of the gones arising from a single hybrid zygote. The second strain is female (*mt−*) and produces four wild-type (streptomycin-sensitive) gones from each back-cross zygote. This indicates the non-chromosomal type of inheritance reported in *C. reinhardtii* by Sager *et al.* (see Chapter 6, this volume), and paradoxically shows that maternal inheritance in *C. eugametos* is through the male (as designated by Gowans; see Introduction to this section). In addition, A.C. McBride (unpublished) isolated two mutants resistant to 250 µg/ml neamine, one resistant to 20 µg/ml spectinomycin, and one which is neamine-dependent. This last mutant, which requires 40 µg/ml neamine, is a female (*mt⁻*) and in crosses does not transmit the neamine-dependence to any of the progeny.

6.2   *Analogues of metabolites*

3-acetyl pyridine, an analogue of nicotinic acid, acts as an antimetabolite of nicotinic acid in *C. eugametos* (the inhibitory effect of 3-acetyl pyridine on growth is reversed by additions of nicotinamide to the medium). Nakamura and Gowans (1964) recovered mutants resistant to 3-acetyl pyridine, and demonstrated that these mutants excreted nicotinic acid into the medium although no excess acid could be detected in the cells. They postulated that such strains are either derepressed or deretroinhibited. Nakamura and Gowans (1965) later reported on genetic and physiological studies of two of these mutants, with mutations located at two independent loci and with different levels of resistance. High-level resistance (*apy-1*) was found to be epistatic to resistance to lower levels of the antimetabolite (*apy-2* = partial resistance). The level of resistance closely paralleled the amount of nicotinic acid excreted. These two resistant mutants were crossed with all of the nicotinamide auxotrophs, for two reasons. First, if allelism between a resistant gene and an auxotrophic gene could be demonstrated, this would be evidence that the mechanism resulting in nicotinic acid excretion was deretroinhibition rather than derepression. Secondly, it was felt that double-mutant strains (*apy-1 nic*) would be helpful in identifying biochemical blocks in the auxotrophs, because they could be expected to excrete excess amounts of accumulated intermediates. The gene of the resistant strain (*apy-1*) was found to be non-allelic with any of the nicotinamide auxotroph genes (*nic-4* through *nic-8*). Because of difficulties in classifying segregants of *apy-2* and wild type, not all of their allelism tests were conclusive; but it seems clear that *apy-2* is not allelic with *nic-4, -5,* or *-8,* and may not be allelic with *nic-6* or *-7*.

Whereas crosses of the partially resistant strain (*apy-2*) produce normally viable double-marker recombinants (*apy-2 nic*), segregants bearing *apy-1* in combination with any of the nicotinamide auxotroph genes are lethal, the double-mutant strains dying within several divisions following meiosis. Thus, an NPD tetrad from a *apy-1* × *nic* cross yields only two viable products, both of which are wild type; a T tetrad produces three viable products, one wild type, one *apy-1*, and one *nic*. The observed lethality of the doubly marked (*apy-1 nic*) cells makes it seem unlikely that *apy-1* is either a derepressed or deretroinhibited mutant, since if so one would have to attribute the lethal effect to autotoxicity of an accumulated intermediate. A more likely hypothesis, suggested by the work of Davis (1963) with *Neurospora* mutants, is that the *apy-1* mutant has a defect in its nicotinamide-transport system. Excretion of nicotinic acid would result from this defect, and might be expected to produce a secondary derepression or deretroinhibition. Resistance would be due to the failure of the mutant to transport the analogue into the cell, and death of the doubly marked strains (*apy-1 nic*) could be attributed to their failure to take in exogenously supplied nicotinamide, superimposed on their inability to synthesize their own supply of the vitamin.

Mutants resistant to another analogue of nicotinic acid, 3-pyridine sulphonic acid, were also obtained (Nakamura & Gowans 1964). The inhibition of wild type by this substance, however, is not reversed by additions of nicotinamide to the media, and the resistant characteristic does not exhibit simple Mendelian segregation in crosses.

Pyrithiamine, an analogue of thiamine, is an extremely effective inhibitor of *C. eugametos*: growth of the wild type is completely inhibited by as little as $1 \cdot 0$ ng per ml (McBride J.C. & Gowans 1967)! Pyrithiamine inhibition is reversible: inhibition by 1000 ng/ml is reversed by 1500 ng/ml thiamine (ratio $= 1 \cdot 5 : 1 \cdot 0$), whereas inhibition by 1 ng/ml requires 10 ng/ml of thiamine to effect reversal (ratio $= 10 : 1$), these different thiamine: pyrithiamine ratios suggesting that there may be at least two separate sites of pyrithiamine inhibition. Several pyrithiamine-resistant mutants have been isolated, and one such mutant (*pt-1*) has been studied genetically and physiologically (McBride J.C. & Gowans 1967). Whereas growth of wild type cells is completely inhibited by pyrithiamine at levels around 1 ng/ml, this mutant grows normally in concentrations as high as 1500 ng/ml, and only at levels exceeding 6000 ng/ml (6 μg/ml) is its growth completely inhibited.

The *pt-1* mutant does not excrete detectable amounts of thiamine, and appears to contain less intracellular thiamine than does the wild type. Cellular uptake of thiamine by the *pt-1* mutant is considerably less than that of wild type. Double-mutant strains, carrying the *pt-1* gene in combination with the thiamine-auxotrophy gene (*thi-2*, actually $=$ *thi-1*, see above), are lethal, paralleling the situation for 3-acetyl pyridine resistance and nicotinamide auxotrophy, as reported above. Resistance to pyrithiamine, like resistance to 3-acetyl pyridine, could be explained by the decreased uptake of this anti-metabolite by the cells. However, *pt* cells do not excrete thiamine, whereas *apy* cells do excrete nicotinic acid, leading J. C. McBride and Gowans (1967) to postulate that pyrithiamine resistance may be due to a cloistering of the thiamine pathway (perhaps by some failure of intracellular transport) rather than to failure in a cell-membrane transport system.

Loeppky (1967) found that certain nicotinamide auxotrophs had an increased resistance to pyrithiamine. Thus, whereas wild type is completely inhibited by 1 ng/ml, 10 ng/ml was required to inhibit *nic-4* or *nic-5* strains.

The phenylalanine analogue 2-amino-3-phenylbutanoic acid (APBA, 3-methylphenylalanine) was also found to be an effective inhibitor of *C. eugametos* (McBride A.C. & Gowans 1969). The reversal of inhibition by APBA, however, is not confined to phenylalanine. Although phenylalanine and alanine were the most effective, all other amino acids tested, and even two alpha-keto acids, phenylpyruvic and para-hydroxyphenylpyruvic, have shown some ability to reverse APBA inhibition (McBride A.C. 1967). Resistant mutants are readily obtainable, since APBA induces both point mutations and chromosome aberrations (see section on Mutation), and three independent genes conferring APBA-resistance have been genetically mapped. One of these (*mpa-2*) confers low-level resistance (less than 200 µg/ml), a second (*mpa-3*) confers resistance to 200 but not to 500 µg APBA/ml, and a third (*mpa-1*) permits growth even on agar media containing more than 500 µg APBA/ml.

The mechanism of resistance to APBA is not known. Neither *mpa-1* nor *mpa-2* strains excrete or accumulate excess phenylalanine, and phenylalanine uptake is not significantly or consistently different among *mpa-1*, *mpa-2* and wild-type strains. Because of the reversal of APBA inhibition by all tested amino acids, differences in transamination were suspected. However, resistant strains and wild type have comparable glutamate-pyruvate transaminase activity, and phenylalanine-pyruvate transaminase activity was not detectable in either wild type or APBA-resistant strains. No differences were found between wild type and resistant strains in their utilization of ammonium and nitrate-nitrogen sources (McBride A.C. 1967). Strains with the *mpa-1* gene showed a slightly higher tolerance to ethionine (an analogue of methionine) than did either wild type or *mpa-2* cells, but uptake of labelled ethionine was similar in all of these strains.

Five mutants of *C. eugametos* showing different levels of resistance to caffeine were isolated by Thometz (1973): three were shown to segregate 2:2, and the gene-centromere distance for one of the mutants (*caf-1*) was estimated to be 21·5 map units. Cells of these mutants generally accumulated more ^{14}C from labelled caffeine than did the wild type. These mutants were found to be more resistant to UV irradiation, both with and without photo-reactivation (see Mutation section for discussions of dark repair in wild type).

Two mutants resistant to L-methionine-DL-sulfoxamine, one sucrose-dependent mutant and one tryptophan-sensitive mutant have also been isolated (McBride J.C. unpublished) but not further studied.

7 OBLIGATE MIXOTROPHS

Although *C. moewusii* and *C. eugametos* are obligate phototrophs (J. Lewin 1950, Wetherell 1958), mutants have been recovered which require a carbon source in addition to light. Lewin (1953) reported such an obligately mixotrophic mutant of *C. moewusii* which failed to grow appreciably on minimal medium, but which grew well on media supplemented with citrate, fumarate, succinate, pyruvate or malate, and less well on media supplemented with acetate (Lewin & Frenkel 1960). He found no significant differences from wild type in the pigments of the mutant, which appeared capable of carrying out photosynthesis. The gene segregated in a Mendelian fashion, although there was considerable reduction in viability among the sexual progeny. Zygotes homozygous for the carbon-source requirement required long periods of illumination for maturation, but ultimately appeared to be normal.

Chu-Der (1971), Der and Gowans (1972) isolated ten unstable mutants of *C. eugametos* which likewise were obligate mixotrophs, although even in media supplemented with potassium acetate they grew more slowly than the wild type, and cultures reverted or died out more readily, necessitating more frequent subculturing. The chlorophyll *a*:chlorophyll *b* ratios in the different acetate-requiring mutants ranged from ratios of 4·37 to 19·67, whereas in the wild type it is 2·92. Total chlorophyll was less than wild-type in some, but greater in others. The Mehler reaction (Mehler 1951) was demonstrated in all of the four mutants tested and the Hill reaction in all of the seven tested. Of the five obligate mixotrophs tested for specific activities of the TCA enzymes, three lacked succinate dehydrogenase activity and one lacked citrate synthase activity; the fifth showed activity for all of the TCA enzymes.

These acetate-requiring mutants produced relatively few active gametes, and, among the few back-cross zygotes formed, germination was low, ranging from 12 to 74% in various crosses. Nevertheless, Chu-Der and Gowans succeeded in showing that five of six mutants tested showed a 2:2 segregation of the acetate requirement. The sixth mutant, a female (*mt⁻*), failed to transmit the acetate requirement to any of the sexual progeny, which might indicate 'non-chromosomal' inheritance if one could exclude the possibility that the rare gametes formed in suspensions of this strain were in fact revertants.

8 MODIFIER AND SUPPRESSOR MUTANTS

Lewin (1974) recovered a gene from his paralysed mutant stock M. 74 which partially alleviates paralysis, allowing a genetically paralysed strain to swim in an erratic fashion. Segregants from zygotes homozygous for the paralysis gene showed that this partial suppressor segregated 2:2. In one cross it showed false linkage (see Section 10). Lewin also found indications of the existence of two complementary paralysis-suppressor genes associated with the M. 82 paralysed strain. Nakamura and Gowans (1965) reported a modified gene (*mod-1*), found in two different nicotinamide auxotrophs, which inhibited the utilization of nicotinic acid and, in the cases of *nic-4* and *nic-6*, also of quinolinic acid. (Cells with *nic-4* and *nic-6* are the only nicotinamide auxotrophs found that can use quinolinic acid: see above, section on Auxotrophic mutants.) The *mod-1* gene had no evident effect on the utilization of nicotinamide. In crosses the character segregated 2:2, and genetic data indicated that the two independently recovered modifier genes were allelic (or closely linked) and independent of the auxotrophic genes whose action they modified. Like Lewin's partial suppressor of paralysis, *mod-1* showed false linkage in certain crosses. Evidence was also presented that the defect produced by the *mod-1* gene could be remedied by the addition of excess cations (K, NH_4, Ca or Mg) to the medium (Nakamura & Gowans 1967). The remedial effect was clearly due to the cations themselves and not to an increased ionic pressure in the media or to the presence of additional anions. Actual measurements of the comparative uptake of both quinolinic acid and calcium by strains with and without the *mod-1* gene have revealed no difference in the uptake of quinolinic acid and only a slight difference in the uptake of calcium (Nakamura & Gowans 1972). If this is a transport difficulty it evidently does not occur at the cell membrane, but rather at some intracellular transport site. Calcium (and other cations), by altering such an intracellular system, may facilitate the entry of endogenous quinolinic and nicotinic acid into a normally cloistered pathway.

9   MUTATION

9.1   *Response to irradiation*

Survival of *C. eugametos* following X-irradiation was non-exponential, indicating that more than one hit is required to produce inactivation (Nybom 1953; Godward 1962). From comparative irradiation studies of haploid and diploid *C. reinhardtii*, and wild type and mutants of *C. eugametos* which showed differential responses to supplements, Wetherell and Krauss (1957) concluded that *C. eugametos* had a high degree of duplication of genetic information, though less than would be found in the diploid state. Alternatively, these observations could be explained if *C. eugametos* and *C. moewusii* become autopolypoloid in high light intensities, as indicated by Buffaloe (1958), who found uniformly small cells in cultures illuminated at 1000 lux and a direct relationship between light intensity and cell size at higher intensities. Chromosome preparations of the large cells showed polyploidy; and when such large cells were placed in the dark they underwent successive bipartitions until the haploid level was restored. The light intensities employed by Nybom (1953) were not given in his paper; Wetherell and Krauss (1957) used ca. 2500 lux.

Most of the cells killed by X-irradiation are end-point killed cells (i.e. the cells can still divide, but the progeny die), whereas death induced by U.V. irradiation is primarily from zero-point killing (death without post-irradiation cell division) (Nybom 1953). End-point death is generally associated with nuclear damage, and zero-point death with cytoplasmic damage (Godward 1962). Nybom demonstrated that after UV irradiation the potential of *C. eugametos* for photoreactivation was extremely high, and that photoreactivation was primarily of the zero-point killed cells. Thometz (1973) confirmed the high potential for photoreactivation in *C. eugametos*, and also observed two components in the photoreactivation curves which indicated the presence of two types of cells, perhaps end-point and zero-point killed cells or, possibly, haploid and polyploid cells.

Thometz (1973) who studied the survival of cells of *C. eugametos* and *C. reinhardtii* after UV irradiation, found that when cells were held in a caffeine solution in the dark survival was less than in the controls, and concluded that *C. eugametos* had a dark-repair mechanism which could be inhibited by caffeine.

9.2   *Simultaneous mutation*

Lewin (1974) reported three cases of simultaneous mutation in *C. moewusii* and indications of a fourth. M. 53 yielded independently segregating genes for sterility and for paralysis. M. 85 carried two paralysis mutations which were linked, 28 map units apart. M. 87 carried genes for paralysis and slow growth. The fourth case occurred in a study of UV-induced reversion of paralysis genes. One such revertant, of the paralysed strain M. 02, simultaneously regained motility and lost its sexual fertility; however, it was impossible to test the involvement of two genes in this simultaneous change since the resulting strain was sterile. Similarly, a paba-less mutant (Lewin 1953) originated simultaneously with a volutin marker, and a thiamine-less mutant with a paralysis mutation. Lewin discussed the occurrence of simultaneous mutations (Lewin 1974) and pointed out that a similar observation had been made in studies of *C. reinhardtii* (McVittie 1972). If the cultures exposed to irradiation in mutant recovery experiments consisted of mixtures of haploid and polyploid cells, one would expect a high frequency of such simultaneous mutations. Gowans (unpublished) has indications that the frequency of mutant recovery is higher in experiments where cells are irradiated after a 12-hour dark period than in experiments where the irradiated cells were grown in continuous light.

9.3   *Reversion, back-mutation, and chemical mutagens*

Wetherell and Krauss (1957) reported a high frequency of spontaneous reversion among their strains of *C. eugametos* which showed differential responses to supplements. Thus, out of 47 isolated, 30 eventually reverted to the normal phenotype. Since in most cases the phenotypic differences observed in these strains were not established as genetically segregating characteristics, these observations are difficult to interpret. Lewin (1974) observed relatively high frequencies of reversions of two independently derived, partially paralysed strains (M. 11 and M. 38), attributed to allelic (or closely linked) genes. He also studied UV-induced revertants of these two unstable strains, as well as two such revertants of the stable, inseparably sex-linked gene of M. 02. One revertant of M. 36, one of M. 02 and two of M. 75 completely regained wild-type motility. In addition a revertant of M. 75

was found which had partially restored motility, and a revertant of M. 02 which had simultaneously regained normal motility and lost the ability to agglutinate in a sexual reaction (see Simultaneous mutation section). This reversion of M. 02 is of particular interest because of the absolute linkage of the flagellar paralysis gene (*pf*) to the mating-type locus. M. 02 may actually bear a *pf* allele of mating type, and the kind of motility controlled at this locus may be a pleiotropic manifestation of normal mating-type gene function. Alternatively, the *pf* gene of M. 02 may be located along with the mating-type locus on a non-homologous section of the sex chromosome (as an X-linked or Y-linked gene), or the M. 02 stock may have an inversion in the mating-type or sex chromosome which effectively suppresses crossing-over between *mt* and *pf*. The high incidence of lethality among sexual progeny of M. 02 adds support to the latter hypothesis (see Chromosome aberration section).

Loeppky (1967) conducted extensive reversion experiments with two nicotinamide auxotrophs (*nic-4* and *nic-5*) of *C. eugametos* after treatment with chemical mutagens. No reversions at either locus were found after treatment of 10^9 cells with 2-amino purine, 10^9 cells with 5-bromodeoxyuridine (plus aminopterin), 10^7 cells with hydroxylamine, or 10^7 cells with 2-methoxy-6-chloro-9 (-3(ethyl-2-chloroethyl)-aminopropylamino) acridine dihydrochloride (ICR 170), nor did reversions occur among a total of 10^{10} untreated cells used as controls in these experiments. This correlates with the observation that no spontaneous reversions of any of the nicotinamide auxotrophs have been noted during their routine subculture (see, however, Mutator gene section). In contrast, the *thi-2* (*thi-1*) gene (strain Mo27) showed rare spontaneous reversion in the laboratory, and Loeppky was able to increase the rate of reversion to thiamine independence by a factor of 10,000 (from 10^{-9} to 10^{-5}) by treatment of mutant cells with either 2-amino purine or 5-bromodeoxyuridine (plus aminopterin). Three of these revertants, selected at random, were crossed with wild type to determine whether reversion was due to a suppressor mutation or to back-mutation at the *thi-1* locus. None of the segregants from these crosses required thiamine, indicating that the reversions were true back-mutations (or, just possible, closely linked suppressor mutations).

Loeppky also found that ICR 170 increased mutation from wild type to pyrithiamine resistance 1,000-fold (from 10^{-6} to 10^{-3}).

A.C. McBride and Gowans (1969) demonstrated by fluctuation tests that the phenylalanine analogue 2-amino-3-phenylbutanoic acid (APBA. 3-methylphenylalanine) induces gene mutation, as does streptomycin in *C. reinhardtii* (Sager & Tsubo 1962), and also chromosome aberration. The mutagenic effect of APBA was first discovered during experiments designed to find mutants resistant to APBA, but subsequent studies demonstrated that APBA could also induce mutants resistant to neamine and to streptomycin. The 'point' mutants induced by APBA segregate 2:2 in crosses and are, on this basis, chromosomal genes rather than 'non-chromosomal' genes (see chapter by Gillham and Boynton in this volume). The evidence for aberration induction was on the basis of lethality patterns among gonal progeny (see Chromosome aberration section). McBride and Gowans proposed four hypotheses to account for the action of APBA in inducing mutations and aberrations: (1) incorporation of the analogue into chromosome-associated proteins, (2) donation of methyl groups of APBA to DNA bases, (3) incorporation of APBA into DNA polymerase, or (4) induction of controlling elements similar to those observed in maize (McClintock 1965). The mode of action of APBA on *C. eugametos* and that of streptomycin on *C. reinhardtii* may be basically the same, since streptomycin is known to inhibit the incorporation of phenylalanine into peptide linkages in the presence of the synthetic messenger polyuridylic acid (Kirk & Tilney-Bassett 1967).

There were indications in the work of Thometz (1973) on caffeine-resistant mutants that caffeine, too, might act as a mutagen as well as a selective agent; but fluctuation analyses were not carried out.

9.4   *Chromosome aberrations*

Perkins (1966, 1974) summarized the various lethality patterns to be expected in tetrad analysis of zygotes heterozygous for different chromosomal aberrations.

The best evidence for aberrations, in the absence of cytological confirmation, derives from the patterns expected from reciprocal and insertional translocations, since each predicts an equality of two different viable:lethal ratios among the four products from different zygotes. In a cross involving meiotic mother cells heterozygous for a *reciprocal* translocation, the meioses with no chromatid exchanges will fall into two classes of equal frequency: (1) those resulting from adjacent centromere segregations (0 viable:4 lethal), and (2) those resulting from alternate centromere segregations (4 viable:0 lethal). Meioses with one chromatid exchange will yield 2 viable:2 lethal, whether the centromere segregation is alternate or adjacent. In a cross involving meiotic mother cells heterozygous for an *insertional* translocation, the meioses with no chromatid exchanges will fall into two classes of equal frequency: (1) those which after first division have one nucleus with two normal chromosomes and the other nucleus with an interstitially translocated chromosome and a chromosome with a deletion (4 viable:0 lethal), and (2) those which after first division have one nucleus with one normal chromosome and one chromosome with a deletion, and the other nucleus with an interstitially translocated chromosome and a normal chromosome (2 viable:2 lethal). Meioses with one chromatid exchange will yield 3 viable:1 lethal.

A.C. McBride and Gowans (1969) reported that crosses involving five out of 31 different strains isolated after exposure to APBA gave lethality patterns indicating segregation of a heterozygous reciprocal translocation. These putative translocations were not necessarily associated with APBA resistance: three of the five strains were resistant to APBA, two were not, and of these two one was resistant to streptomycin. Furthermore, the identified APBA-resistance genes (*map-1, -2, -3*) are borne in stocks which show no evidence of chromosomal aberration. After exposure to APBA four other strains showed, in crosses, lethality patterns which could be attributed to heterozygous inversions or to more complex chromosomal aberrations.

In his analysis of flagellar-activity mutants of *C. moewusii*, Lewin (1974) observed progeny lethality ranging from 18 to 56% of the progeny in different crosses, which could be due to chromosomal aberrations. Of particular interest is the mutant M. 02, in which *pf* is apparently inseparable from the mating-type locus. (This was the strain involved in the interspecific cross between *C. eugametos* and *C. moewusii* analysed by Gowans (1963).) Such absolute linkage could be attributable to an inversion. Unfortunately, no conclusions can be drawn from the genetic analysis of 37 tetrads from the M. 02×wild-type cross, since the lethality pattern did not fit any of the patterns expected from a single simple chromosome aberration.

Crosses involving one mutant (M. 43) gave a pattern which seems (to C.S.G.) indicative of a pericentric inversion, long enough to include frequent double exchanges, with one break at a dispensible chromosome tip.

9.5 *Mutator gene*

Wang (1972) recovered a mutator gene (*m*) of *C. eugametos*, which arose spontaneously in a haploid segregant of a vegetative putative diploid stock, capable of inducing reversion at the loci *nic-4* and *nic-5*, which had not previously been observed to revert either spontaneously or after treatment with chemical mutagens (Loeppky 1967). The *m* gene segregates 2:2 in crosses, and may be linked to the *nic-5* gene. Wang presented evidence that the gene was indeed a mutator and not a suppressor, since he isolated no auxotrophic segregates from crosses of strains with a reverted *nic-4* gene and wild type. The action of the *m* gene appears to be specific, since it did not increase the rate of mutation to either streptomycin resistance or neamine resistance.

9.6 *Limited mutation spectrum*

The paucity of types of auxotrophic mutants has been discussed for *Chlamydomonas* (Ebersold 1962) and for photosynthetic forms in general (Li *et al.* 1967). No amino-acid-requiring mutants of *C. moewusii* or *C. eugametos* have been recovered, and arginine-requirers are the only reported amino-acid auxotrophs of *C. reinhardtii*. This was noted early in the mutation experiments of Gowans (1956), but extensive searches for such mutants in *C. eugametos*, on agar media supplemented with various mixtures of amino acids or with single amino acids (particularly arginine), proved fruitless. The simplest explanation for these results is obligate cellular autonomy (Redei

1967), and the simplest type of obligate autonomy would be determined by a failure of membrane uptake of an exogenous supply of the metabolite. However, some amino-acid analogues are inhibitory to *C. eugametos* and *C. moewusii* (McBride A.C. & Gowans 1970); *C. eugametos* can grow on media with either histidine or tryptophan as its sole nitrogen source; *C. eugametos* has been shown to incorporate ^{14}C into cells supplied with labelled leucine, glycine, tryptophan, histidine or phenylalanine; and the label from histidine or phenylalanine has been shown to be incorporated into the soluble proteins (Wang 1972). Nevertheless, special efforts to recover phenylalanine auxotrophs (McBride A.C. & Gowans 1970) and histidine auxotrophs (Wang 1972) have failed.

The limited mutant spectrum could be attributable to some kind of genetic redundancy: euploidy or aneuploidy, gene duplication in a haploid complement, or the duplication of nuclear genetic abilities by the DNA of the chloroplast or other organelles. Irradiation experiments give evidence for redundancy (Nybom 1953) and cytological observations indicate a kind of reversible autopolyploidy (Buffaloe 1958). Simple autopolyploidy, however, could not be the sole explanation, since dark periods result in haploid cells in which mutations hidden by ploidy would be expected to segregate out. Redundancy involving duplication of nuclear-determined synthetic abilities by organelle-determined synthetic abilities might be detectable in experiments involving metabolites known to be synthesized by different pathways in different organisms. One could expect, for instance, that synthesis of nicotinamide might proceed via the tryptophan pathway in the nuclear-gene sequence, and via the glycerol-succinate pathway in some organellar-gene sequence—although in fact only the tryptophan pathway is present in *C. eugametos* (Uhlik & Gowans 1974). This would not be a critical test of nuclear-organelle redundancy, however, because nicotinamide auxotrophs can be recovered and therefore redundancy is not suspected for this pathway. A better test would be the study of the synthesis of lysine, which is produced by two separate biochemical sequences in different organisms (Lingens *et al.* 1966), and for which a mutant requirement has never been recovered in *C. eugametos* or *C. moewusii*.

Wang (1972) sought to obtain histidine mutants by repeated sequential irradiations of the same cells, in hopes of mutating all of the replicates of one of the biosynthetic genes. His lack of success is understandable. If the average mutation rate is 10^{-6}, the compound probability law makes recovery of mutants by this approach highly unlikely, even if such a gene is merely duplicated and not multiply redundant. Another approach currently being tried is the isolation of mutants hypersensitive to antimetabolites, since these may represent cells with a reduction in wild-type gene dosage through the occurrence of loss mutations. The results obtained so far have been negative.

10   FORMAL GENETICS

Lewin (1953) found that a mutant gene of *C. moewusii* controlling the para-aminobenzoic-acid requirement was linked to mating type. He later (1974) identified seven linkage groups on the basis of crosses of mutants with mutational defects involving flagellar activity. Two pairs of linked genes were identified, and one paralysis gene (M. 02) was shown to be inseparably linked or allelic with the mating-type locus.

Gowans (1960) tested 20 of the 36 possible gene-gene relationships between nine different genes of *C. eugametos* and concluded that none showed linkage (although one case of 'false linkage' was found). Likewise Nakamura and Gowans (1965) tested 23 of the 36 possible relationships in this alga without detecting linkage (though here, too, one case of 'false linkage' was found). Linkage between *nic-5* and *m* (= mutator gene) was reported for this species by Wang (1972). The cases of linkage and certain gene-centromere distances of non-linked genes in *C. moewusii* and *C. eugametos* are summarized in Fig. 7.1.

An interesting situation was presented by Lewin (1974). The mutant genes in M. 85-1 and M. 82 were found to be alleles (PD:NPD:T ratio = 20:0:0). However, whereas the *pf* gene in mutant M. 82 is, by the criterion of PD:NPD ratio, linked to that of M. 21 (25 PD:6 NPD:23 T), the allele of this gene in M. 85-1 is, by the same criterion, independent of the M. 21 gene (22 PD:28 NPD:81 T). In the relationship of these two alleles with the M. 21 gene, the tetratype frequencies are strikingly different (0·426 for the M. 82 by M. 21 relationship, and 0·618 for the M. 85-1 by M. 21 relationship). Although by the criterion of PD:NPD ratio the M. 21 gene is linked to that of M. 82 with a probability greater than 99%, on the basis of the NPD:T ratio the two genes seem to segregate independently since this ratio (6:23) barely exceeds the theoretical 1:4 ratio limit (Perkins 1955). It should be pointed out, however, that Lewin (in his Table 3) reported a 34% lethality among progeny from M. 82×M. 85-1 crosses; perhaps too few NPD tetrads were recovered because of lethal crossing-over resulting from a chromosome aberration in one of the parental stocks.

10.1   *False linkage and negative linkage*

In tetrad analysis, linkage can be detected by a significant excess of parental ditype (PD) over non-parental ditype (NPD) tetrads (Perkins 1955). If the genes segregate independently the PD:NPD ratio should approximate 1:1. Significant excesses of NPD tetrads over PD tetrads are not expected; when they occur the situation might be termed 'negative linkage'. Gowans (1960) reported a case of excess NPDs in *C. eugametos* crosses involving *pab-1* and *mt*, and also discussed two cases of negative linkage reported for *C. reinhardtii*

(Hartshorne 1955; Sager 1955). False linkage refers to cases where evidence for linkage is statistically significant only in particular crosses, but is not consistently so. False linkage (Longley 1945) and negative linkage (Schult & Lindegren 1956) could both be explained by invoking some kind of affinity between non-homologous chromosomes.

Since in *Chlamydomonas* all cases of negative linkage and false linkage are barely significant at or near the 5% level, perhaps they should all be dismissed as random variations. However, certain correlations indicate that further observations may be warranted. Firstly, there is often an association between false linkage or true linkage and negative linkage. Thus from pooled data of many crosses Gowans (1960) found evidence for negative linkage of *pab-1* with *mt*. In some individual crosses it was very striking; and in certain crosses *pab-1* showed negative linkage to *nic-1* and at the same time false linkage to *thi-1*. Likewise Wang (1972), who reported linkage of *nic-5* and *m*, from the same cross obtained data indicating negative linkage of *nic-4* and *nic-5*. Lewin (1974) reported complete linkage between the M. 02 *pf* gene and *mt*, and examination of Table 2 in his paper shows three cases of negative linkage of the M. 02 gene to other paralysis genes (M. 22, M. 74 and M. 17), each being significant at the 5% level. Secondly, there is a frequent involvement of the mating type (*mt*) gene. Thus, *mod-1* shows false linkage to *mt* in *C. eugametos* (Nakamura & Gowans 1965), *pab-1* shows negative linkage to *mt* (Gowans 1960), *pab* shows linkage to *mt* in *C. moewusii* (Lewin 1953), *yl* shows negative linkage to *mt* in *C. reinhardtii* (Sager 1955), and the *pf* of M. 02, which is inseparably linked to *mt* in *C. moewusii*, shows negative linkage to some other markers (Lewin 1974).

10.2   *High tetratype-tetrad frequencies*

Perkins (1955) has pointed out that tetratype tetrad frequencies which exceed 2/3 can be used as evidence that sister-strand exchanges and non-sister-strand exchanges do not affect chiasma interference to the same extent. Figure 1 shows that at least four genes (*nic-3* and *nic(42)* may be allelic) have exchange frequencies in excess of the theoretical upper limit of 66·7%. The most likely explanation for these high tetratype frequencies would be a nearly obligate exchange between such a gene and its corresponding centromere combined with a very strong chiasma interference, although, as Perkins points out, the same result could theoretically be obtained if centromeres were usually postreduced for one bivalent and prereduced for the other.

The *nic* genes of *C. eugametos* do not appear to be randomly distributed with regard to their centromeres. Thus at least four of the loci are within four map units of their respective centromeres, and at least two segregate independently of their centromeres. No *nic* genes have been found to be located in regions of their linkage groups intermediate between these extremes.

10.3   Temperature and crossing-over

Gowans (1960) tested the germination of zygotes from a single cross at six different temperatures (5, 13, 20, 25, 28, and 30°C). At the two extreme temperatures (5° and 30°) they failed to germinate. In the 13–28° range the tetratype frequencies between *pab-1* and *mt* were found to vary inexplicably: 31·0% at 13°, 12·5% at 20°C, 29·8% at 25°C and 13·0% at 28°C.

10.4   'Non-chromosomal' genes

As already mentioned in section 6.1 on Antibiotics, A.C. McBride (unpublished) recovered two streptomycin-resistant strains and one neamine-dependent strain of *C. eugametos* which do not show segregation of the traits in crosses; Chu-Der (1971) reported an acetate-requiring strain of a similar type. All of these strains, except one streptomycin-resistant mutant, were female, and did not transmit the mutant character to sexual progeny. Correspondingly the one male (*mt+*) streptomycin-resistant strain transmitted streptomycin resistance to all four products of each zygote.

11   CONCLUSION

A considerable body of knowledge concerning the genetics and basic physiology of *C. moewusii* and *C. eugametos* has been accumulated, providing opportunities for research into a variety of areas of fundamental interest to the biologist. We who work with these species recognize, nonetheless, the value of extensive pioneering work of W.T. Ebersold, R.P. Levine, R. Sager and others who have established *C. reinhardtii* as the organism of choice for most kinds of investigation in chlamydomonad genetics. Thus, *C. eugametos* and *C. moewusii* assume a secondary position in genetic research, similar to those of *Neurospora sitophila* and *Drosophila pseudoobscura*. Research on such organisms serves to support, or question, generalizations which tend to be made when concerted efforts are concentrated on a single species. Differences in morphology or physiology may provide special opportunities for investigations which are not possible with the 'organism of choice'. The existence of the tandem stage in *C. moewusii* allows complementation in heterokaryons. The obligate phototrophy of *C. moewusii* and *C. eugametos*, and the recovery of obligately mixotrophic mutants, offer promise of a better understanding of the special functions of light in chlamydomonad physiology and of the metabolic interrelationships between photosynthetic and exogenously-supplied carbon compounds. Although 2-amino-3-phenylbutanoic acid has no apparent effect on *C. reinhardtii*, it has been found to induce both gene mutations and chromosome aberrations in *C. eugametos*. It would be interesting to know what biochemical features underly such differences.

12 REFERENCES

BEADLE G.W., MITCHELL H.K. & NYC J.F. (1947) Kynurenine as an intermediate in the formation of nicotinic acid from tryptophane by Neurospora. *Proc. Nat. Acad. Sci. U.S.* **33**, 155–8.

BERNSTEIN E. & JAHN T.L. (1955) Certain aspects of the sexuality of two species of Chlamydomonas. *J. Protozool.* **2**, 81–5.

BROKAW C.J. (1960) Decreased adenosine triphosphate activity of flagella from a paralyzed mutant of *Chlamydomonas moewusii*. *Exptl. Cell Res.* **19**, 430–2.

BROWN R.M., JOHNSON C. & BOLD H.C. (1968) Electron and phase-contrast microscopy of sexual reproduction in *Chlamydomonas moewusii*. *J. Phycology*, **4**, 100–20.

BROWN R.M. & WALNE P.L. (1967) Comparative immunology of selected wild types, varieties, and mutants of Chlamydomonas. *J. Protozool.* **14**, 365–73.

BUFFALOE N.D. (1958) A comparative cytological study of four species of Chlamydomonas. *Bull. Torrey Bot. Club.* **85**, 157–78.

CHU-DER O.M. (1971) Acetate mutants and obligate photoautotrophy in *Chlamydomonas eugametos*. Ph.D. dissertation, University of Missouri-Columbia.

COLEMAN A.W. (1962) Sexuality. *in* Physiology and Biochemistry of Algae, ed. Lewin R.A. Academic Press, N.Y.

CZURDA V. (1935) Über die 'Variabilität' von *Chlamydomonas eugametos* Moewus. *Beih. Bot. Centralbl.* **53A**, 133–57.

DAVIS R.H. (1963) Lethality of Neurospora arginine mutants associated with a factor from wild type. *Neurospora Newsletter* **4**, 5 (Cited by permission).

DEMPSEY L.T. & DELCARPIO J.B. (1971) Electrophoretic analysis of enzymes from three species of Chlamydomonas. *Amer. J. Botany*, **58**, 716–20.

DER O.C. & GOWANS C.S. (1972) TCA enzymes in obligate- and facultative-autotrophic species, and in acetate mutants of Chlamydomonas. *Can. J. Gen. Cytol.* **14**, 724.

EBERSOLD W.T. (1962) Biochemical Genetics. *in* Physiology and Biochemistry of Algae, ed. Lewin R.A. Academic Press, N.Y.

GERLOFF J. (1940) Beiträge zur Kenntnis der Variabilität und Systematik der Gattung Chlamydomonas. *Arch. f. Protist.* **94**, 311.

GIBBS S.P., LEWIN R.A. & PHILPOTT D.E. (1958) The fine structure of the flagellar apparatus of *Chlamydomonas moewusii*. *Exptl. Cell Res.* **15**, 619–22.

GODWARD M.B.E. (1962) Invisible Radiations, *in* Physiology and Biochemistry of Algae, ed. Lewin R.A. Academic Press, N.Y.

GOWANS C.S. (1956) Genetic investigations on *Chlamydomonas eugametos*. Ph.D. dissertation, Stanford University.

GOWANS C.S. (1960) Some genetic investigations on *Chlamydomonas eugametos*. *Zeit. f. Indukt. Abstam.- u. Vererb.-lehre*, **91**, 63–73.

GOWANS C.S. (1963) The conspecificity of *Chlamydomonas eugametos* and *Chlamydomonas moewusii*: An experimental approach. *Phycologia*, **3**, 37–44.

GOWANS C.S. (1965) Tetrad analysis. *Taiwania*, **11**, 1–19.

GUILLARD R.R.L. (1960) A mutant of *Chlamydomonas moewusii* lacking contractile vacuoles. *J. Protozool.* **7**, 262–8.

HARTSHORNE J.N. (1953) The function of the eyespot in *Chlamydomonas*. *New Phytologist* **52**, 292–7.

HARTSHORNE J.N. (1955) Multiple mutation in *Chlamydomonas reinhardii*. *Heredity*, **91**, 239–48.

HENDERSON L.M., RAMASARMA G.B. & JOHNSON B.C. (1949) Quinolinic acid metabolism, IV. Urinary excretion by man and other mammals as affected by the ingestion of tryptophane. *J. Biol. Chem.* **181**, 731–8.

HENDERSON L.M., SOMEROSKI J.F., RAO D.R., WU P.H.L., GRIFFITH T. & BYERRUM R.U. (1959). Lack of a tryptophan-niacin relationship in corn and tobacco. *J. Biol. Chem.* **234**, 93–5.

HUTNER S.H. & PROVASOLI L. (1951) The Phytoflagellates. *in* Biochemistry and Physiology of Protozoa, ed. Lwoff A. Academic Press, N.Y.

JONES B.R. (1968) Genetic and physiological studies of thiamine requiring mutants of *Chlamydomonas eugametos*. Ph.D. dissertation, University of Missouri-Columbia, 1968.

KIRK J.T.O. & TILNEY-BASSETT R.A.E. (1967) The Plastids, their Chemistry, Structure, Growth and Inheritance. W.H. Freeman Press, San Francisco.

LEETE E., MARION L. & SPENSER I.D. (1955) The biosynthesis of damascenine and trigonelline. *Can. J. Chem.* **33**, 405–10.

LEWIN J.C. (1950) Obligate autotrophy in *Chlamydomonas Moewusii* Gerloff. *Science*, **112**, 652–3.

LEWIN R.A. (1952a) Studies on the flagella of algae. I. General observations on *Chlamydomonas moewusii Gerloff. Biol. Bull.* **103**, 74–9.

LEWIN R.A. (1952b) Ultraviolet induced mutations in *Chlamydomonas moewusii Gerloff. J. Gen. Microbiol.* **6**, 233–48.

LEWIN R.A. (1953) The genetics of *Chlamydomonas moewusii Gerloff. J. Genet.* **51**, 543–60.

LEWIN R.A. (1954) Mutants of *Chlamydomonas moewusii* with impaired motility. *J. Gen. Microbiol.* **11**, 358–63.

LEWIN R.A. (1956) Control of sexual activity in Chlamydomonas by light. *J. Gen. Microbiol.* **15**, 170–85.

LEWIN R.A. (1960) A device for obtaining mutants with impaired motility. *Can. J. Microbiol.* **6**, 21–5.

LEWIN R.A. (1974). Genetic control of flagellar activity in *Chlamydomonas moewusii. Phycologia* **13**, 45–55.

LEWIN R.A. & FRENKEL A.W. (1960) Difektita aŭtotrofo de mutaciita *Chlamydomonas. Plant and Cell Physiol.* **1**, 327–30.

LEWIN R.A. & MEINHART J.O. (1953) Studies on the flagella of algae. III. Electron micrographs of *Chlamydomonas moewusii. Can. J. Botany* **31**, 711–7.

LI S.L., REDEI G.P. & GOWANS C.S. (1967) A phylogenetic comparison of mutation spectra. *Molec. Gen. Genetics* **100**, 77–83.

LINGENS F., VOLLPRECHT P. & GILDEMEISTER V. (1966) Zur biosynthese der nicotinsaure in Xanthomonas and Pseudomonasarten, *Mycobacterium phlei* und rotalgen. *Biochem. Zeit.* **344**, 462–77.

LOEPPKY C.B. (1967) Chemical mutagenesis of *Chlamydomonas eugametos*. M.A. thesis, University of Missouri-Columbia.

LONGLEY A.E. (1945) Abnormal segregation during megasporogenesis in maize. *Genetics*, **30**, 100–13.

MCBRIDE A.C. (1967) Genetic and physiological effects of an amino acid analog in *Chlamydomonas eugametos*. Ph.D. dissertation University of Missouri-Columbia.

MCBRIDE A.C. & GOWANS C.S. (1969) The induction of gene mutation and chromosome aberration in *Chlamydomonas eugametos* by a phenylalanine analog. *Genet. Res., Camb.* **141**, 121–6.

MCBRIDE A.C. & GOWANS C.S. (1970) Comparative sensitivity of three species of Chlamydomonas to analogs of metabolites. *J. Phycology*, **6**, 54–6.

MCBRIDE J.C. & GOWANS C.S. (1967) Pyrithiamine resistance in *Chlamydomonas eugametos. Genetics*, **56**, 405–12.

MCCLINTOCK B. (1965) The control of gene action in maize. *Brookhaven Symp. Biol.* **18**, 162–84.

MCVITTIE A. (1972) Genetic studies on flagellar mutants of *Chlamydomonas reinhardi. Genet. Res., Camb.* **9**, 157–64.

172 CHAPTER 7

MEHLER A.H. (1951) Studies on reactions of illuminated chloroplasts. I. Mechanism of the reduction of oxygen and other Hill reagents. *Arch. Biochm.* **33**, 65–77.

MOEWUS F. (1931) Neue Chlamydomonaden. *Archiv. f. Protist.* **75**, 284–96.

NAKAMURA K. & GOWANS C.S. (1964) Nicotinic acid-excreting mutants in Chlamydomonas. *Nature*, **202**, 826–7.

NAKAMURA K. & GOWANS C.S. (1965) Genetic control of nicotinic acid metabolism in *Chlamydomonas eugametos. Genetics*, **51**, 931–45.

NAKAMURA K. & GOWANS C.S. (1967) Ionic remediability of a mutational transport defect in Chlamydomonas. *J. Bacteriol.* **93**, 1185–7.

NAKAMURA K. & GOWANS C.S. (1972) A modifier of a *nic* gene in Chlamydomonas. *Can. J. Gen. Cytol.* **14**, 733.

NISHIZUKA Y. & HAYAISHI, O. (1963) Studies on the biosynthesis of nicotinamide adenine dinucleotide. I. Enzymic synthesis of niacin ribonucleotides from 3-hydroxyanthranlilic acid in mammalian tissues. *J. Biol. Chem.* **238**, 3369–77.

NYBOM N. (1953) Some experiences from mutation experiments in Chlamydomonas. *Hereditas*, **39**, 317–24.

ORTEGA M.V. & BROWN G.M. (1960) Precursors of nicotinic acid in *Escherichia coli. J. Biol. Chem.* **235**, 2939–45.

PERKINS D.D. (1955) Tetrads and crossing-over. *J. Cell. & Comp. Physiol.* **45**, *Suppl. 2*, 119–49.

PERKINS D.D. (1966) Preliminary characterization of chromosome rearrangement using shot asci. *Neurospora Newsletter*, **9**, 10–11 (Cited by permission).

PERKINS D.D. (1974) The manifestation of chromosome rearrangements in unordered asci of Neurospora. *Genetics*, in press.

REDEI G.P. (1967) Genetic estimate of cellular autarky. *Experientia*, **23**, 584.

SAGER R. (1955) Inheritance in the green alga *Chlamydomonas reinhardi. Genetics*, **40**, 476–89.

SAGER R. & GRANICK S. (1953) Nutritional studies with *Chlamydomonas reinhardi. Ann. N.Y. Acad. Sci.* **56**, 831–8.

SAGER R. & TSUBO Y. (1962) Mutagenic effects of streptomycin in Chlamydomonas. *Archiv. f. Mikrobiol.* **42**, 159–75.

SCHULT E.E. & LINDEGREN C.C. (1956) Mapping methods in tetrad analysis. I. Provisional arrangement and ordering of loci preliminary to map construction by analysis of tetrad distribution. *Genetica*, **28**, 165–76.

STARR R.C. (1964) The culture collection of algae at Indiana University. *Am. J. Botany*, **51**, 67–86.

SYRETT P.J. (1962) Nitrogen Assimilation. *in* Physiology and Biochemistry of Algae, ed. Lewin R.A. Academic Press, N.Y.

THOMETZ D.S. (1973) Effects of caffeine on *Chlamydomonas eugametos*. M.A. thesis, University of Missouri-Columbia.

TRAINOR F.R. (1959) A comparative study of sexual reproduction in four species of Chlamydomonas. *Amer. J. Botany*, **46**, 65–70.

UHLIK D.J. & GOWANS C.S. (1972) The pathway of synthesis of nicotinamide in *Chlamydomonas eugametos. Can. J. Gen. Cytol.* **14**, 739.

UHLIK D.J. & GOWANS C.S. (1974) The pathway of synthesis of nicotinamide in *Chlamydomonas eugametos. Intl. J. Biochem.* **5**, 79–84.

WALLER G.R., YANG K.S., GHOLSON R.K. & HADWIGER L.A. (1966) The pyridine nucleotide cycle and its role in the biosynthesis of ricine by *Ricinus communis* L. *J. Biol. Chem.* **241**, 4411–8.

WANG W. (1972) Mutational and physiological studies of *Chlamydomonas eugametos*. Ph.D. dissertation, University of Missouri-Columbia.

WETHERELL D.F. (1958) Obligate phototrophy in *Chlamydomonas eugametos. Physiol. Plantarum*, **11**, 260–74.

WETHERELL D.F. & KRAUSS R.W. (1957) X-ray induced mutations in *Chlamydomonas eugametos*. *Am. J. Botany*, **44**, 609–19.

WIESE L. (1969) Algae. *in* Fertilization. Comparative Morphology Biochemistry and Immunology, eds. Metz C.B. & Monroy A. Academic Press, N.Y.

WIESE L. & HAYWARD P.C. (1972) On sexual agglutination and mating-type substances in isogamous dioecious Chlamydomonads. III. The sensitivity of sex cell contact to various enzymes. *Amer. J. Botany*, **59**, 530–6.

WILSON R.G. & HENDERSON L.M. (1960) Niacin biosynthesis in the developing chick embryo. *J. Biol. Chem.* **235**, 2099–2102.

CHAPTER 8

GENETIC ASPECTS OF SEXUALITY IN VOLVOCALES

LUTZ WIESE

Department of Biological Science,
Florida State University, Tallahassee, U.S.A.

1 Introduction 174

2 Mating systems and sex determination 174

3 Gametogenesis, fertilization and
meiosis 178

4 Sexual incompatibility, syngens,
and speciation 181

5 Conclusions 192

6 References 193

1 INTRODUCTION

The genetic aspects of sex phenomena in the *Volvocales* have been insufficiently investigated. Their haplontic life cycles, their experimental amenability, and their apparently simple sexual differentiation systems render many algae in this order exceptionally suitable for systematic analysis of mating systems, modes of sex determination, and the genetic control of sexual differentiation. Much of the recent research in the *Volvocales* has concentrated on mechanisms controlling reproductive isolation between and within species, sex determination and the differentiation process at gametogenesis, and fertilization. This review deals primarily with the genetic bases of these phenomena.

2 MATING SYSTEMS AND SEX DETERMINATION

The mating system of an organism is determined by its modes of sex distribution, sex determination, and fertilization. Being predominantly haplonts, members of the *Volvocales* exhibit either haplogenetic sex determination, by meiotic segregation of sex or mating-type alleles at the sex or mating-type locus (*mt*), or a physiological (modificatory) type of sex determination in which maleness and femaleness arise by differentiation processes within a single genotype.

Many phycologists who would rather not use the word 'sex' for bipolar isogamy prefer to speak of mating types. According to Grell (1963) this is an unjustifiable extension of the term, which was coined for a special situation in which there is bi- or multipolarity of gamete-producing parental stages, as in *Neurospora* or *Paramecium*, superimposed on a sexual bipolarity of the gametes or nuclei. Such mating types, which control outbreeding in monoecious organisms, are not known in *Volvocales*, although they may exist in some diatoms. However, since the term 'mating type' is now in general use in the phycological literature, it is employed in its broader sense in this review.

In cases of isogamy, it is customary to designate the mating types arbitrarily as *mt*⁺ and *mt*⁻ when one cannot distinguish between male and female (Lewin 1954b). In related but incompatible species one may be able to homologize gamete types, as in three isogamous heterothallic taxa of *Chlamydomonas*, with similarities in gamete behaviour and biochemistry. In *C. eugametos*, and in *C. moewusii syngen I* (Provasoli strains) and *syngen II* (Lewin strains), one gamete type takes over the locomotion of the prezygotic pairs (Lewin 1952b; Bernstein & Jahn 1955; Wiese & Metz 1969). The sexual agglutination factor of this mating type is insensitive to certain proteases but sensitive to concanavalin A and α-mannosidase, whereas that of its partner is sensitive to the proteases but not to the other two agents (Wiese & Hayward 1972; Wiese 1974; McLean & Brown 1974; Wiese & Wiese 1975).

Where sex determination is haplogenetic, the diplophase is generally sexually indifferent. An important exception was detected in *C. reinhardtii* by Ebersold (1967). In some crosses approximately 4% of the zygotes divide mitotically instead of developing into a zygospore, thereby establishing stable diploid clones heterozygous for the mating-type alleles. In all such diploid clones the cells exhibit mating activity, which has been shown to be exclusively of the *mt*⁻ type since they copulate only with haploid *mt*⁺ cells. Evidently here *mt*⁻ dominates over *mt*⁺. Such a situation might provide the starting point for the evolution of a progressive sexual differentiation of the diplophase, as encountered in certain systems of mating types *sensu stricto* or in diplogenetic sex determination.

Haplogenetic sex determination is necessarily linked to heterothallism, and two sexually different clones have to be combined in order to assure fertilization and zygote formation. In homothallic species, on the other hand, sex can only be physiologically determined. The presence of both male and female colonies in single clones of certain oogamous *Volvocales*, e.g. *Eudorina conradi* and *E. californica* (Goldstein 1964) and *Volvox aureus* (Darden 1966) reveals that dioecism is not necessarily associated with haplogenetic sex determination. In several unicellular and colonial *Volvocales* there is a kind

of subdioecism, with two sexes and haplogenetic sex determination, although
some cells in predominantly male clones exhibit a certain degree of female
differentiation, while some female individuals may exhibit male characteristics. Such clones are capable of self-fertilization, as found in *Dunaliella
salina* (Lerche 1937), *Volvox africanus* (Starr 1971b) and certain polyploid
or aneuploid strains of *Eudorina* arising from interspecific hybridization
(Goldstein 1964).

Some of the statements of early authors on modes of sex differentiation
and sex determination in a given species have to be re-evaluated in the light
of the intraspecific variability of these traits which has been more recently
detected. *Pandorina morum*, for instance, was described by Schreiber (1925)
as heterothallic with haplogenetic sex determination; but comprehensive
investigation of this species revealed several homothallic strains with physiological sex determination (Coleman 1959).

Haplogenetic sex determination can be unequivocally demonstrated by
tetrad analysis, the sexes being distributed in a 1:1 ratio among the four gone
cells liberated after zygote germination. Tetrad analysis demonstrating
genetic segregation was first performed in *Chlamydomonas* crosses by Pascher
(1916, 1918). In *Gonium*, the two mt^+ and two mt^- gone cells formed after
meiosis remain united for one generation in a common four-celled colony,
and unisexual clones segregate only after the first asexual reproduction
(Schreiber 1925). Haplogenetic sex determination is often inferred merely
from the fact that clonal cultures are demonstrably either male or female, or
mt^+ or mt^-. In the *Volvocales* there seems to be an evolutionary trend involving a reduction in the number of surviving gone cells from four, as in all
unicellular forms and in *Gonium* and *Stephanosphaera* (Cohn & Wichura
1857), to one, as in all other colonial forms, where three of the four meiotic
nuclei die. In the latter case one can demonstrate haplogenetic sex determination by a 1:1 sex ratio among the gone cells indicating equal chances of
survival of the mt^+ or mt^- progeny. (Another trend, from isogamy via anisogamy (heterogamy) to oogamy, is not considered in this review; it is
described by Kalmus (1932), Scudo (1967), Parker, G. A. *et al.* (1972), and
Parker, B. C. (1972).)

Haplogenetic sex determination has been shown to occur in dioecious
strains of the following *Volvocales*: *Chlamydomonas* spp. (Pascher 1918;
Gerloff 1940), *C. reinhardtii* (Smith & Regnery 1950; Sager 1955), *C. moewusii*
(Gerloff 1940; Lewin 1953), *C. eugametos* (Moewus 1938; Foerster & Wiese
1954a; Gowans 1960), *Chlorogonium euchlorum* (Schulze 1927), *C. elongatum*
and *C. leiostracum* (Strehlow 1929), *Dunaliella salina* (Lerche 1937). *Gonium
pectorale* (Schreiber 1925; Stein 1958b), *Pandorina morum* (Schreiber 1925;
Coleman 1959), *P. unicocca* (Rayburn & Starr 1974), *Eudorina elegans*
(Schreiber 1925; Goldstein 1964; Mishra & Threlkeld 1968), *Volvulina steinii*
(Carefoot 1966), *V. pringsheimii* (Starr 1962), *Platydorina caudata* (Harris &
Starr 1969), *Astrephomene gubernaculifera* (Brooks 1966), *Volvox gigas* (Van

de Berg & Starr 1971), *V. rousseletii* (McCracken & Starr 1970), *V. carteri* f. *nagariensis* (Starr 1969) and *V. carteri* f. *weismannia* (Kochert 1968).

Apart from their decisive role in sex determination, little is known of the actual nature of the sex alleles. According to Moewus (1936, 1938), the sex-determining factors in *C. eugametos* are not alleles but represent different genes on homologous chromosomes. In his matings of $mt^+ \times mt^-$ strains a certain percentage of genes appeared, he reported, which gave rise to homo-thallic clones, and he attributed their origin to crossing-over between non-allelic factors (see Gowans, p. 317). In spite of intensive efforts by many other workers, Moewus' statements were never confirmed, neither in *C. reinhardtii* (Smith & Regnery 1950, Sager 1955, 1972) nor in *C. moewusii* (Lewin 1953) or Moewus' strains of *C. eugametos* (Foerster & Wiese 1954a; Gowans 1960). Homothallic strains could conceivably arise from a heterothallic species as a result of mutations not involving the sex loci themselves, which could account for the appearance of both homothallic and heterothallic strains in several species.

Linkage between the mating-type locus and a gene involved in flagella movement was detected in *C. moewusii* (Lewin 1954a). The sex-determining gene in *C. reinhardtii* was mapped, in relation to various other kinds of somatic genes, by Sager (1955, 1972), and those of *C. eugametos* and *C. moewusii* were similarly located by Gowans (1960) and Lewin (1974).

In *C. reinhardtii* a mutation was reported which changes sexual reactivity of the gametes from mt^+ to mt^- (Sager 1956). Although the allelism of the sex-determining factors is generally assumed, Gillham (1969) indicated on the basis of experiments on uniparental inheritance that the mating-type factors in *C. reinhardtii* are not in fact simple alleles but are different genes or gene clusters with distinct functions.

A sex allele determines the haploid genotype as male or female, or as mt^+ or mt^-, possibly by acting as a specific repressor for the phenotypic expression of the opposite sex in a potentially bisexual organism. Nothing is known of additional genes which may also be involved in the expression of sex. The phase of expression of genetically determined sexuality differs from species to species, as do the various genetic and environmental conditions responsible for the induction of sexual differentiation. In colonial *Volvocales* such as *Eudorina* and *Volvox*, sexual competence may be restricted to special generative cells. In some of these, notably *Eudorina* (Szostak *et al.* 1973) and *Volvox* species (Darden 1970; Starr 1971a; Starr & Jaenicke 1974), specific inducer substances are synthesized which trigger sexual differentiation. The analysis of the action of such a biochemical inducer has enabled Starr (1972) to formulate a theory for the biochemical basis of certain aspects of sexual differentiation in *Volvox* (p. 191). Such analyses of sex determination and differentiation in related homothallic and heterothallic species can be expected to elucidate the special roles of the sex alleles and their interrelations with the genotype.

Whenever maleness and femaleness are expressed in the same colony or within one clone, sex determination is modificatory, e.g. in *Polytoma uvella* (Pringsheim & Ondratschek 1939), *Chlamydomonas philotes* (Lewin 1957a), *C. gymnogama* (Deason 1967), *Golenkinia minutissima* (Starr 1963), *Volvox globator* (Smith 1944), and *V. aureus* (Darden 1966). As in heterothallic species, sexual differentiation as such must be under precise genetic control. The conditions which trigger sexual differentiation in a homothallic strain may be identical with those which elicit sexualization in heterothallic species.

In unicellular species, one may assume that some sex determination takes place in all species with intraclonal copulation, being obvious in anisogamous or oogamous species such as *Chlamydomonas suboogama* (Tschermak-Woess 1959, 1962). However, in isogamous homothallic species, with no recognizable morphological or physiological differences, one may question whether the gametes have a hidden mt^+ or mt^- bipolarity, as postulated by Hartmann (1956), or whether each gamete has the potential to copulate with any other gamete (Pringsheim & Ondratschek 1939; Pringsheim 1963).

Using the 'Restgameten' method, developed by Haemmerling (1934) to demonstrate bipolar sexual differentiation among the isogametes of *Acetabularia*, Lerche (1937) apparently revealed the existence of mt^+ and mt^- gametes in a homothallic isogamous species, *Haematococcus pluvialis* (Hartmann 1956; Hartmann, pers. communication). She also noted that N and P depletion interfered with normal gametogenesis and induced the production of gametes of one sex only.

As in heterothallic species, sexual differentiation in homothallic *Volvocales* is apparently triggered by the effect of a certain factor or factors on individual cells, which have to be competent for induction. Intimately connected with the initiation of sexual differentiation is the sexual determination of genetically identical cells as male or female. The discovery and isolation of sex-inducing substances in *Volvox* (Darden 1966; Kochert 1968; Starr 1969) and the investigation of their modes of action in homothallic species (Darden 1966, 1970) have provided some insight into the complexity of the phenomena involved and have led to the identification of certain essential steps (p. 188).

3 GAMETOGENESIS, FERTILIZATION, AND MEIOSIS

Investigations of gametogenesis and fertilization in *Volvocales* have concerned chiefly the morphology, physiology and biochemistry of cells involved in these phenomena. Algal geneticists have concentrated their research on *Chlamydomonas*, which in gametogenesis represents an outstanding model for an intracellular gene-dependent differentiation, and on *Volvox*, in which sexual differentiation is more complex and displays an integrated succession of steps.

In *Chlamydomonas* gametogenesis may be induced by depriving vegetative cells of available nitrogen (Sager & Granick 1954); details are somewhat species specific and depend largely on the state of the vegetative cells to be induced. When agar cultures are flooded, cells first become flagellated and motile and then acquire the capacity to copulate, for which process flagella are essential. Both the formation of flagella and their motility are subject to mutation: flagellaless and flagellate but paralysed mutants have been studied by Lewin (1952a, 1953, 1954a, 1974) and by Randall and Starling (see Chapter 3). The capacity to copulate also requires the synthesis of mating-type substances (Gerloff 1940; Hutner & Provasoli 1951; Smith 1946; Lewin 1950, 1952b; Foerster & Wiese 1954b). These gamete agglutinins are constituents of the cell surface, and are located at the flagella tips, as in most isogamous *Chlamydomonas* species; along the entire flagella, as in *C. chlamydogama* (Bold 1949) and *Chloromonas saprophila* (Tschermak-Woess 1963); or on the surface of the gamete cell bodies, as in *Chlamydomonas suboogama* (Tschermak-Woess 1959, 1962).

The transformation of vegetative cells into gametes seems to involve several distinct gene-controlled steps at the nuclear, cytoplasmic and ultrastructural levels, connected with profound changes in metabolic activities. It is, however, questionable whether all of the alterations in cell structure and metabolism observed after N-depletion actually reflect features of gametogenesis or whether some merely represent reactions to sudden N-deprivation.

The use of synchronous liquid cultures (Bernstein 1960) permits the analysis of individual component steps in gametogenesis, and allows one to correlate gametogenesis and the competence for its induction with certain stages in the cell cycle (Kates & Jones 1964; Jones 1970; Siersma & Chiang 1971; Surzycki 1971). Under standardized conditions, with a 12-h light/12-h dark cycle in a complete nutrient medium, vegetative cells of *C. reinhardtii* divide every 24 h into four vegetative cells. Cell enlargement occurs in the light; DNA replication, nuclear and cell division take place in the last half of the dark period; and four daughter cells are released when the light is turned on. In exponentially growing cultures, the DNA replicates in the semiconservative mode (Sueoka 1960; Sueoka *et al.* 1967). Vegetative cells deprived of N likewise divide twice to produce four gametes (Kates & Jones 1964; Sueoka *et al.* 1967; Kates *et al.* 1968; Siersma & Chiang 1971). When induced after 6 h in the light phase, the cells have the greatest gametogenic potential (Kates & Jones 1964; Schmeisser *et al.* 1973), and some 17 h later respond with 100% gamete formation. At an early stage gametogenesis can be inhibited by the addition of ammonium or nitrate salts; these ions also cause differentiated gametes to revert to the vegetative condition (Sager & Granick 1954; Kates & Jones 1964). The two gametogenic divisions ensue at about the same time as the two vegetative mitoses, preceded by two rounds of DNA replication late in the dark phase. DNA replication can be uncoupled from the nuclear divisions (Kates *et al.* 1968). In synchronized cultures, the G 1 phase lasts about 15 hours, and the S period and the division period about four hours each; there is no pronounced G 2 phase (Kates *et al.* 1968; Surzycki 1971; Schmeisser *et al.* 1973). Net protein synthesis ceases under N-deprivation, but some protein turnover continues (Jones *et al.* 1968).

Mating efficiency (percent pairing under optimum conditions) varies among different strains of one species, and may be reduced in subcultures (Chiang *et al.* 1970).

Although there is no appreciable pool of precursors for DNA synthesis, the nuclear DNA content per cell is unaffected by the N-depletion (Sueoka *et al.* 1967). During gametogenesis Siersma and Chiang (1971) observed an extensive degradation (80–90%) of cytoplasmic and chloroplast ribosomes, and a simultaneous degradation of ribosomal RNA, whereas in vegetative reproduction ribosomes are conserved. Isotope-transfer studies have shown that about 60% of the RNA label, freed in the first phase of gametogenesis, is later incorporated in newly synthesized gamete DNA. Siersma and Chiang therefore suggested that gamete production may be synchronized by the synchronous provision of DNA precursors from degraded ribosomal RNA due to an increase of ribonuclease activity under gametogenic conditions. The interrelation between ribosome content and gametic activity was also indicated by experiments of Honeycutt and Margulies (1972), who obtained equal growth of *C. reinhardtii* in mineral media with either NH_4^+ or arginine as N source, but noted that cells grown with arginine possess less than half as many ribosomes as cells grown in NH_4^+ media, and up to 36% may function as gametes. Transfer of NH_4^+-grown cells into arginine medium results in a reduction of ribosome content; conversely, transfer of arginine-grown cells into NH_4^+ medium results in an increase in ribosomes. (The effects of such transfers on gametic potency were not determined.) Cells of the arginine-requiring mutant (*arg*-1) possess a ribosome content equal to that of wild-type NH_4^+-grown cells, and are not able to copulate. Because of the obvious repression of gametogenesis in wild-type cells and the known dedifferentiation of gametes caused by NH_4^+ (Sager & Granick 1954; Kates & Jones 1964), Honeycutt and Margulies ascribed to the ammonium ion a specific repressor function within the genetic control mechanism involved in the biosynthesis of proteins essential for the gametic phase. (Simultaneously, NH_4^+ is supposed to repress a system which degrades ribosomes.) This concept has much in common with that developed by Starr for sexual differentiation in *Volvox* (see p. 191).

Experiments with synchronized cell divisions in liquid cultures have helped to elucidate one of the pathways leading to gametogenesis (Siersma & Chiang 1971). Genetically identical cells, identically induced, may nevertheless enter different modes of gametogenesis in experiments starting from asynchronous liquid cultures or from cells in the late growth phase on agar slants (Schmeisser *et al.* 1973). As in other potentially hologamous *Chlamydomonas* species (*C. eugametos*, *C. moewusii*), palmelloid cells in *C. reinhardtii*

can transform directly into gametes within about an hour, without gametogenic divisions, indicating that gametogenesis is not essentially predicated on
cell division. The actual connection of certain observed metabolic changes
with gametogenesis is therefore questionable (Schmeisser *et al.* 1973). Finally,
the attainment and maintenance of the gametic state in mt^+ strains of *C.
eugametos* and *C. moewusii* (Provasoli strains) are specifically dependent on
light, with an action spectrum different from the photosynthetic spectrum
(Foerster & Wiese 1954b; Lewin 1956; Foerster 1957, 1959; Stifter 1959;
Lorch & Karlander 1973).

In certain *Volvocales* the actual gamete contact may be preceded by a
chemotactic attraction between sexually different gametes mediated by an
erotactin, as demonstrated for *Chlorogonium oogamum* (Pascher 1931) and
Chlamydomonas moewusii f. *rotunda* (Tsubo 1961). Ultimately contact
between the sex cells triggers a succession of steps leading to the formation
of the zygote. In *C. reinhardtii*, the process of agglutination seems to activate
a lytic enzyme in the cells, which dissolves their cell envelopes (Claes 1971)
and permits the gametes to come together by means of a copulation tube
(Friedmann *et al.* 1968).

The fertilization process itself has not been studied genetically, and until
recently zygote formation and germination have only been investigated
cytologically and physiologically (e.g. by Lewin 1957b). Latterly Sueoka *et
al.* (1967) and Chiu and Hastings (1973) have begun a genetic analysis of the
processes of meiosis.

4 SEXUAL INCOMPATIBILITY, SYNGENS, AND SPECIATION

Several instances of sexual isolation among various strains of a morphologically characterized species have been found in the *Volvocales*. Often such
reports on sexual incompatibility phenomena are incomplete and demand
further studies with genetically defined material under controlled experimental conditions. Within certain species, however, sexual isolation and the
existence of distinct syngens (sexually compatible pairs of strains) have been
neatly demonstrated by systematic combinations of clones isolated from the
same site or from various geographical locations. Strains have been observed
to differ in their responsiveness to various intracellular or environmental
conditions which may elicit the sexual phase, in the various levels of their
sexual differentiation or compatibility during fertilization, and in zygote
formation and zygote germination. Such differences may account for sexual
isolation and, by preventing sexual reproduction, may exclude genetic recombination. On the other hand several successful crosses between morphologically distinct species have been reported.

I

182 CHAPTER 8

[I once tried to create a kind of artificial sexual differentiation in *C. dysosmos*, working on the hypothesis that in this isogamous, homothallic species each clone—perhaps even each cell—has the combined potentialities of both sexes (or mating-types), and that one or the other might be inactivated by mutation. In normal clones, abundant zygotes are formed which become orange as they mature. I irradiated vegetative cells with UV light to kill about 99%, streaked them on agar, and among 1050 surviving clones found 7 which for some reason failed to produce such zygospores. Then I paired these clones in all possible ways, hoping to find some complementary combinations in which zygotes would be formed. Although I failed in this attempt, I still feel that an approach of this sort may ultimately prove fruitful in elucidating primitive cases of sexual differentiation. See Lewin R.A. (1954c) Unsuccessful attempt to produce a heterothallic mechanism in a homothallic organism. *Microbial Genetics Bulletin* **10**, 18–19.]

In *Chlamydomonas* sexual isolation between heterothallic isogamous species can often be traced to the absence of initial contact between sexually different gametes. In the most extensively investigated species gametogenesis is induced by certain specific inner and environmental conditions; in contrast to the situation in *Pandorina* and many *Volvox* species (see below), it does not require prior mutual or unilateral induction between their sexes. In the mating-type reaction, contact is followed by an almost instantaneous agglutinative adhesion between the flagella of compatible cells; activated gametes of incompatible strains do not agglutinate in this way. *C. eugametos*, two of the four known syngens of *C. moewusii*, *C. mexicana*, *C. chlamydogama*, and *C. reinhardtii* are all sexually isolated in this manner (cf. Wiese 1974).

The capacity and specificity of such flagellar agglutination reside in special mating-type substances, which are not present on flagella of vegetative cells, and which are apparently synthesized at gametogenesis (Smith 1946; Hutner & Provasoli 1951; Lewin 1950, 1952b, 1956; Sager & Granick 1954; Foerster & Wiese 1954b; Foerster 1957, 1959; Stifter 1959). They can be isolated (Moewus 1933; Foerster & Wiese 1954b, 1955; Wiese 1965, 1969, 1974) and shown to interact selectively with flagella of the complementary gamete type, causing iso-agglutination between gametes of identical sex.

The mating-type substances are evidently glycoprotein complexes (Foerster & Wiese 1954b; Foerster *et al.* 1956; Wiese 1965, 1969, 1974), and apparently each heterothallic species or syngen has its own specific pair of complementary sexual agglutinins (Wiese 1969, 1974). The synthesis of such components is necessarily gene-controlled, and is correspondingly sensitive to actinomycin D (Greenwood 1973). The mating-type activity is inherent to certain special membrane vesicles described by McLean *et al.* (1974).

The sexual agglutinins in several species of *Chlamydomonas* have been shown to be differentially sensitive to specific enzymes and lectins (Wiese & Metz 1969; Wiese & Shoemaker 1970, Wiese & Hayward 1972, Wiese 1974). In *C. reinhardtii* and *C. chlamydogama*, the mating-type activity of both *mt*$^+$ and *mt*$^-$ gametes is sensitive to proteases and insensitive to concanavalin A. A striking pattern appears in *C. eugametos*, *C. mexicana*, and in two syngens of *C. moewusii*. In all four taxa, one gamete type is resistant to pronase, trypsin and subtilisin, while the other is sensitive to these enzymes. The contact capacity of mt^+ gametes of the two *C. moewusii* syngens is evidently determined by the presence of terminal mannose residues in α-glycosidic linkage since it is sensitive to concanavalin A and α-mannosidase. The strict incompatibility between these two syngens must therefore be based on some additional difference between their respective mt^+ mating-type substances, or on a different exposure of the mannose residues, permitting selective recognition by their homologous mt^- gametes. One major function of the sex-specific mt^- substances, their sugar-binding capacity, must likewise be syngen-specific since the mt^+ gametes can apparently discriminate between the mt^- substances of compatible and incompatible gametes, selectively adsorbing only the former (Wiese & Wiese 1975).

The analyses of the mating-type reaction in these two syngens and in related species suggest a simple explanation for syngen formation and speciation, which would simultaneously account for both the observed general existence of sterile strains and some aspects of the geographical distribution of syngens in other genera. The species' genotype must code for the functional structure of the mating-type substances and their complementarity. A minor mutation in the functional structure of the mating-type substances might interfere with the agglutinative capacity or its specificity and thereby incapacitate the sexual activity of one gamete type. Such a mutant, excluded from sexual reproduction with its complementary parental strain, could survive only by asexual reproduction; on isolation, it would appear sterile; and it could resume sexual reproduction only if it back-mutated or if a complementary mutation occurred in its erstwhile mating partner. In the latter case, a new syngen would be constituted.

The possibilities of genetic change affecting the sexual glycoproteins are presumably limited, certain alterations being more likely to occur than others. Within a species sexual isolation into separated syngens, and their geographic distribution, would have to be understood on the basis of frequent, highly specific mutations of the contact mechanism, such as are known to occur, for instance, in virus-host attachment mechanisms (cf. Lindberg 1973).

In some cases two heterothallic species can successfully be crossed in only one way, but not in the reciprocal combination. In the crossing of two heterothallic *Chlamydomonas* species, the mt^+ gametes of *C. paradoxa* mate with mt^- gametes of *C. botryoides* whereas the reciprocal combination produces no such interactions (Strehlow 1929). There is complete interfertility between *C. reinhardtii* and *C. smithii*, two taxa which differ in some morphological and fine-structural traits (Hoshaw & Ettl 1966), as was earlier reported between the other unidentified *Chlamydomonas* species with which Pascher (1916) opened the era of algal genetics (see Appendix A).

A comprehensive study revealed the existence of both homothallic and heterothallic strains in the isogamous species *Pandorina morum* (Coleman 1959). Gametogenesis evidently depends on a system of mutual induction between the two sexes. Induction seems to involve soluble substances since completion of gametogenesis can be effected by cell-free filtrates. Sexual isolation between *Pandorina* syngens is closely related to this induction system, any disharmony preventing the formation of gametes. Incompatibility between different syngens may be complete or partial, permitting sexual reproduction between two syngens in one combination but not in the reciprocal combination, as in the crossing of two syngens from Iowa and Indiana which seem to have one sex in common (Coleman 1959). Different syngens may occur at the same location; conversely, cross reactions may occur between strains from locations geographically far apart.

In extensive studies of sexual isolation among clones of the isogamous *Gonium pectorale* from Canada, the United States and Great Britain, Stein (1958b, 1965, 1966) detected only one homothallic population; all other isolated clones belonged to heterothallic strains. Although in *Gonium*, according to Stein, sexual isolation is the exception rather than the rule, her data exhibited a very informative incompatibility pattern. Two pairs of mt^+ and mt^- strains, one from Massachusetts and one from Indiana, neither interbred nor mated with any other clones, and thus represent true syngens. Some of the other strains interbred with most others, certain combinations showing only a restricted compatibility. For each combination, compatibility was assessed on the basis of zygote formation observed some weeks after mixing the two clones. Such a technique, however, does not reveal whether a case of incompatibility involves impairment of gametogenesis, mating-type reaction, pairing or zygote formation. As in *Pandorina*, different syngens were found to coexist in one pond; on the other hand, compatibility was observed among N. American clones from diverse sites in British Columbia, Indiana, California, Minnesota, and Texas, and even two clones originating from Great Britain. The two British clones, respectively from Scotland and Cambridge, characteristically failed to produce zygotes, although of opposite sex. The distribution pattern of the compatible and incompatible strains was explained by Stein (1965) in terms of geographical isolation with reference to the geological history of the isolation sites, possibilities of spore distribution by migrant waterfowl, communicating waterways, wind, etc. Stein (1966) also demonstrated the existence of physiological races in *Gonium pectorale* differing in the nutritional factors affecting their growth and in the readiness of the cells to enter the sexual phase. In various populations sexualization evidently may depend differently on the N source, vitamin B_{12} concentration and/or temperature, which may thus indirectly affect sexual isolation among clones.

In *Volvulina steinii* three syngens were described, all with a haploid chromosome number of 7 (Carefoot 1966). Two of these syngens, one originating from Texas and one from California, consisted of only one pair of sexes. The third syngen included several pairs of sexually compatible clones from different geographic locations (Texas, Indiana, Iowa) which were only partially cross-compatible, crossing with certain other pairs of strains only in one direction. There was complete sexual isolation between two pairs of strains from Texas, although both pairs interacted with other strains of the syngen. Certain combinations produced only a few zygotes, and three clones showed no sexual response at all.

The system of sexual isolation in the heterothallic species *Astrephomene gubernaculifera* is cytogenetically interesting (Stein 1958a; Brooks 1966). In confirmation of the original report by Cave and Pocock (1956), Stein found clones differing in chromosome number (n = 4, 6, or, most commonly, 7); in certain strains with 7 chromosomes, some clones contained colonies with 6 and 7, or 7 and 8, or 6, 7, and 8 chromosomes, the abnormal chromosome numbers presumably attributable to non-disjunction and chromosome breakage. Clones with 7 chromosomes were reported to produce zygotes in any combination. Stein's clones with 4 chromosomes did not produce zygotes within the clones or in any of the combinations tested, but mated with cultures of the higher chromosome number. Two of the 4-chromosome clones reacted with gametes of both sexes in test clones (Stein 1958a). With additional data, Brooks (1966) demonstrated the existence of at least five true syngens, each with two sexes, not all having the same chromosome number. Within the syngens sexual compatibility was always complete and, once a mating reaction was initiated, it always led to zygote formation.

In a systematic study of sex distribution and determination and of sexual isolation in *Eudorina*, Goldstein (1964) reorganized the taxonomy of the genus on the bases of morphological, embryonic, and physiological features. New species were established and old ones were re-evaluated. The genus includes both homothallic and heterothallic taxa, some exhibiting subdioecy and parthenogenesis. Among 44 investigated populations, Goldstein recognized 12 homothallic strains, 24 pairs of heterothallic clones, eight parthenosporic strains and a few presumably heterothallic clones without a sexual partner, and assigned them to six species and three varieties. The study illustrates the dilemma and the justification of such attempts at classification. Goldstein observed sexual isolation within, and compatibility between, certain of his established species. In various clone combinations, five groups of compatible taxa emerged; only a few combinations between these syngens led to zygote formation (Table 8.1). In one case a female clone of *E. unicocca* (1F) mated with any of several male clones of three other syngens belonging taxonomically to *E. elegans*, *E. cylindrica*, and *E. illinoisensis*, though the hybrid zygotes produced did not give rise to viable offspring. (Remarkably, the reciprocal crosses did not produce any zygotes.) Intrasyngenic combinations of this clone 1F with male clones of *E. unicocca* var. *peripheralis* resulted in normal zygotes. *E. elegans* appears to be composed of four syngens, one of which, syngen IV, includes also strains of *E. cylindrica* and *E. illinoisensis*, which leads one to question the status of these taxa as separate species. However, *E. illinoisensis* was found to have some genetic stability, since in intrasyngenic crosses with *E. elegans* the zygote progeny exhibited reduced viability and included diploid or aneuploid cells, 'male' and 'female' clones capable of selfing, and homothallic forms. Syngen V, comprising

Table 8.1. Intercrossing of heterothallic strains of *Eudorina*. (From Goldstein 1964.)
■ = Zygotes of a heterothallic pair; X = Intercross zygotes; · = No zygotes; ▲ = Clone not tested.

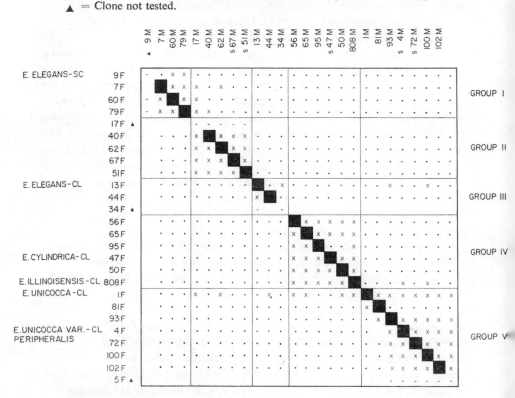

heterothallic strains of *E. unicocca* and *E. unicocca* var. *peripheralis*, proved to comprise two separate groups of male and female clones interconnected merely by one unilateral intercross.

In *Platydorina caudata* six pairs of strains originating from Iowa and Kansas exhibited no sexual isolation (Harris & Starr 1969).

In *Volvox*, likewise, there are homothallic and heterothallic species, including both monoecious and dioecious strains. The pattern of sexual isolation between existing species and varities appears to be more stable than in the genera described above (Table 8.2) (Smith 1944; Starr 1968, 1971a). The species are characterized by morphological and embryonic traits and by their modes of reproduction and sexual differentiation.

There is a trend in dioecious *Volvox* species to reduce the role of cells in

Table 8.2. Sexual types of *Volvox* species. (From Starr 1968)

Name of species	Strain designation	Origin	Homo- thallic	Hetero- thallic	Monoe- cious	Dioe- cious	Partheno- sporic
V. africanus	Mo	Missouri	×			×	
V. africanus	Darra	Australia		×		×	
V. aureus	M-5	Michigan	×			×	
V. aureus	65-98	Mississippi					×
V. barberi	IU 804	California	×		×		
V. carteri f. *nagariensis*	HK	Japan		×		×	
V. carteri f. *weismannia*	NB	Nebraska		×		×	
V. carteri f. *weismannia*	Wf	Australia		×		×	
V. dissipatrix	Marburg	Australia	×		×	×	
V. gigas	K25	South Africa		×		×	
V. globator	IU 955	Massachusetts	×		×	×	
V. obversus	Wd	Australia		×		×	
V. perglobator	HP	Indiana		×		×	
V. powersii	17-10	Nebraska	×			×	
V. rousseletii	K32	South Africa		×		×	
V. spermatosphaera	17-14-4	Nebraska	×			×	
V. spermatosphaera (small form)	H	Missouri	×			×	
V. tertius	IU 132	England	×			×	
V. sp. (*Euvolvox*)	62-22	Mexico			×	males	

Homothallic, sexual reproduction occurring within a clonal population.
Heterothallic, sexual reproduction occurring only between two different clonal populations.
Monoecious, both eggs and sperm formed in the same individual.
Dioecious, eggs and sperm formed in separate individuals.
Parthenosporic, forming resistant spores without fertilization.

male colonies to sperm production, winding up with dwarf males which may or may not possess somatic cells. Thus, *V. rousseletii* and *V. dissipatrix* have full-sized male colonies; in *V. tertius* (Pocock 1938) and *V. carteri* f. *weismannia* (Kochert 1968) the male colonies have reduced numbers of somatic cells; in *V. carteri* f. *nagariensis* (Starr 1969) they have equal numbers of spermatogenous and somatic cells; while in *V. spermatosphaera* (Powers 1908) and *V. pocockiae* (Starr 1970) all of the cells in the male colonies ultimately become spermatozoids.

In many species induction of sexual activity involves specific chemical inducers acting upon one or both sexes and affecting susceptible embryonic stages which differ from taxon to taxon (Darden 1970; Starr 1971a). In view of the variety of sexual systems in the genus, species of *Volvox* are of exceptional interest not only for the genetic analysis of sex determination, sexual differentiation and sexual isolation, but also for the study of morphogenesis. The genetic control of these phenomena is apparent from comparative studies of the various species (Smith 1944; Starr 1968, 1971a) and of mutants isolated from nature (Darden 1968) or appearing spontaneously in clonal cultures (Starr 1971a, 1972; Van de Berg & Starr 1971). A systematic study of the genetic bases of asexual reproduction and morphogenesis by means of chemically induced mutants has been initiated by Sessoms and Huskey (1973).

In general, a species is characterized as dioecious, with separate male and female colonies, or monoecious, with eggs and sperm produced on the same individual (*V. globator, V. capensis*). In monoecious species, male and female cells are differentiated in some way from genetically identical cells. In dioecious species, two genetically different clones may be needed for sexual reproduction (heterothallism; e.g., *V. rousseletii, V. perglobator*) or a special sex-determination system may control the production of male and female individuals within one clone (homothallism; e.g., *V. aureus*). The mode of sex determination is normally typical for a given species. However, species well defined by morphological and embryonic characters may comprise strains or syngens with different types of sex differentiation and sex determination. An outstanding example is *V. africanus* in which Starr (1971b) isolated dioecious, heterothallic strains from Australia; a dioecious, homothallic strain from Missouri; monoecious clones from South Africa; and a clone from India which produced monoecious and male colonies, but no purely female colonies.

In various species of *Volvox*, Starr and his colleagues have revealed an extraordinary variety in the pathways of differentiation into somatic and reproductive cells, and in the determination of certain reproductive cells to form male or female gametes. Details in the embryonic development and organization of the colonies have been revealed by studying various mutations induced by ethyl methane sulfonate or N-methyl-N-nitro-N-nitroso guanidine (Sessoms & Huskey 1973). Among 68 independently obtained mutations, some 12 different types could be distinguished affecting cellular differentiation (from somatic cells to gonidia), cell-division patterns, morphogenetic movements, or other features of the organization, polarity and stability of the colonies. Some of these mutants are temperature-sensitive, permitting normal development at 25°C and causing specific aberrations at 35°C. Some of these chemically induced mutants resemble spontaneous variants detected earlier by Starr (1971a) and Van de Berg and Starr (1971).

In certain species (e.g., *V. pocockiae*) sexual differentiation in both sexes apparently occurs spontaneously in response to cellular or environmental conditions as yet unidentified (Starr 1970). In others special inducer substances emitted by male cells trigger further sexual differentiation processes. Only in a heterothallic variety of *V. dissipatrix* are there both male and female colonies which produce inducing substances affecting both sexes (cf. Starr 1971a).

Much of the recent *Volvox* research was initiated by the study of sexual differentiation in *V. aureus* strain M5, a homothallic, dioecious species (Darden 1966). The first males in a population, which seem to arise spontaneously, release a specific male-inducing substance (MIS), which induces young undivided gonidia in asexual colonies to differentiate into other male colonies producing packets of spermatozoa. These sperm packets in turn attach to other young colonies freshly released during asexual reproduction, the young undivided gonidia of which function as eggs. Another isolate of *V. aureus*, strain DS, also homothallic and dioecious, forms male colonies only very rarely, generally reproducing by asexual division or by parthenospores. Although cells of this strain can produce MIS, as demonstrated with test cultures of strain M5, they evidently do not respond to it as do those of strain M5 (Darden 1968).

In the heterothallic *V. carteri* f. *weismannia*, Kochert (1968) has described a femaleness-inducing principle which is secreted by male colonies and acts upon young gonidia in undifferentiated colonies of the female clone, inducing them to develop into sexual, egg-producing colonies. In another heterothallic strain, *V. carteri* f. *nagariensis*, sexualization in both male *and* female clones depends on an inducer secreted by male cells, which in response to some environmental conditions release a substance inducing young gonidia of other colonies to differentiate into sperm-producing colonies. In these male spheroids, equal numbers of androgonidia and somatic cells are formed by one differentiating mitosis. Gonidia of colonies in female clones develop into egg-producing spheroids only after exposure to the same inducer substance (Starr 1969).

In other species, corresponding sex substances induce young gonidia to differentiate directly into egg initials or androgonidia. The heterothallic species *V. gigas* and *V. rousseletii* are examples of this sort (Van de Berg & Starr 1971; McCracken & Starr 1970). It appears that the potentially reproductive cells in young colonies pass through a stage of competence for sexual induction, and in the absence of inducer at the susceptible stage they develop asexually. In *V. powersii*, which is homothallic and dioecious, such an induction system has yet to be identified. In a morphological variant of this species somatic cells can develop spontaneously into reproductive cells (Van de Berg & Starr, 1971).

The inducing agents are species-specific substances of high molecular weight, evidently proteins, active in some cases at dilutions as low as 3×10^{-15} M, and in treatments of susceptible cells as short as 5–15 minutes (Darden

Table 8.3. Sexual cross-induction between various clones of *Volvox carteri* (Starr 1971a).

	Female susceptible to induction			
	f. *nagariensis*	f. *weismannia*		
Source of inducer (male)				
		Indian	Australian	Nebraskan
f. *nagariensis*	+	−	−	−
f. *weismannia*				
Indian	−	+	+	+
Australian	−	+	+	−
Nebraskan	−	+	−	+

1970; Starr 1971a). Their specificity is best illustrated in various strains of *V. carteri* (Table 8.3, Fig. 8.1). Their biochemistry is reviewed by Darden (1970), Starr (1971a) and Starr and Jaenicke (1974).

The actual role of the inducers is yet unclear. Because of differences in

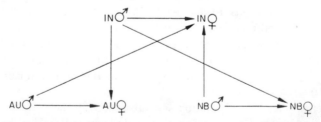

Fig. 8.1. Pattern of cross-induction of female strains by inducing fluid from males. The HK clones are *Volvox carteri* f. *nagariensis* from Japan while the other strains (Australia, India, and Nebraska) belong to *V. carteri* f. *weismannia* (Starr 1971a).

their mode and time of action, McCracken and Starr (1970) concluded that the various inducers have no common biochemical basis of operation. It is debatable whether they are in effect inhibitors, rather than inducers (Darden 1970). In *V. carteri* f. *nagariensis*, quantitative studies under standard conditions indicated that as few as two molecules of inducer are required to initiate sexual development of a competent cell (Pall 1973). Starr (1972) therefore inferred that these highly efficient agents may act close to the gene

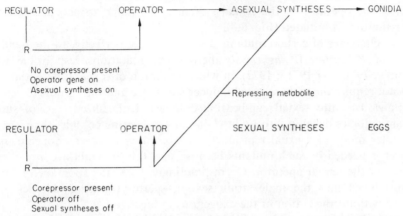

Fig. 8.2. Postulated scheme for regulation of asexual and sexual systems in *Volvox carteri* f. *nagariensis*. (Asexual spheroids with gonidia result when the asexual system is 'on', producing a repressing metabolite which acts as a corepressor in turning off the sexual system.) (Starr 1972.)

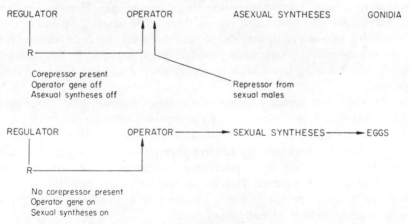

Fig. 8.3. Postulated scheme for action of substance from sexual males in the regulation of the asexual and the sexual systems in *Volvox carteri* f. *nagariensis*. (The chemical substance from sexual male acts as a corepressor, turning off the operator of the asexual system. In the absence of the repressing metabolite from the asexual syntheses (see Fig. 8.2), the operator of the sexual system initiates the sexual syntheses, resulting in a female spheroid with eggs.) (Starr 1972.)

level, and interpreted their action on the basis of the Jacob-Monod operon model. (See Figs. 8.2 and 8.3, illustrating sexual induction in a female culture.) On this hypothesis, asexual development and sexual differentiation represent two interlocking repressible systems. Under conditions favouring asexual development a metabolite is produced which acts as a corepressor which turns off the operator of the sexual system (Fig. 8.2). The inducer substance of the male strain is assumed to act as a corepressor, turning off the operator of the asexual system so that the repressing metabolite is not formed, the operator of the sexual system is permitted to act, and sexual differentiation is initiated (Fig. 8.3).

The character of certain mutants fits this concept perfectly. In the female strain of *V. carteri* f. *nagariensis* there occur mutations responsible for 'spontaneity' (Starr 1971a, 1972), in which initiation of sexual differentiation does not require the male-secreted inducer substance and the gonidia develop spontaneously into sexual egg-bearing colonies. Unfertilized eggs of such mutant colonies develop directly into similar egg-bearing colonies, bypassing any other mode of asexual reproduction. According to Starr's concept, no inducer is needed by such mutants because they fail to synthesize any corepressor of the sexual operator. Correspondingly, when the spontaneity gene is introduced into the male strain sexual, sperm-producing colonies are formed without the action of the exogenous inducer.

A number of mutants have been detected which in various ways modify differential cell divisions in sexual or asexual colonies (Starr 1971a). For instance, in a mutant of the female strain of *V. carteri* f. *nagariensis* the presumptive somatic cells may enlarge and develop into daughter colonies or, if exposed to inducer, into sexual spheroids producing eggs. The same mutant gene, introduced into male cells, causes them all to respond to inducer by differentiating into sexual spheroids, so that not only the androgonidia but also the presumptive somatic cells produce sperm packets. In this mutant, evidently, the normally irreversible determination of somatic *Volvox* cells is suppressed.

Whether the first sexual colonies in a male population of f. *nagariensis* arise by spontaneous mutation, or as a consequence of some combination of environmental factors, is still unresolved. From the evolutionary point of view Starr (1972) questioned the relative survival value of a sexual system that can be initiated only by a spontaneous mutation rather than by some combination of environmental factors. In species other than *V. carteri* f. *nagariensis*, a taxon-specific response to inner or environmental factors is thought to elicit the first appearance of males.

5   CONCLUSIONS

Compared with other areas of algal genetic research the genetic analysis of the complex sexual phenomena in *Volvocales* and in other algae has been generally neglected. The lack of adequate information is especially obvious when one considers the detailed information now available for comparable phenomena in fungi and bacteria. The diversity of sexual differentiation in *Volvocales* provides us with a variety of systems by which we may study its genetic control, especially the nature of the sex-determining factors, their special role within the entire control system, and the nature of the sex-determining switch in organisms with modificatory sex determination. The genetic analysis of sexuality in *Volvocales* and other haploid organisms with a strict alternation of haplo- and diplophase by fertilization and meiosis will provide information on the evolution of sex and on sexuality as such from its expression in bacteria to its manifestation in haplodiplonts and diplonts.

6   REFERENCES

BERNSTEIN E. (1960) Synchronous division in *Chlamydomonas moewusii*. *Science*, **131**, 1528–9.

BERNSTEIN E. & JAHN T.L. (1955) Certain aspects of the sexuality of two species of *Chlamydomonas*. *J. Protozool.* **2**, 81–5.

BOLD H.C. (1949) The morphology of *Chlamydomonas chlamydogama*, sp. nov. *Bull. Torrey Botan. Club*, **76**, 101–8.

BROOKS A.E. (1966) The sexual cycle and intercrossing in the genus *Astrephomene*. *J. Protozool.* **13**, 368–75.

CAVE M.S. & POCOCK M.A. (1956) The variable chromosome number in *Astrephomene gubernaculifera*. *Amer. J. Bot.* **43**, 122–34.

CAREFOOT R.J. (1966) Sexual reproduction and intercrossing in *Volvulina steinii*. *J. Phycol.* **2**, 150–6.

CHIANG K.-S., KATES J.R., JONES R.F. & SUEOKA N. (1970) On the formation of a homogeneous zygotic population in *Chlamydomonas reinhardtii*. *Develop. Biol.* **22**, 655–69.

CHIU S.M. & HASTINGS P.J. (1973) Pre-meiotic DNA synthesis and recombination in *Chlamydomonas reinhardi*. *Genetics*, **73**, 29–43.

CLAES H. (1971) Autolyse der Zellwand bei den Gameten von *Chlamydomonas reinhardii*. *Arch. Mikrobiol.* **78**, 180–8.

COHN, F. & WICHURA M. (1857) Ueber *Stephanosphaera pluvialis*. *Acta Acad. Caes. Leop. Cur. Halle*, **26**, 1–32.

COLEMAN A.W. (1959) Sexual isolation in *Pandorina morum*. *J. Protozool.* **6**, 249–64.

DARDEN W.H. (1966) Sexual differentiation in *Volvox aureus*. *J. Protozool.* **13**, 239–55.

DARDEN W.H. (1968) Production of a male-inducing hormone by a parthenosporic *Volvox aureus*. *J. Protozool.* **15**, 412–14.

DARDEN W.H. (1970) Hormonal control of sexuality in the genus *Volvox*. *Ann. New York Acad. Sci.* **175**, 757–63.

DEASON T.R. (1967) *Chlamydomonas gymnogama*, a new homothallic species with naked gametes. *J. Phycol.* **3**, 109–12.

EBERSOLD W.T. (1967) *Chlamydomonas reinhardi*: Heterozygous diploid strains. *Science*, **157**, 447–9.

FOERSTER H. (1957) Das Wirkungsspektrum der Kopulation von *Chlamydomonas eugametos*. *Z. Naturforsch.* **12b**, 765–70.

FOERSTER H. (1959) Die Wirkungsstaerken einiger Wellenlaengen zum Ausloesen der Kopulation von *Chlamydomonas moewusii*. *Z. Naturforsch.* **14b**, 479–80.

FOERSTER H. & WIESE L. (1954a) Untersuchungen zur Kopulationsfaehigkeit von *Chlamydomonas eugametos*. *Z. Naturforsch.* **9b**, 470–1.

FOERSTER H. & WIESE L. (1954b) Gamonwirkungen bei *Chlamydomonas eugametos*. *Z. Naturforsch.* **9b**, 548–50.

FOERSTER H. & WIESE L. (1955) Gamonwirkung bei *Chlamydomonas reinhardti*. *Z. Naturforsch.* **10b**, 91–2.

FOERSTER H., WIESE L. & BRAUNITZER G. (1956) Ueber das agglutinierend wirkende Gynogamon von *Chlamydomonas eugametos*. *Z. Naturforsch.* **11b**, 315–17.

FRIEDMANN I., COLWIN A.L. & COLWIN L.A. (1968) Fine structural aspects of fertilization in *Chlamydomonas reinhardi*. *J. Cell Sci.* **3**, 115–28.

GERLOFF J. (1940) Beitrage zur Kenntnis der Variabilitaet und Systematik der Gattung *Chlamydomonas*. *Arch. Protistenk.* **94**, 311–502.

GILLHAM N.W. (1969) Uniparental inheritance in *Chlamydomonas reinhardi*. *Amer. Naturalist*, **103**, 355–88.

GOLDSTEIN M. (1964) Speciation and mating behaviour in *Eudorina*. *J. Protozool.* **11**, 317–44.

GOWANS C.S. (1960) Some genetic investigations on *Chlamydomonas eugametos*. *Zeits. Indukt. Abstamm. Vererbungslehre*, **91**, 63–73.

GREENWOOD H.L. (1973) The effects of actinomycin D, gibberellin A7 and cyclic AMP on mating of *Chlamydomonas reinhardtii*. *J. Protozool.* **20**, 305–7.

GRELL K.G. (1963) Morphologie und Fortpflanzung der Protozen (einschliesslich Entwicklungsphysiologie und Genetik.) *Fortsch. Zool.* **14**, 1–85.

HAEMMERLING J. (1934) Ueber die Geschlechtsverhaeltnisse von *Acetabularia mediterranea* und *Acetabularia wettsteinii*. *Arch. Protistenk.* **83**, 57–97.

HARRIS D.O. & STARR R.C. (1969) Life history and physiology of reproduction of *Platydorina caudata* Kofoid. *Arch. Protistenk.* **111**, 138–55.

HARTMANN M. (1956) Die Sexualitaet. Gustav Fischer Verlag, Stuttgart, Germany.

HOSHAW R.W. & ETTL H. (1966) *Chlamydomonas smithii* sp. nov.—a chlamydomonad interfertile with *Chlamydomonas reinhardtii*. *J. Phycol.* **2**, 93–6.

HONEYCUTT R.C. & MARGULIES M.M. (1972) Control of ribosome content, gamete formation and amino acid uptake in wild-type and ARG-1 *Chlamydomonas reinhardti*. *Biochim. biophys. Acta*, **281**, 399–405.

HUTNER S.H. & PROVASOLI L. (1951) The phytoflagellates. In *Biochemistry and Physiology of Protozoa*, ed. Lwoff A., vol. 1, pp. 27–128. Academic Press, New York.

JONES R.F. (1970) Physiological and biochemical aspects of growth and gametogenesis in *Chlamydomonas reinhardtii*. *Ann. New York Acad. Sci.* **175**, 648–59.

JONES R.F., KATES J.R. & KELLER S.J. (1968) Protein turn over and macromolecular synthesis during growth and gametic differentiation in *Chlamydomonas reinhardtii*. *Biochim. biophys. Acta*, **157**, 589–98.

KALMUS H. (1932) Ueber den Erhaltswert der phaenotypischen (morphologischen) Anisogamie und die Entstehung der ersten Geschlechtsunterschiede. *Biol. Zentralbl.* **52**, 716–20.

KATES J.R. & JONES R.F. (1964) The control of gametic differentiation in liquid cultures of *Chlamydomonas*. *J. Cell. Comp. Physiol.* **63**, 151–64.

KATES J.R., CHIANG K.-S. & JONES R.F. (1968) Studies on DNA replication during synchronized vegetative growth and gametic differentiation in *Chlamydomonas reinhardtii*. *Exp. Cell Res.* **49**, 121–35.

KOCHERT G. (1968) Differentiation of reproductive cells in *Volvox carteri*. *J. Protozool.* **15**, 438–52.

LERCHE W. (1937) Untersuchungen ueber Entwicklung und Fortpflanzung in der Gattung *Dunaliella*. *Arch. Protistenk.* **88**, 236–68.

LEWIN R.A. (1950) Gamete behaviour in *Chlamydomonas*. *Nature*, **166**, 76.

LEWIN R.A. (1952a) Ultraviolet induced mutations in *Chlamydomonas moewusii* Gerloff. *J. Gen. Microbiol.* **6**, 233–48.

LEWIN R.A. (1952b) Studies on the flagella of algae. I. General observations on *Chlamydomonas moewusii* Gerloff. *Biol. Bull.* **102**, 74–9.

LEWIN R.A. (1953) The genetics of *Chlamydomonas moewusii* Gerloff. *J. Genetics*, **51**, 543–60.

LEWIN R.A. (1954a) Mutants of *Chlamydomonas* with impaired motility. *J. gen. Microbiol.* **11**, 358–63.

LEWIN R.A. (1954b) Sex in unicellular algae. In *Sex in Microorganisms*, ed. D. H. Wenrich, pp. 100–33. Amer. Assoc. Adv. Science, Washington, D.C.

LEWIN R.A. (1954c) Unsuccessful attempt to produce a heterothallic mechanism in a homothallic organism. *Microbial Genetics Bulletin*, **10**, 18–19.

LEWIN R.A. (1956) Control of sexual activity in *Chlamydomonas* by light. *J. gen. Microbiol.* **15**, 170–85.

LEWIN R.A. (1957a) Four new species of *Chlamydomonas. Canad. J. Bot.* **35**, 321–6.

LEWIN R.A. (1957b) The zygote of *Chlamydomonas moewusii. Canad. J. Bot.* **35**, 795–804.

LEWIN R.A. (1974) Genetic control of flagellar activity in *Chlamydomonas moewusii* (Chlorophyta, Volvocales). *Phycologia*, **13**, 45–55.

LINDBERG A.A. (1973) Bacteriophage receptors. *Ann. Rev. Microbiol.* **27**, 205–41.

LORCH S.K. & KARLANDER E.P. (1973) Gametogenesis in *Chlamydomonas eugametos*: I. Light requirements. *Plant Physiol.* **51**, 1046–50.

McCRACKEN M.D. & STARR R.C. (1970) Induction and development of reproductive cells in the K-32 strains of *Volvox rousseletii. Arch. Protistenk.* **112**, 262–82.

McLEAN R.J. & BROWN R.M. (1974) Cell surface differentiation of *Chlamydomonas* I. *Developm. Biol.* **36**, 279–85.

McLEAN R.J., LAURENDI C. & BROWN R.M. (1974) The relationship of gamone to the mating reaction in *Chlamydomonas moewusii. Proc. Nat. Acad. Sci. USA.* **71**, 2610–2613.

MISHRA N.C. & THRELKELD S.F.H. (1968) Genetic studies in *Eudorina. Genet. Res.* **11**, 21–31.

MOEWUS F. (1933) Untersuchungen ueber die Sexualitaet und Entwicklung von Chlorophyceen. *Arch. Protistenk.* **80**, 469–520.

MOEWUS F. (1936) Faktorenaustausch, insbesondere der Realisatoren bei *Chlamydomonas*-Kreuzungen. *Ber. deutsch. bot. Ges.* **54**, 45–57.

MOEWUS F. (1938) Vererbung des Geschlechts bei *Chlamydomonas eugametos* und verwandten Arten. *Biol. Zentralbl.* **58**, 516–36.

PALL M. (1973) Sexual induction in *Volvox carteri*: a quantitative study. *J. Cell. Biol.* **59**, 238–41.

PARKER B.C. (1972) On the evolution of isogamy to oogamy. In *Contributions in Phycology*, eds. Parker B.C. & Brown R.M. pp. 59–66.

PARKER G.A., BAKER R.R. & SMITH V.G.F. (1972) The origin and evolution of gamete dimorphism and the male-female phenomenon. *J. theor. Biol.* **36**, 529–53.

PASCHER A. (1916) Ueber die Kreuzung einzelliger haploider Organismen: *Chlamydomonas. Ber. deutsch. bot. Ges.* **34**, 228–42.

PASCHER A. (1918) Ueber die Beziehung der Reduktionsteilung zur Mendelschen Spaltung. *Ber. deutsch. bot. Ges.* **36**, 163–8.

PASCHER A. (1931) Ueber einen neuen einzelligen und einkernigen Organismus mit Eibefruchtung. *Beih. Bot. Centralbl. A.* **48**, 466–80.

POCOCK M.A. (1938) *Volvox tertius* Meyer. With notes on two other British species of *Volvox. J. Quekett Microsc. Club*, **1**, 1–25 (cited in Starr, R.C. 1970).

POWERS J.H. (1908) Further studies in *Volvox*, with descriptions of three new species. *Trans. Amer. Microsc. Soc.* **28**, 141–75.

PRINGSHEIM E.G. (1963) *Farblose Algen.* Gustav Fischer Verlag, Stuttgart.
PRINGSHEIM E.G. & ONDRATSCHEK K. (1939) Untersuchungen ueber die Geschlechts-
vorgange bei *Polytoma. Beih. Bot. Zentralb. A* **59,** 118–72.
RAYBURN W.R. & STARR R.C. (1974) Morphology and nutrition of *Pandorina unicocca,*
sp. nov. *J. Phycol.* **10,** 42–50.
SAGER R. (1955) Inheritance in the green alga *Chlamydomonas reinhardi. Genetics,* **40,**
476–89.
SAGER R. (1956) Inheritance of streptomycin resistance: further studies. *Microbial. Genet.
Bull.* **13,** 28.
SAGER R. (1972) *Cytoplasmic genes and organelles.* Academic Press, New York & London.
SAGER R. & GRANICK S. (1954) Nutritional control of sexuality in *Chlamydomonas rein-
hardti. J. gen. Physiol.* **37,** 729–42.
SCHMEISSER E.T., BAUMGARTEL D.M. & HOWELL S.H. (1973) Gametic differentiation in
Chlamydomonas reinhardi: Cell cycle dependency and rates in attainment of mating
competency. *Developm. Biol.* **31,** 31–7.
SCHREIBER E. (1925) Zur Kenntnis der Physiologie und der Sexualitaet hoeherer *Volvocales.
Z. Botanik,* **17,** 337–76.
SCHULZE B. (1927) Zur Kenntnis einiger *Volvocales. Arch. Protistenk.* **58,** 508–76.
SCUDO F.M. (1967) The adaptive value of sexual dimorphism I. Anisogamy. *Evolution,*
21, 285–91.
SESSOMS A.H. & HUSKEY R.J. (1973) Genetic control of development in *Volvox:* Isolation
and characterization of morphogenetic mutants. *Proc. Nat. Acad. Sci. U.S.* **70,** 1335–8.
SIERSMA P.W. & CHIANG K.S. (1971) Conservation and degradation of cytoplasmic and
chloroplast ribosomes in *Chlamydomonas reinhardtii. J. Mol. Biol.* **58,** 167–85.
SMITH G.M. (1944) A comparative study of the species of *Volvox. Trans. Amer. Microsc.
Soc.* **63,** 265–310.
SMITH G.M. (1946) The nature of sexuality in *Chlamydomonas. Amer. J. Bot.* **33,** 625–30.
SMITH G.M. & REGNERY D.C. (1950) Inheritance of sexuality in *Chlamydomonas reinhardi.
Proc. nat. Acad. Sci. U.S.* **36,** 246–8.
STARR R.C. (1962) A new species of *Volvulina* Playfair. *Arch. Microbiol.* **42,** 130–7.
STARR R.C. (1963) Homothallism in *Golenkinia minutissima.* In *Studies on micro-algae and
photosynthetic bacteria,* pp. 3–6. Ed. Japan. Soc. Plant Physiologists, Univ. Tokyo Press,
Tokyo.
STARR R.C. (1968) Cellular differentiation in *Volvox. Proc. Nat. Acad. Sci. U.S.* **59,** 1082–8.
STARR R.C. (1969) Structure, reproduction, and differentiation in *Volvox carteri f. nagari-
ensis* Iyengar, strains HK 9 and 10. *Arch. Protistenk.* **111,** 204–22.
STARR R.C. (1970) *Volvox pocockiae,* a new species with dwarf males. *J. Phycol.* **6,** 234–9.
STARR R.C. (1971a) Control of differentiation in *Volvox.* In: Changing synthesis in Develop-
ment. Symposium Soc. Developm. Biol., 1970, ed. Runner M. *Developm. Biol. Suppl.*
4, 59–100.
STARR R.C. (1971b) Sexual reproduction in *Volvox africanus.* In *Contributions in Phycology,*
eds. Parker B.C. & Brown R.M. pp. 59–66.
STARR R.C. (1972) A working model for the control of differentiation during development
of the embryo of *Volvox carteri f. nagariensis. Memoires Soc. bot. France,* 175–82.
STARR R.C. & JAENICKE L. (1974) Purification and characterization of the hormone
initiating sexual morphogenesis in *Volvox carteri f. nagariensis. Proc. nat. Acad. Sci.
U.S.* **71,** 1050–4.
STEIN J.R. (1958a) A morphological study of *Astrephomene gubernaculifera* and *Volvulina
steinii, Amer. J. Bot.* **45,** 388–96.
STEIN J.R. (1958b) A morphological and genetic study of *Gonium pectorale. Amer. J. Bot.*
45, 664–72.
STEIN J.R. (1965) Sexual populations of *Gonium pectorale (Volvocales). Amer. J. Bot.* **52,**
379–88.

STEIN J.R. (1966) Growth and mating of *Gonium pectorale* in defined media. *J. Phycol.* **2,** 23–8.

STIFTER I. (1959) Untersuchungen ueber einge Zusammenhaenge zwischen Stoffwechsel und Sexualphysiologie an dem Flagellaten *Chlamydomonas eugametos*. *Arch. Protistenk.* **104,** 364–88.

STREHLOW K. (1929) Ueber die Sexualitaet einiger *Volvocales*. *Z. Botanik* **21,** 625–92.

SUEOKA N. (1960) Mitotic replication of deoxyribonucleic acid in *Chlamydomonas reinhardti*. *Proc. nat. Acad. Sci. U.S.* **46,** 83–91.

SUEOKA N., CHIANG K.-S. & KATES J.R. (1967) Deoxyribonucleic acid replication in meiosis of *Chlamydomonas reinhardi*. *J. Mol. Biol.* **25,** 47–66.

SURZYCKI S. (1971) Synchronously grown cultures of *Chlamydomonas reinhardi*. *Methods Enzymol.* **23,** 66–73.

SZOSTAK J.W., SPARKUHL J. & GOLDSTEIN M.E. (1973) Sexual induction in *Eudorina*. *J. Phycol.* **9,** 215–18.

TSCHERMAK-WOESS E. (1959) Extreme Anisogamie und ein bemerkenswerter Fall der Geschlechtsbestimmung bei einer neuen *Chlamydomonas* Art. *Planta*, **52,** 606–22.

TSCHERMAK-WOESS E. (1962) Zur Kenntnis von *Chlamydomonas suboogama*. *Planta*, **59,** 68–76.

TSCHERMAK-WOESS E. (1963) Das eigenartige Kopulationsverhalten von *Chloromonas saprophila*, einer neuen Chlamydomonacee. *Oesterr. bot. Z.* **110,** 294–307.

TSUBO Y. (1961) Chemotaxis and sexual behaviour in *Chlamydomonas*. *J. Protozool.* **8,** 114–21.

VAN DE BERG W.J. & STARR R.C. (1971) Structure, reproduction and differentiation in *Volvox gigas* and *Volvox powersii*. *Arch. Protistenk.* **113,** 195–219.

WIESE L. (1965) On sexual agglutination and mating type substances (gamones) in isogamous heterothallic Chlamydomonads. I. Evidence of the identity of the gamones with the surface components responsible for sexual flagellar contact. *J. Phycol.* **1,** 46–54.

WIESE L. (1969) Algae. In *Fertilization. Comparative morphology, biochemistry, and immunology*, eds. Metz C.B. & Monroy A. Vol. 2: 135–88. Academic Press, New York.

WIESE L. (1974) Nature of sex specific glycoprotein agglutinins in *Chalydomonas*. Ann. New York Acad. Sci. **234,** 383–95.

WIESE L. & HAYWARD P.C. (1972) III. The sensitivity of sex cell contact to various enzymes. *Amer. J. Bot.* **59,** 530–6.

WIESE L. & METZ C.B. (1969) On the trypsin sensitivity of fertilization as studied with living gametes in *Chlamydomonas*. *Biol. Bull.* **136,** 483–93.

WIESE L. & SHOEMAKER D.W. (1970) On sexual agglutination and mating type substances in isogamous heterothallic Chlamydomonads. II. The effect of concanavalin A upon the mating type reaction. *Biol. Bull.* **138,** 88–95.

WIESE L. & WIESE W. (1975) IV. Unilateral inactivation of the sex contact capacity in compatible and incompatible taxa by α-mannosidase and snake venom protease. *Devel. Biol.* **43,** 264–76.

CHAPTER 9

GENETICS OF ZYGNEMATALES

PAUL BIEBEL

Department of Biology, Dickinson College,
Carlisle, Pennsylvania, U.S.A.

1   Introduction   198

2   Life cycles   199

3   Inheritance of traits   201
   3.1   Mating types   201
   3.2   Lethals   202
   3.3   Zygospore morphology   202

3.4   Chloroplasts   203
3.5   Cell morphology and size
     203

4   Discussion   207

5   References   208

1   INTRODUCTION

Members of the Zygnematales have great potential value as subjects for genetic investigations. Desirable features found in members of this order include: (1) a wealth of morphological diversity, particularly in cell shape, cell wall markings, and chloroplast form, (2) relative ease of cultivation and manipulation of life-cycle stages, often in chemically defined media, (3) life cycles amenable to particular kinds of genetic analysis. Their uniqueness among green algae is evident from the complete absence of flagellated cells throughout the order and their peculiar method of sexual union. Non-flagellate, naked gametes fuse in the characteristic and familiar process of conjugation which is responsible for the older name of the order, Conjugales.

There are at least three distinct morphological series within the order. Genera such as *Spirogyra* and *Zygnema*, which are filamentous and have smooth, undecorated cell walls in the vegetative stage, are members of Zygnemataceae. Saccoderm desmids, the Mesotaeniaceae, have cell walls similar to those of the Zygnemataceae, but they tend to be unicellular especially just prior to conjugation. The Desmidiaceae, or placoderm desmids, make up the third and largest series with the richest morphological variety comprising both unicellular and filamentous forms: the cell walls of their vegetative cells are porose and are often elaborately decorated with granules and spines.

2 LIFE CYCLES

Representatives of all three series have been grown in simple synthetic media, providing the opportunity for experimental manipulation of the life cycles and the possibility of biochemical genetic analysis. *Zygnema* (Gauch 1966) and saccoderm desmids in the genera *Netrium* and *Cylindrocystis* (Biebel 1972) have been grown axenically in defined mineral media. Some placoderm desmids, including species of *Cosmarium* and *Closterium* (Tassigny 1971), have also been grown in mineral media or in synthetic media containing vitamins.

Reproductive cycles have been controlled in both defined and undefined media. The first demonstration that nitrogen deprivation induces gamete formation in algae, a phenomenon now widely known and used in genetic crosses, was in *Spirogyra communis* (Benecke 1908). Pringsheim (1919) induced conjugation and germination of zygospores in *Cylindrocystis brebissonii* by varying the concentration of nitrogen compounds. The ease of inducing gamete fusion by this means has been confirmed in several strains of *Cylindrocystis brebissonii* and *C. crassa*, but germination of the resulting zygospores has proven to be more difficult to induce (Biebel 1975). Gauch (1966) obtained the complete sexual cycle of *Zygnema circumcarinatum* and another (unidentified) species on chemically defined medium by altering salt concentrations to obtain different stages. In two strains of *Netrium digitus* a complete sexual cycle was induced in axenic culture on a chemically undefined medium, a salt solution supplemented with soil extract (Biebel 1964). Allen (1958) reliably controlled the complete sexual cycle of *Spirogyra* in unialgal cultures (containing bacteria). Complete sexual cycles were induced under similar conditions in the saccoderm desmid *Spirotaenia condensata* (Hoshaw & Hilton 1966) and the placoderm desmids *Cosmarium* (Starr 1954, Brandham & Godward 1965) and *Closterium* (Lippert 1967). In most cases, zygospore germination could only be induced after a dormant period of two or more months, so that the sexual generation time in experimental crosses of Zygnematales generally requires approximately three months. One striking exception is the spontaneous germination of large numbers of *Spirotaenia* zygotes, observed only 22 days after compatible mating types were mixed (Hilton 1970).

In common with most other Chlorophyta, Zygnematales have a haploid haplobiontic life cycle—the dormant zygote is the only diploid cell of the sexual cycle, with meiosis occurring upon germination. Karyogamy occurs early in the ripening of zygospores in *Spirotaenia condensata* (Hilton 1970). In *Cosmarium turpinii* (Starr 1955) and *Netrium digitus* (Biebel 1964) the gamete nuclei remain distinct throughout the dormant period in the zygospore in a kind of heterokaryon relationship, and they fuse just prior to meiosis. The number of meiotic nuclear products which survive and are incorporated into distinct germling cells varies not only from one genus to another but even within the same strain. In most species of *Spirogyra* (Allen 1958), *Zygnema* (Gauch 1966), and *Sirogonium* (Hoshaw 1968) only one product survives. *Cosmarium* (Starr 1955; Brandham & Godward 1965), *Closterium* (Lippert 1967), and heterothallic strains of *Netrium digitus* (Biebel 1964) regularly produced two germlings. There is evidence (Starr 1954, Korn 1970) that these cases involve a selective abortion of nuclei which permits genetic analysis of what could be called 'incomplete ordered tetrads'. Complete but unordered tetrads are formed in *Spirotaenia condensata* (Hoshaw & Hilton 1966) and *Cylindrocystis brebissonii* (Biebel 1975).

There is a great range in chromosome numbers among members of the order. Brandham (1965a) cited haploid numbers among desmids from nine to 592. Among strains whose sexual cycle has been completed in culture, the haploid number of chromosomes is 12 ± 1 for *Spirotaenia* (Hilton 1970), approximately 30 for *Cosmarium turpinii* (Starr 1954), 20 for another *Cosmarium* species (Brandham 1965a), approximately 70 for *Closterium moniliferum* (Lippert 1967), and approximately 150 for *Netrium digitus* (Biebel 1964).

There are few plastids per vegetative cell in Zygnematales, and these may be elaborate and of a number and shape characteristic for each genus. Continuity of plastids can often be traced through the zygote stage (Starr 1955), though during maturation of *Cosmarium* zygospores some of the chloroplasts disintegrate, and only those contributed by one of the mating types survive (Korn 1969). The plastids disappear completely in mature *Spirotaenia* zygospores and reappear during germination (Hilton 1970). In *Cylindrocystis brebissonii*, which has two plastids per gamete, portions of all plastids in the zygospore remain visible and identifiable throughout dormancy and germination (Biebel 1975).

Sexuality and mating systems in these algae are relatively uncomplicated. With few exceptions the desmids are all isogamous; cultured strains are asexual, homothallic or heterothallic (with two complementary mating types). Since dormant zygospores serve for species survival through adverse environmental conditions, the significance of sexuality may be independent of other genetic evolutionary functions such as gene mixing. Indeed, for some desmids there is evidence that very close inbreeding is the rule, and cells only rarely fuse with unrelated cells. Regular conjugation of sister cells shortly after cell division occurs in *Netrium digitus* (Biebel 1964), and Brandham (1965b) cited obligate conjugation of sister cells in *Closterium siliqua* and *Staurastrum denticulatum*. Both homothallic and heterothallic strains are known to occur within a single morphological species of desmids (Brandham & Godward 1965). Sexual clones and non-conjugating clones otherwise indistinguishable from them have been isolated from the same population (Brandham & Godward 1965). Starr (1959) isolated a large number of sexual *Cosmarium* strains which were referred to five species. Among these, most were shown to be heterothallic, and two morphologically distinct varieties were found to be interfertile.

3 INHERITANCE OF TRAITS

3.1 *Mating types*

The first definitive analysis of mating type inheritance in a species of the Zygnematales was done by Starr (1954) with *Cosmarium turpinii*. Among 105 germlings, or 'gones', considered at random, he found 1:1 segregation of the two mating types. No more than two gones per zygospore were produced, but from 23 germinating zygospores, both gones survived and could be genetically analysed. Among these pairs of survivors the two gones were of different mating types in 22 cases, whereas in the remaining pair both cells gave rise to clones of the same mating type. Starr interpreted these results as confirmation of cytological evidence that the 'two nuclei which survive are non-sister-nuclei of the four formed as a result of the two divisions' of meiosis. The majority of pairs thus resulted from first-division segregation of Mendelian mating-type alleles, while the exceptional pair was the result of crossing-over between the mating-type locus and the centromere.

The apparent determination of mating type by a single pair of Mendelian alleles has also been demonstrated in *Cosmarium formosulum* and *C. biretum* (Starr 1959), *Closterium moniliferum* and *Cl. ehrenbergii* (Lippert 1967), *Netrium digitus* (Biebel & Reid 1965), *Spirotaenia condensata* (Hilton 1970), and *Cosmarium botrytis* (Brandham & Godward 1965). In the last mentioned case, segregation ratios appeared to be modified by lethality (see below).

In all of these cases except *Spirotaenia*, germinating zygospores regularly produced only two viable meiotic products, the other two nuclei becoming pycnotic and disintegrating as observed in *Cosmarium turpinii*. If Starr's hypothesis is correct, and the surviving nuclei are always non-sisters of the second meiotic division, then among the germinating heterozygous zygospores the frequency of intact pairs with the same genotype would represent the map distance, between the locus and its centromere. If, on the other hand, nuclear survival were at random, a frequency of 33% pairs with the same genotype would be expected. As already mentioned, Starr (1954) found a frequency of 4% for *Cosmarium turpinii* in 23 intact pairs, while Korn (1970) who studied the same strains obtained 12 pairs in which the two gones were of the same mating type out of 89 intact pairs analysed. Combining the results from Starr and Korn, one can calculate a crossover frequency of approximately 12%. Out of 32 survivor pairs in *Cosmarium formosulum*, Starr (1959) obtained a frequency of 3% pairs with the same genotype. Biebel and Reid (1965) found no pairs with the same mating type among 10 pairs of *Netrium digitus* gones. Lippert (1967) analysed larger numbers of survivor pairs and obtained a frequency of 4% among 97 in *Closterium moniliferum* and 24% among 58 in *Cl. ehrenbergii*. Taken together, these data tend to indicate a general ordered or non-random survival of nuclei in desmids.

3.2 *Lethals*

A gene which in homozygous condition causes lysis of immature zygospores though it has no apparent effect on haploid vegetative cells was discovered in *Cosmarium turpinii* by Starr (1954). Lysis occurs before the thickened spore wall develops, i.e., at some stage well before karyogamy. The normal allele, when in a functional heterozygote, permits survival and full development of zygospores, apparently exerting its 'dominance' while there are still separate gametic nuclei in a heterokaryon. Starr (1959) demonstrated that the lethal gene occurred in the natural population, and was able to isolate clones bearing it from three separate collections within a period of 18 months. In each case it was found in the same mating type, although backcrosses showed that the lethal gene and its allele segregated independently of *mt*. Out of 23 pairs of gones surviving only two pairs were of the same genotype, i.e., 9%. In later studies with the same strains Korn (1970) obtained a corresponding frequency of 10% among 89 pairs.

Brandham and Godward (1965) reported an unusual case of apparent lethality in *Cosmarium botrytis*. Matings between any of three *mt*-clones isolated from the same locality at Radlett, Herts, with any of four clones from other localities in England produced offspring which were all *mt⁻*. When two of the F_1 clones were backcrossed, the zygotes yielded mating types in an approximate 1:1 ratio. Evidently post-meiotic abortion of *mt⁺* nuclei was governed by a factor originally present in the Radlett *mt⁻* clones; but since only a small number of zygospore products were tested, Brandham and Godward could not determine whether it was cytoplasmic or nuclear.

3.3 *Zygospore morphology*

A pair of genes which affect zygospore wall structure in *Netrium digitus* var. *lamellosum* was reported by Biebel and Reid (1965). The case was similar to the lethal gene reported by Starr in *Cosmarium turpinii* in that a dominant gene in a heterokaryon determined the wild-type phenotype, which in the wild-type *Netrium* is an elongated zygospore with a wall composed of three layers. Zygospores homozygous for the recessive allele were spherical, and had a double-layered wall which lacked the normal middle layer. Viability of homozygous mutant zygospores was not determined, but from hetero-zygotes the mutant gene segregated independently from the *mt* locus. Among nine survivor pairs from such germinating zygospores, two pairs had the same genotype for wall morphology.

3.4   Chloroplasts

Korn (1969) attributed certain features in the plastid constitution of *Cosmarium turpinii* cells to nuclear influences. From indirect evidence he concluded that plastid survival through the zygospore stage was determined by mating type, and that surviving plastids were those contributed by the mt^+ gamete. This implies that nuclear genes change plastids, since these plastids contributed by the mt^-, which regularly disintegrate, have been inherited from a mt^+ parent. This situation was compared with that in species of *Spirogyra* and *Zygnema* in which the plastids of the passive or female gamete survive in the zygote, while those of the active or male gamete disintegrate. In none of these studies were morphological phenotypic differences between plastids found.

In *Spirogyra* inheritance of the number of chloroplasts per cell was studied by Allen (1958) in a strain which gave rise to relatively stable variants in culture. The original cells contained only a single chloroplast. Among the variations was a clone with two or more plastids per cell, and a larger average cell volume, apparently determined by nuclear ploidy (see below). The increased plastid condition was inherited in succeeding vegetative cell generations, indicating that the plastids are self-reproducing organelles. Inheritance of plastid number through the zygospore was associated with cell volume rather than with the plastid number of the parental strain. Gauch (1966) also recorded inheritance of increased plastid numbers in presumed diploids of *Zygnema*.

3.5   Cell morphology and size

The rich variety of cell forms among members of the Zygnematales, particularly placoderm desmids, would seem to offer unique opportunities for studying the genetic control of morphogenesis. The genus *Micrasterias*, with its large and geometrically symmetrical cells seemed to offer special promise, and Waris and Kallio (1964) conducted detailed studies of morphogenetic problems in this genus for more than a decade. Because their work has been thoroughly discussed in an excellent review (Waris & Kallio 1964), only a brief summary will be given here. Some variations in cell form occurred spontaneously; others were induced by cold shock, heat, continuous illumination, treatment with chemicals, or centrifugation. Variant forms were manifested mostly in the amount of 'radiation', or the number of wings extending laterally from the longitudinal cell axis. Normal wild-type cells of *Micrasterias* with two such wings, are termed biradiate: uniradiate, triradiate, polyradiate and even aradiate (consisting only of the central axis) cells have also been obtained. In many cases these abnormal conditions have been associated with changes in nuclear ploidy, both euploid and aneuploid changes having been obtained. As an explanation of these morphogenetic phenomena Waris

and Kallio proposed a 'cytoplasmic framework' theory according to which
the characteristic symmetry and basic form of the cells is determined by three
cytoplasmic axes which transmit these features to newly forming semicells.
The inheritance of these features is thought to be independent of the nucleus,
but the degree of differentiation superimposed on the basic pattern is postulated to be under nuclear control. Kiermayer (1970), on the other hand,
attributed morphogenetic effects in *Micrasterias* to reorganization of cortical
protoplasm rather than to changes in the axial structures. Since characteristic
forms can be maintained in clonal cultures with different frequencies of reversion to wild type, it is clear that heritable phenomena are involved. The lack
of sexuality in the *Micrasterias* strains studied has precluded genic analysis,
and their mechanisms of form inheritance remain in considerable doubt.

In *Cosmarium turpinii* inheritance of the triradiate condition was investigated by Starr (1958). Two types of triradiate cells arose in clonal
cultures, large, apparently diploid cells and small, haploid cells. Each of these
could be maintained as relatively stable clones though with some reversion
to the biradiate condition. The strain studied was heterothallic, and both
diploid and haploid triradiate cells retained the ability to form zygospores
when paired with haploid biradiate cells of the complementary mating type.
Zygospores formed between haploid triradiate and biradiate cells failed to
germinate. Zygospores formed between diploid triradiate and haploid biradiate cells germinated but produced no viable progeny: cytological preparations revealed some univalent and trivalent chromosomes during meiosis,
which may account for the mortality of the products.

Haploid triradiate cells were isolated by Brandham and Godward (1964)
from clones of both homothallic and heterothallic *Cosmarium botrytis*
strains. In each case they had arisen spontaneously from germinating zygospores: no such abnormal cells were seen in vegetatively reproducing clonal
cultures of the biradiate wild type. (It is a curious fact that the haploid triradiate clone of *C. turpinii* isolated by Starr (1958) also originated from a
germinating zygospore, although his diploid triradiate clones arose spontaneously in biradiate cultures reproducing asexually.) Brandham and
Godward crossed their triradiate form with biradiate cells of the complementary mating type, obtaining 13 biradiate and two triradiate gones
from eight zygospores. Among the gones recovered from two crosses between
biradiate clones of the same strain, 22 biradiate and three triradiate gones
were isolated. Thus it can be seen that not only is the triradiate character
not inherited in a Mendelian manner, but a cross between one triradiate
parent and one biradiate cell produces no more triradiate offspring than
between two biradiate cells. Brandham and Godward attributed determination of radiation to the cytoplasmic organization which occurs at the time of
germination. Germling cells of desmids normally have a juvenile morphology,
and the shape characteristic of each genus does not arise until the first division
following zygospore germination.

Diploid triradiate cells were found in a heterothallic strain of *Cosmarium botrytis*, and diploid multiradiate cells were isolated from a homothallic strain of *Staurastrum denticulatum* and from a non-sexual clone of *Staurastrum dilatatum* (Brandham 1965b). The triradiate *Cosmarium* clone appeared, following ultraviolet irradiation, as biradiate diploid cells much larger than the haploid biradiate cells from which they arose. Then, as the cells in the culture multiplied, the triradiates arose, and increased in proportion until they comprised 97% of the cell population. Diploid cells had an increased number of nucleoli as well as approximately twice the chromosome number. They fused with haploid cells of the complementary mating type to form triploid zygospores which had 1·5 times the volume of diploid zygospores, and which ultimately germinated, although no viable products survived.

The diploid *Staurastrum denticulatum* clone arose spontaneously from a germinating zygospore. At first it consisted of triradiate cells characteristic of the genus but larger than the haploid cells, but after a period of time the cells of the clone became almost entirely quadriradiate. The diploid cells retained the ability to conjugate and were homothallic, as was the parent clone. Tetraploid zygospores produced from fusion of such diploid gametes had twice the volume of diploid zygospores, but could not be induced to germinate.

The asexual clone of *Staurastrum dilatatum* had both triradiate and quadriradiate cells. When larger cells appeared, some of these were isolated giving rise to clones which contained mostly pentaradiate and a few hexaradiate cells which, by comparison with similar results in *Cosmarium* and *Staurastrum denticulatum*, Brandham assumed to be diploid.

A homothallic strain of *Closterium siliqua* likewise produced diploid clones (Brandham 1965b) which could be recognized not only by their larger cells but also an increase of the nucleolar-organizing chromosomes from two to four. Since each semicell of *Closterium* has a radial symmetry around the longitudinal axis, in contrast to the bilateral symmetry of *Cosmarium* and *Staurastrum* cells, no increase in radiation could be expected. The diploid cells conjugated and produced mature tetraploid zygospores, which did not germinate.

A polyploid series of strains arose in cultures, originating apparently in single steps from a homothallic clone, of *Spirogyra* studied by Allen (1958). Three different kinds of subcultures were distinguished according to cell diameters, with the following ranges: 15·0 to 19·5, 21·0 to 25·5, and 27·0 to 33·0 μm. Chromosome counts for cells of the three respective groups were: 11 to 15, 26 to 30, and 56 to 60, apparently in a polyploid series of n, 2n, and 4n. Conjugation within a group and between cells from different groups was obtained, and zygospores germinated to produce one germling per zygospore. Variation among germling cultures was greater than in cells of clonal cultures, and there was some indication that partial reversions occurred by loss of chromosomes leading to various aneuploid conditions. It is of some interest that the ratios of the volumes of haploid, diploid and tetraploid cells were approximately 1:4:16, rather than 1:2:4 as found in desmids by Brandham (1965b).

In occasional filaments of *Zygnema* Gauch (1966) observed abrupt changes in cell diameter, somewhat like those of Allen for *Spirogyra*, but he was unable to confirm corresponding increase in ploidy.

Morphological mutants of *Cosmarium turpinii* induced by ultraviolet irradiation were investigated by Korn (1970), who obtained 47 different cell-shape mutants. Two of these, termed 'bonnet' and 'gaunt', were studied through the sexual cycle. The nature of the cell division process in placoderm desmids permits a kind of strand analysis, since each daughter cell receives the old, mature, highly characteristic cell wall of one parental semicell and forms a newly synthesized wall on the other semicell. The time when a change in cell-wall pattern occurs can thus be detected easily. Korn disposed irradiated cells on agar plates, and for four asexual generations after treatment he separated the daughter cells on the surface, arranging them in such a way that lineages could be traced and the origin of mutations thereby correlated with post-irradiation cell divisions. All but one of the mutations appeared during the first three divisions, the phenotypic lag period. According to Korn, his set of data 'supports the ideas . . . that there is one double helix for DNA in the chromosome at G_1 and two double helices at the G_2 stage of the mitotic cycle'.

The *bonnet* mutant cells differed in shape from those of wild type, and no reversions were seen in culture. The *gaunt* mutant differed from the wild type and from *bonnet* not only in cell shape but also in polymorphism, in that it produced uniradiate and aradiate cells as well as biradiate cells. Unlike those of the *Micrasterias* mentioned above, even the aradiate cells of *C. turpinii* tended to revert to the biradiate condition. The *gaunt* clones, on the other hand, did not produce any reversions to wild type. Since the original strains were the same as those used by Starr (1954), genes for mating-type and zygote lethality could be included as markers in crosses involving Korn's morphological mutants; they always segregated in the expected 1:1 Mendelian ratio. Both the *bonnet* and the *gaunt* genes segregated independently of *mt* and the zygote-lethal locus. In three crosses, wild-type and *bonnet* segregated in an approximate 2:1 ratio: wild-type and *gaunt* segregated similarly. In a cross involving both mutants, wild-type recombinants were three times as frequent as single mutants or double mutants. In all crosses there was an excess of wild-type gones over expectations on the basis of Mendelian inheritance, possibly attributable to reduced penetrance or reduced viability of the mutants. Since only two of the four nuclear products of meiosis in the zygospore are normally viable, Korn concluded that the wild type genes confer an advantage in the competition to survive, whereas the mutants evidently produced lethality at some stage. Further evidence of the detrimental nature of the *bonnet* gene is the fact that in crosses involving an *mt⁻ bonnet* gamete, zygote viability is reduced.

4 DISCUSSION

It should be clear from the above account that the potential of the Zygnematales for contribution to our genetic knowledge has hardly been realized. Mating-type inheritance seems to follow the pattern well established for other algae. The effect of mating type on the survival of plastids and other cytoplasmic structures may be of special interest, and should be further investigated. The strange lethal effect of mating type cited by Brandham and Godward (1965), though the evidence is based on scant data, should be sought in other populations. The occurrence of such lethal effects may help explain the infrequent occurrence of desmid zygospores in nature, and the frequent observation that isolates from some localities are apparently asexual while clones of similar cells from other localities can be made to conjugate by relatively simple manipulation of physical or chemical conditions. Many natural populations may consist of only a single mating type.

It is a curious fact that, with the exception of mating-type alleles, the first two observed cases of typical Mendelian segregation in Zygnematales were expressed in a heterokaryotic, 'functional diploid' condition in the zygospore, and could be detected only in a backcross. Since recessive genes were involved, which had no apparent effect on the predominantly haploid vegetative cells it might be profitable to look more closely at characteristics of zygospores in seeking other gene differences. It would not be surprising to find that, in Nature, such recessive genes accumulate in the vegetative phase, where they face little or no selective pressure.

One feature of this order which has attracted much attention and stimulated genetic study is the complex and diverse cell morphology of placoderm desmids, particularly in the genera *Micrasterias*, *Staurastrum*, and *Cosmarium*. Although variant forms have been obtained and studied in culture, no satisfactory explanations have been found for the mode of inheritance or the mechanism of gene action on morphogenesis. Variations apparently are perpetuated almost exclusively by asexual means. For the most part, they have been studied in strains which do not conjugate or in which the traits do not persist through the zygospore stage. In the two cases studied by Korn (1970), in which modified cell morphology was transmitted through the zygospore, the numbers of surviving mutant genes were below expectations. Modification of whole lobes and the degree of 'radiation', cited by Waris and Kallio (1964), Starr (1958) and Brandham and Godward (1964), may be so drastic that they could only be attributed to alterations in chromosome ploidy and hence in the balance of genes controlling the basic cell architecture. While such abnormal conditions can be transmitted asexually, the complex reorganization which accompanies sexual reproduction may preclude their transmission through the zygote. Perhaps more subtle variations in form, similar to those studied by Korn, may prove to be more profitable subjects for investigation.

The potential for studies of plastid inheritance, gene mapping in 'ordered incomplete tetrads', strand-analysis and effects of ploidy seems to have been demonstrated in this group.

Given the potential for classical genetic research, why has it, too, not been realized in this group? It would seem that not all of the desirable features of Zygnematales have been exploited. For example, despite the fact that many species have now been cultured axenically in chemically defined media (Tassigny 1971), not a single biochemical mutant is known. Furthermore, although life cycles can usually be manipulated more reliably under defined conditions (Biebel 1975), most of the genetic studies have been done in media containing soil extracts of highly variable composition and often in the presence of contaminating bacteria. Insufficient attention has been devoted to research on critical phases of the life cycles of some of the more promising organisms. The physiology of zygospore germination is still generally a refractory problem in this group. For example, the heterothallic *Cylindrocystis brebissonii* which grows rapidly and conjugates readily in inorganic defined media, has visibly persistent chloroplasts throughout the entire sexual cycle, and eventually produces a complete tetrad of meiotic products (Biebel 1975). The one aspect of its life cycle for which we lack adequate control is the induction of reliable germination of zygospores in a period of less than three months. If one could overcome this obstacle, this might prove a very suitable species for genetic research. Another case is a heterothallic *Spirotaenia condensata* which produces a complete tetrad of meiotic products and has a sexual generation time of only 22 days (Hilton 1970): the information lacking here is a chemically defined medium for growth and gametogenesis in this species.

The great potential for genetic research has not been realized in the Zygnematales because for this order there has not yet been that favourable conjunction of desirable features in one organism and the investigator ready to recognize and exploit them. It may not be too much to hope that the promise will soon be fulfilled.

5 REFERENCES

ALLEN M.A. (1958) The Biology of a Species Complex in *Spirogyra*. Ph.D. thesis. 240 pp. Indiana University. Bloomington.
BENECKE W. (1908) Über die Ursachen der Periodizität im Auftreten der Algen, auf Grund von Versuchen über die Bedingungen der Zygotenbildung bei *Spirogyra communis*. *Internat. Rev. d. ges. Hydrobiol. u. Hydrographie.* **1,** 533–52.
BIEBEL P. (1964) The sexual cycle of *Netrium digitus*. *Amer. J. Bot.* **51,** 697–704.
BIEBEL P. (1975) Morphology and life cycles of saccoderm desmids in culture. *Beih. Nova Hedwigia.* **42,** 39–57.

BIEBEL P. & REID R. (1965) Inheritance of mating types and zygospore morphology in *Netrium digitus* var. *lamellosum*. *Proc. Pennsylvania Acad. Sci.* **39**, 134–7.

BRANDHAM P.E. (1965a) Some new chromosome counts in the desmids. *Brit. phycol. Bull.* **2**, 451–5.

BRANDHAM P.E. (1965b) Polyploidy in desmids. *Canad. J. Bot.* **43**, 405–17.

BRANDHAM P.E. & GODWARD M.B.E. (1964) The production and inheritance of the haploid triradiate form in *Cosmarium botrytis*. *Phycologia*, **4**, 75–83.

BRANDHAM P.E. & GODWARD, M.B.E. (1965) The inheritance of mating type in desmids. *New Phytol.* **64**, 428–35.

GAUCH H.G. (1966) Studies on the Life Cycle and Genetics of *Zygnema*. M.S. thesis. 91 pp. Cornell University. Ithaca.

HILTON R. (1970) A cytological and morphological study of the alga *Spirotaenia condensata* Brébisson. Ph.D. thesis. 87 pp. University of Arizona, Tucson.

HOSHAW R.W. (1968) Biology of the filamentous conjugating algae. In *Algae, Man and the Environment*, ed. Jackson D.F. pp. 135–84. Syracuse University Press, Syracuse, N.Y.

HOSHAW R.W. & HILTON R.L. Jr. (1966) Observations on the sexual cycle of the saccoderm desmid *Spirotaenia condensata*. *J. Arizona Acad. Sci.* **4**, 88–92.

KIERMAYER O. (1970) Causal aspects of cytomorphogenesis in *Micrasterias*. *Ann. New York Acad. Sci.* **175**, 686–701.

KORN R.W. (1969) Chloroplast inheritance in *Cosmarium turpinii* Bréb. *J. Phycol.* **5**, 332–6.

KORN R.W. (1970) Induction and inheritance of morphological mutations in *Cosmarium turpinii* Bréb. *Genetics*, **65**, 41–9.

LIPPERT B.E. (1967) Sexual reproduction in *Closterium monoliferum* and *Closterium ehrenbergii*. *J. Phycol.* **3**, 182–98.

PRINGSHEIM E.G. (1919) Die Kultur der Desmidiaceen. *Ber. deut. botan. Ges.* **36**, 482–5.

STARR R.C. (1954) Inheritance of mating type and a lethal factor in *Cosmarium botrytis* var. *subtumidum* Wittr. *Proc. Nat. Acad. Sci. U.S.* **40**, 1060–3.

STARR R.C. (1955) Zygospore germination in *Cosmarium botyris* var. *subtumidum*. *Amer. J. Bot.* **42**, 577–81.

STARR R.C. (1958) The production and inheritance of the triradiate form in *Cosmarium turpinii*. *Amer. J. Bot.* **45**, 243–8.

STARR R.C. (1959) Sexual reproduction in certain species of *Cosmarium*. *Arch. f. Protistenk.* **104**, 155–64.

TASSIGNY M. (1971) Observations sur les besoins en vitamines des desmidiées (Chlorophycées-Zygnematales). *J. Phycol.* **7**, 213–5.

WARIS H. & KALLIO P. (1964) Morphogenesis in *Micrasterias*. *Adv. in Morphogenesis*. **4**, 45–80.

CHAPTER 10

GENETICS OF CHAROPHYTA

VERNON W. PROCTOR

Department of Biology, Texas Tech University,
Lubbock, Texas 79409, U.S.A.

1 Introduction 210

2 Controlled crosses 211
2.1 Crosses between dioecious clones 211
2.2 Mechanical emasculation 211
2.3 Male-sterile hybrids 212

2.4 Radiation-induced mutants 214

3 Taxonomic significance of morphological features 215

4 References 217

1 INTRODUCTION

Although charophytes today flourish in a wide variety of aquatic habitats, as apparently they have for the past 350 million years, there is no indication that members of the group play any vital role in the lives of most geneticists. Far less is known about the genetics of Charophyta than of most vascular groups. The basic difficulty has been one of ensuring controlled crosses between monoecious, i.e. bisexual, clones. No such difficulties hinder attempted crosses of dioecious, i.e. unisexual, clones, or of a dioecious female fertilized by sperm of a monoecious plant, which need only be cultured in a common container. However, most charophytes are monoecious, with small and conjoined gametangia. The challenge arises when one attempts to fertilize female gametes of such plants by sperm of a second, be it monoecious or dioecious.

All monoecious charophytes yet examined have proven fully autogamous. To ensure controlled crosses, the monoecious sprigs destined as the maternal parent must be emasculated prior to the attempted union, and must bear only unfertilized eggs. Failure of either contingency could result in selfed, as well as cross-fertilized, oospores. Only within the past decade have techniques become available for successfully emasculating monoecious plants and hereby ensuring controlled crosses. The respective procedures of three such methods and the preliminary results obtained will now be considered.

2 CONTROLLED CROSSES

2.1 *Crosses between dioecious clones*

Controlled crosses between Charophyta were first reported by McDonald and Hotchkiss (1956). They placed male sprigs of *Chara australis* in a common container with female plants (identified as *Protochara australis*) and recovered dark oospores, which appeared normal but which could not be induced to germinate. Attempts at the reciprocal cross, i.e. *C. australis*/'*P. australis*', yielded no oospores. The potential interfertility of these morphologically distinct forms therefore remains an open question. Wood (1965) incorrectly reported that this study had involved the successful hybridization of dioecious and monoecious charophytes. No monoecious species were involved.

The first attempts to cross dioecious and monoecious charophytes were by McCracken *et al.* (1966), who reported that eggs on female sprigs of dioecious *C. rusbyana* could not be fertilized by sperm of monoecious *C. zeylanica*, *C. haitensis*, *C. foliolosa*, or *C. martiana*. Reciprocal crosses were equally unsuccessful. These five morphologically similar species of *Chara*, all in subsection *Willdenowia*, have been characterized by McGuire (1971).

2.2 *Mechanical emasculation*

In the unsuccessful attempts, reported above, to fertilize eggs of monoecious *Chara* spp. with sperm of the dioecious *C. rusbyana*, the initially bisexual sprigs were emasculated by surgical removal of the antheridia, one by one, under low magnification. With care, most of the associated oogonia could be retained undamaged.

Over the past decade, utilizing such mechanically emasculated sprigs we have attempted controlled crosses between species in most of the major groups or 'species complexes' within the genus *Chara*. All have been uniformly unsuccessful (Proctor, unpublished). Such well known and morphologically distinct species as *C. zeylanica*, *C. globularis* (= *C. fragilis*), the *C. vulgaris+ contraria* complex, *C. braunii*, and *C. evoluta* are reproductively isolated; not even hybrid oospores are produced. Reports of hybrids *between* pairs of any of these species must therefore be viewed with considerable scepticism. Such few hybrids as have been experimentally obtained to date have always been *within* these major complexes, e.g. *C. connivens**/*C. globularis*, *C. vulgaris**/ *C. contraria*.

Although emasculation permits one to make controlled crosses between compatible clones, the method is subject to three major limitations. First, it is both tedious and time-consuming. Second, emasculation is not permanent, newly developed shoots of dioecious species being, of course, bisexual.

---

* Maternal parent is conventionally listed first.

Finally, it is often impossible to ensure that every antheridium has been removed during emasculation, and that fertilization of some of the eggs has not already occurred prior to emasculation.

2.3   *Male-sterile (M-S) hybrids*

These limitations were partially overcome by use of male-sterile strains. It has been found that a substantial fraction of the hybrid offspring from inter-populational conspecific crosses are often male-sterile but female-fertile (Proctor 1971a; Proctor & Wiman 1971). Such plants, many of which are vegetatively normal or near-normal, bear both antheridia and oogonia, but only the eggs are functional; the antheridia either fail to open or liberate non-motile sperm. (To some extent the terms male-sterile (M-S) and self-sterile (S-S) are interchangeable. They are used here to the degree of precision known. Where self-sterility is known to result from nonfunctional sperm the plant is designated M-S, otherwise S-S. So far, every instance of self-sterility among charophyte hybrids appears to result from male-sterility. The symbols S-F (self-fertile), F-F (female-fertile), F-S (female-sterile) and M-F (male-fertile) are used in similar fashion. Partial fertility or sterility is indicated by the prefix P. No distinction is drawn between P. S-S and P. S-F.) Such M-S clones, which can be vegetatively propagated, remain permanently male-sterile. Several dozen have been in continuous culture for more than five years without reversion to the S-F state.

The frequency of self-sterility observed among the offspring of four clones of *C. zeylanica sensu stricto*, crossed in all possible combinations, is shown in Table 1a. The four clones, collected initially in Texas, Oklahoma, California and Yucatan, each bore tetrascutate antheridia; in each clone the haploid chromosome number was 28. Sprigs were carefully trimmed, emasculated, and crossed to yield 12 lots of cross-fertilized oospores and four of selfed. After a short rest period, the 16 lots of oospores were allowed to germinate. Where possible, isolated progeny were reared to maturity and scored as S-F or S-S. Offspring involving only the parental clones 415 (Texas), 256 (Okla.), and X-187 (Calif.) were fully S-F, with one exception. (A single isolate from the cross 415 (Texas)/256 (Okla.) was only 2·5% S-F. Selfed oospores were collected and germinated for two additional generations; all resulting off-spring were still <5% S-F.) However, crosses involving X-008 (Yuc.), regard-less of whether this clone served as maternal or paternal parent, gave quite different results. A substantial fraction of their offspring were either partially or fully S-S. Oospores of selfed X-008 and from the cross X-008/X-397 (N. Car.) yielded only S-F offspring.

Two vegetatively normal but S-S isolates from the cross 415/X-008 (Table 10.1), and two from the reciprocal cross, were then selected for back-crossing to each of the parental clones, viz., 415 and X-008. An examination of their antheridial filaments at the moment of sperm release revealed no

Table 10.1. Fertility of offspring from crosses involving four N. American clones of *C. zeylanica sensu stricto*. All four clones produce monecious-conjoined gametangia and tetrascutate antheridia, and all have 28 chromosomes.

♀	♂	SF	SS	Died	Total isolated
256	256	11	0	0	11
(Oklahoma)	415	12	0	0	12
	X-008	5	6	1	12
	X-187	15	0	0	15
415	256	14*	0	0	14
(Texas)	415	12	0	0	12
	X-008	8	7†	0	15
	X-187	15	0	0	15
X-008	256	2	10	2	14
(Yucatan)	415	7	7¶	0	14
	X-008	11	0	0	11
	X-187	1	1	1	3
X-187	256	15	0	0	15
(California)	415	15	0	0	15
	X-008	8	13	0	21
	X-187	15	0	0	15

* One clone<5% SF. † Sources of SS isolates K, L (see Table 10.2). ¶ Sources of SS isolates D, E (see Table 10.2).

motile cells; but since the oogonia darkened when placed with either parental clone, the eggs were assumed to be functional. In due course the eight lots of resulting oospores germinated and the germlings were isolated and later scored as S-F, P. S-F, or S-S (Table 10.2). Most were vegetatively normal or near-normal, but many proved partially or completely S-S. M-S offspring

Table 10.2. Fertility of offspring resulting from backcrossing self-sterile strains D, E, K, and L (see Table 10.1) to parental lines 415 and X-008.

♀	♂	SF	PSF	SS	Died	Total isolated
D	415	10	8	38	2	58
(X-008/415)	X-008	2	5	18	1	26
E	415	5	2	19	2	28
(X-008/415)	X-008	22	0	2	0	24
K	415	15	5	13	0	33
(415/X-008)	X-008	4	4	26	3	37
L	415	4	2	35	0	41
(415/X-008)	X-008	8	1	28	0	37

K

activity of their genes (Fjeld 1971b). Although in the germling the cell number increases exponentially with time, the proportion of *pr* chimaeras induced by UV-irradiation at different times after the two-cell stage is found to be constant, suggesting that *pr* genes are normally active in a particular population of cells which does not increase in number during development but which can later give rise to daughter cells with different developmental fates (as in the HB-cells of Fig. 11.3).

Many mutants with disoriented division planes have been isolated, but only one has been described. Plants of this mutant (*lumpy*) grow as aggregates of undifferentiated cells. Under the electron microscope the fine structure of the walls appears different from that of the wild type (Bryhni 1974), suggesting some causal connection between the pattern of cell division and cell-wall formation. (In the wild type, during filamentous growth the external wall expands preferentially in one direction without any appreciable change in thickness, while the wall separating two daughter cells increases in thickness but not in area. During the growth of the blade, external and internal walls expand their areas uniformly and retain a constant thickness, whereas the wall between two daughter cells increases first in thickness and then in area. Local differences in wall expansion are especially conspicuous during differentiation of the elongated rhizoid cells and the giant stem cells with their tubular outgrowths. Whether the developmental programme is to be expressed or not may thus depend on the biochemical machinery for wall synthesis.)

A temperature-sensitive mutant, *tm*-2, recently isolated by Fjeld (unpublished), grows abnormally at 15°C; at 23°C it exhibits a normal phenotype. When a properly organized mutant plant raised at 23°C is transferred to 15°C, warts of unorganized cells develop on the thallus. Possibly at 15°C this mutant cannot synthesize a normal wall, the cells being presumably unable to participate in the expansion of the walls already present and being therefore extruded from the organized thallus. Such temperature-conditional mutants may ultimately prove as useful for an analysis of the genetic control of morphogenesis in multicellular organisms as for the analysis of phage morphogenesis. (It is interesting to note that cold-sensitive mutations in certain microorganisms are believed to be concerned with an assembly process; Guthrie *et al.* 1969.)

These results suggest that *Ulva* is an experimental organism suitable for a genetic approach to such fundamental problems in developmental biology as those concerning:

(1) determination, and the differential fate of two daughter cells;

(2) dedifferentiation of cells, and the events which make the full developmental programme again available in reproductive cells;

(3) the role of the extracellular matrix in determination and differentiation.

6 CONCLUDING REMARKS

We have considered here the necessity of a genetical approach to solve specific phycological problems, and the possibility of using multicellular algae as models for a genetic approach to general biological problems. The sporulation systems of plants like *Ulva* can also offer other research possibilities of general interest. Since thallus cells tend to sporulate synchronously, synchronized mass cultures, as developed by Nordby and Hoxmark (1972), permit one to make biochemical comparisons between the meiotic formation of zoospores and the mitotic formation of gametes. Because the differentiation of gametes and of zoospores is so similar, subtle processes controlling meiosis may be indentified in this way. On the basis of such analysis, Nordby and Hoxmark have reached the surprising conclusion that each meiotic chromosome consists of at least four strands of DNA, a result with grave implications for existing theories on chromosome organization, mutation mechanisms and cross-over processes.

Although the DNA content of cells generally increases with the size and complexity of the organism, there is no significant difference in this respect between a unicellular form, *Chlamydomonas*, with a haploid DNA content of 0·13 pg (Chiang & Sueoka 1967) and a multicellular form, *Ulva*, with 0·12 pg DNA per haploid cell (Aaneby unpublished). Nevertheless, *Ulva* presents us with the same general problems of determination, cellular differentiation, differential growth and apical dominance as plants and animals at a much higher level of organization. It is worth considering that, in the evolution of multicellular life, substances like growth hormones, originally synthesized by a unicellular ancestor, might be used for new purposes as a multicellular form develops. Furthermore, the regular pattern of early division generally adopted by multicellular organisms in itself introduces differences which could serve as bases for cellular differentiation: e.g., during filamentous growth, the two cells in the middle of a row of four have a somewhat different environment from the end cells. It is therefore theoretically possible that, at least at the organizational level of algae, multicellularity may entail only a small increase in genetic complexity. In practice it may prove difficult to separate genes involved in the multicellular organization itself from genes concerned with general metabolism and common to both unicellular and multicellular forms. The ultimate goal of the current research in our laboratory is to identify some of the former type of genes, and our approach is based upon the hope that there are not too many of them!

7 REFERENCES

BLIDING C. (1968) A critical survey of European taxa in Ulvales, II. *Ulva, Ulvaria, Monostroma, Kornmannia. Bot. Notiser, Lund,* **121,** 535–629.

BRITTEN R.J. & DAVIDSON E.H. (1969) Gene regulation for higher cells: A theory. *Science*, **165**, 349–57.

BRYHNI E. (1974) Genetic control of morphogenesis in the multicellular alga *Ulva mutabilis*. Defect in cell wall production. *Devl. Biol.* **37**, 273–9.

BRÅTEN T. (1971) The ultrastructure of fertilization and zygote formation in the green alga *Ulva mutabilis* Føyn. *J. Cell Sci.* **9**, 621–35.

BRÅTEN T. (1973) Autoradiographic evidence for the rapid disintegration of one chloroplast in the zygote of the green alga *Ulva mutabilis*. *J. Cell Sci.* **12**, 385–9.

CHAPMAN A.R.O. (1974) The genetic basis of morphological differentiation in some *Laminaria* populations. *Marine Biology*, **24**, 85–91.

CHIANG K.-S. & SUEOKA N. (1967) Replication of chloroplast DNA in *Chlamydomonas reinhardi* during vegetative cell cycle: Its mode and regulation. *Proc. Natl. Acad. Sci. U.S.* **57**, 1506–13.

FANG T.C., YIANG B.Y. & LI Y.Y. (1962) On the inheritance of stipe length in Haidai (*Laminaria japonica* Aresch.) *Acta bot. Sin.* **10**, 327–35.

FANG T.C., WU C.Y., YIANG B.Y., LI Y.Y. & REN K.Z. (1963) The breeding of a new variety of Haidai (*Laminaria japonica* Aresch.) *Scientia Sin.* **12**, 1011–18.

FJELD A. (1970a) A chromosomal factor exerting a predetermining effect on morphogenesis in the multicellular green alga *Ulva mutabilis*. *Genet. Res., Camb.* **15**, 309–16.

FJELD A. (1970b) Mosaic mutants: absence in a eucaryotic organism. *Science*, **168**, 843–4.

FJELD A. (1971a) 1:1 distribution between induced mutants and phenocopies. *Nature New Biol.* **232**, 238–9.

FJELD A. (1971b) Genetic control of cellular differentiation in *Ulva mutabilis*. Induced chimeras. *Hereditas*, **69**, 59–66.

FJELD A. (1971c) Unequal contribution of the two gametes to the zygote in the isogameous multicellular alga *Ulva mutabilis*. *Exptl. Cell Res.* **69**, 449–52.

FJELD A. (1972) Genetic control of cellular differentiation in *Ulva mutabilis*. Gene effects in early development. *Develop. Biol.* **28**, 326–43.

FJELD A. (1975) The spontaneous mutability in *Ulva mutabilis*. Occurrence of unstable strains. *Norw. J. Bot.* **22**, 77–82.

FØYN B. (1934) Lebenszyklus und Sexualität der Chlorophycee *Ulva lactuca*. L. *Arch. Protistenk.* **83**, 154–77.

FØYN B. (1955) Specific differences between northern and southern European populations of the green alga *Ulva lactuca* L. *Pubbl. Staz. Napoli*, **27**, 261–70.

FØYN B. (1958) Über die Sexualität und den Generationswechsel von *Ulva mutabilis* (n.s.). *Arch. Protistenk.* **102**, 473–80.

FØYN B. (1959) Geschlechtskontrollierte Vererbung bei der marinen Grünalge *Ulva mutabilis*. *Arch. Protistenk.* **104**, 236–53.

FØYN B. (1960) Sex-linked inheritance in *Ulva*. *Biol. Bull. Wood's Hole*, **118**, 407–11.

FØYN B. (1961) Globose, a recessive mutant in *Ulva mutabilis*. *Botanica Marina*, **3**, 60–4.

FØYN B. (1962a) Two linked autosomal genes in *Ulva mutabilis*. *Botanica Marina*, **4**, 156–62.

FØYN B. (1962b) Diploid gametes in *Ulva*. *Nature, Lond.* **193**, 300–1.

GUTHRIE C., NASHIMOTO H. & NOMURA M. (1969) Structure and function of *E. coli* ribosomes, VIII. Cold-sensitive mutants defective in ribosome assembly. *Proc. Nat. Acad. Sci. U.S.A.* **63**, 384–91.

HANIC L.A. (1973) Cytology and genetics of *Chondrus crispus* Stackhouse. In *Chondrus crispus*, eds. Harvey M.J. & McLachlan F. pp. 23–41. Nova Scotia Institute of Science, Halifax, Nova Scotia.

HOXMARK R.C. (1975) Experimental analysis of the life cycle of *Ulva mutabilis*. *Botanica Marina*, **18**, 123–9.

HOXMARK R.C. & NORDBY Ø. (1974) Haploid meiosis as a regular phenomenon in the life cycle of *Ulva mutabilis*. *Hereditas*, **76**, 239–50.

KORNMANN P. (1970) Eine mutation bei der siphonalen Grünalge *Derbesia marina. Helgoländer wiss. Meeresunters.* **21**, 1–8.

LEWIN R.A. (1958) Genetics and marine algae. In *Perspectives in marine biology*, ed. Buzzati-Traverso A.A., pp. 547–57. Univ. of Calif. Press, Berkeley & L.A.

LØVLIE A. (1964) Genetic control of division rate and morphogenesis in *Ulva mutabilis* Føyn. *Compt. rend. trav. Lab. Carlsberg*, **34**, 77–168.

LØVLIE A. (1968) On the use of a multicellular alga (*Ulva mutabilis* Føyn) in the study of general aspects of growth and differentiation. *Nytt Magasin for Zoologi*, **16**, 39–49.

McLACHLAN J., CHEN L.C.M. & EDELSTEIN T. (1970) The culture of four species of *Fucus* under laboratory conditions. *Can. J. Bot.* **49**, 1463–9.

MÜLLER D.G. (1972) Studies on reproduction in *Ectocarpus siliculosus. Mémoires Soc. bot. Fr.* 87–98.

NEUMANN K. (1969) Beitrag zur Cytologie und Entwicklung der siphonalen Grünalge *Derbesia marina. Helgoländer wiss. Meeresunters.* **19**, 355–75.

NORDBY Ø. & HOXMARK R.C. (1972) Changes in cellular parameters during synchronous meiosis in *Ulva mutabilis* Føyn. *Exptl. Cell Res.* **75**, 321–8.

PERROT Y. (1972) Les *Ulothrix* marins de Roscoff et le problème de leur cycle de reproductions. *Mémoires Soc. Bot. Fr.* 1972, 67–74.

POLLOCK E.G. (1969) Interzonal transplantations of embryos and mature plants of *Fucus. Proc. Int. Seaweed Symp.* **6**, 345–56.

RUENESS J. (1973) Speciation in *Polysiphonia* (Rhodophyceae, Ceramiales) in view of hybridization experiments: *P. hemisphaerica* and *P. boldii. Phycologia*, **12**, 107–9.

RUENESS J. & RUENESS M. (1973) Life history and nuclear phases of *Antithamnion tenuissimum*, with special reference to plants bearing both tetrasporangia and spermatangia. *Norw. J. Bot.* **20**, 205–10.

RUENESS J. & RUENESS M. (1975) Genetic control of morphogenesis in two varieties of *Antithamnion plumula* (Rhodophyceae, Ceramiales). *Phycologia*, **14**, 81–5.

RUSSELL G. & MORRIS O.P. (1970) Copper tolerance in the marine fouling alga *Ectocarpus siliculosus. Nature, Lond.* **228**, 288–9.

RUSSELL G. & MORRIS O.P. (1973) Ship-fouling as an evolutionary process. *Proc. 3rd Int. Congress on marine corrosion and fouling*, pp. 719–30. Northwestern University Press.

SAITO Y. (1972) On the effects of environmental factors on morphological characteristics of *Undaria pinnatifida* and the breeding of hybrids in the genus Undarier. In *Contributions to the systematics of benthic marine algae of the North Pacific*, eds. Abbot J.A. & Kurogi M., pp. 117–30. Japanese Society of Phycology, Kobe, Japan.

SCHREIBER E. (1930) Untersuchungen über Parthenogenesis, Geschlechtsbestimmung und Bastardierungsvermögen bei Laminarien. *Planta* 12, 331–53.

DE SILVA M.W.R.N. & BURROWS E.M. (1973) An experimental assessment of the status of the species *Enteromorpha intestinalis* (L.) Link and *Enteromorpha compressa* (L.) *J. mar. biol. Ass. U.K.* **53**, 895–904.

SUNDENE O. (1958) Interfertility between forms of *Laminaria digitata. Nytt Mag. Bot.* **6**, 121–6.

SVENDSEN P. & KAIN J.A. (1971) The taxonomic status, distribution, and morphology of *Laminaria cucullata* Sensu Jorde and Klavestad. *Sarsia*, **46**, 1–21.

WADDINGTON C.H. (1969) Gene regulation in higher cells. *Science*, **166**, 639–40.

WIK-SJØSTEDT A. (1970) Cytogenetic investigations in *Cladophora. Hereditas*, **66**, 233–62.

CHAPTER 12

APPROACHES TO THE
GENETICS OF *ACETABULARIA*

BEVERLEY R. GREEN

Botany Department, University of British Columbia,
Vancouver, B.C. V6T 1W5, Canada.

1 Introduction   236

2 Life-cycle and sexuality   236

3 Why have no mutants of *Acetabu-laria* been reported?   244

4 Gene expression: classical experiments   246

5 Gene expression: modern experiments and concepts   247

6 How autonomous are *Acetabularia* chloroplasts?   250

7 References   253

1 INTRODUCTION

Acetabularia is a member of the Dasycladales, the order of siphonaceous green algae which remain unicellular and uninucleate throughout most of their life-cycle, even while differentiating elaborate reproductive structures (Valet 1968, 1969). Members of at least three genera, *Acetabularia*, *Batophora* and *Neomeris*, are capable of regenerating a whole plant from a small nucleate portion, while even an anucleate portion can continue to grow and to develop a sterile reproductive structure (Hämmerling 1934; Dao 1958; Puiseux-Dao 1962). *Acetabularia* has been extensively used for studies of differentiation and nucleocytoplasmic interactions, beginning with the work of Hämmerling (1934). Much of the earlier work has been reviewed by Hämmerling (1953, 1963) and by others (Richter 1962; Brachet 1965; Werz 1965; Puiseux-Dao 1970).

2 LIFE-CYCLE AND SEXUALITY

The life-cycle of *Acetabularia* is probably familiar to most readers of this book (see Fig. 12.1). In spite of the large amount of work which has been done on it, however, we still do not know some of the basic facts of its life history, in particular:

(1) when meiosis occurs.
(2) when DNA replicates in the cyst.
(3) when DNA replicates in the primary nucleus.
(4) the time and nature of the chromosome divisions which precede the breakdown of the primary nucleus to form 1,000 or more small secondary nuclei.

It is therefore not surprising that the genetics of *Acetabularia* is largely unexplored.

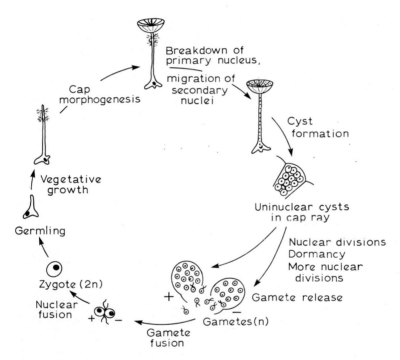

Fig. 12.1.   Life-cycle of *Acetabularia mediterranea*. Not drawn to scale. Gametes are about 10 μm diameter; mature cell is 25–30 mm long with a cap of diameter 6–10 mm; cysts are 100–200 μm diameter.

If an organism is to be studied genetically, it must be possible to make controlled crosses. In *Acetabularia*, the cyst is the only stage which can be manipulated easily, as the gametes are short-lived and apparently cannot be cultured further to give rise to haploid clones. For this reason, it is important to know whether meiosis occurs before or after cyst formation. Hämmerling (1934, 1944) reported that gametes are of two mating types and showed that all the gametes from any one cyst are of one or the other type. This implies that meiosis occurs *before* cyst formation. However, he subsequently stated (1963) that meiosis occurs *after* cyst formation, as suggested by the cytological observations of Schulze (1939). If the original single cyst nucleus were diploid, and if meiosis took place within the cyst, gametes of *both* mating types would be produced. This would mean that gametes from the same cyst could fuse, and the high probability of selfing would complicate analyses of genetic crosses.

A very simple test showed that cysts do indeed release gametes which cannot fuse with each other (Green 1973). A number of tubes were set up containing one cyst in several ml of sea-water medium. Control tubes had 2, 4, 30 or more cysts. After illumination for a suitable period, they were examined for evidence of growth. The data are given in Table 12.1. In three separate experiments using *A. mediterranea*, all but one tube out of the 240 containing a single cyst were completely lacking in germlings. (The single tube containing germlings had such a large number that it probably resulted from an experimental error.) In tubes with two cysts, where the probability of the cysts being of opposite mating-type is 1/2, 11 out of 40 tubes contained germlings. This is the proportion expected if 75% of the cysts released gametes within a certain time period, a figure actually higher than would be predicted from published observations (Woodcock & Miller 1973a). Tubes with more than two cysts produced proportionately more germlings. Similar results have been obtained with *A. (Polyphysa) cliftonii* (Green unpublished).

Table 12.1. Fraction of tubes containing germlings among tubes with various numbers of cysts (Green 1973).

Expt. No.	Number of cysts/tube				
	1	2	4	12	30
1	0/60	—	—	—	5/5
2	1/100	—	3/5	9/10	5/5
3	0/80	11/40	10/16	—	1/1

The only model consistent with these data is one in which meiosis, with consequent segregation of the sex loci, occurs before the enclosure of the single secondary nucleus by the cyst wall, with a number of subsequent mitotic divisions occurring before and after the dormant period (Woodcock & Miller 1973a). In other words, all gametes are haploid, and all the gametes from any one cyst are of the same mating type. The results in Table 12.1 do not agree with Puiseux-Dao's suggestion (1970) that the gametes could be diploid and the zygote tetraploid.

Koop (1975) has shown by quantitative microspectrophotometry that a

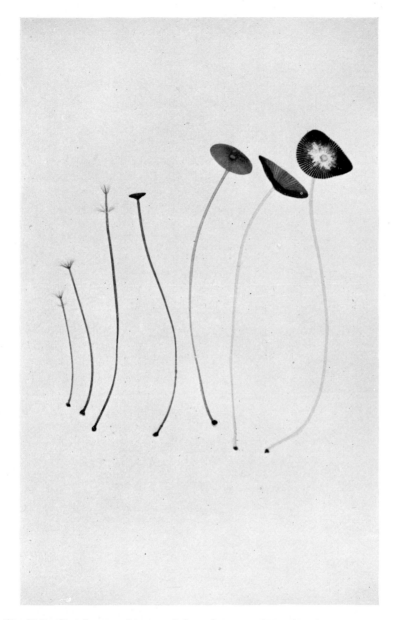

Fig. 12.2. Developmental stages of *A. mediterranea*. (Magnification about ×2.2.)

Fig. 12.3. Fully grown cells of *A. major* and *A. mediterranea*. (Magnification about ×1.6.) (Photographs courtesy of Dr Sigrid Berger, Max-Planck-Institut für Zellbiologie, Wilhelmshaven.)

cyst nucleus contains the same amount of DNA as a gamete nucleus, and half the amount of a zygote nucleus. This confirms my conclusion that meiosis occurs before cyst formation. He also showed that most of the secondary nuclei in the rhizoid, newly produced from the fragmenting primary nucleus, contained the haploid amount of DNA, although 15–20% had diploid or intermediate amounts. He concluded that meiosis probably occurs during the breakdown of the primary nucleus.

There has been some controversy over the possibility that the biflagellate swarmers referred to as gametes are actually diploid zoospores which lose their flagella and germinate directly. Arguments in favour of this have been presented by Puiseux-Dao (Dao 1957; Puiseux-Dao 1962, 1970), and refuted by Hämmerling (1964). Unfortunately, no quantitative criteria were given for distinguishing zoospores from gametes, and the results were presented in a descriptive fashion. Valet (1969) claimed to have observed both gametes and zoospores in *A. clavata* and *A. exigua* and quoted different size ranges for each, but the cell shapes of both were variable and the size ranges overlapped to a considerable extent. The criteria for distinguishing zoospores from gametes have not been well defined for any species of *Acetabularia*.*

Woodcock and Miller (1973a) have studied in detail the ultrastructural features of cyst development. After the cyst wall has formed, the single nucleus divides several times, with a reduction in nuclear diameter from 15 to 5 μm. No further changes occur in the light, but after a period of storage in cold and dark conditions, which seems to be necessary to break dormancy, an increased number of nuclei are observed, implying that more divisions have occurred. Since the nuclear divisions are asynchronous, it has not been possible to quantitate this process (Puiseux-Dao personal communication). Gametogenesis itself is very rapid, and because gamete release cannot be predicted more than a few hours in advance it has not been studied.† However, it probably involves more nuclear divisions, since on the average some 1,800

---

* Puiseux-Dao now suggests (personal communication) that some of her early results which indicated asexual reproduction may have been due to the limiting conditions (pure sea-water, low light) under which the algae were cultivated at that time. In addition, the strain of *A. mediterranea* which she used came from a brackish pond, and did not grow as well as those from the Bay of Naples from which most of the current laboratory cultures are descended, including those used in my experiments. It may well be that asexual reproduction, either by means of zoospores or by direct germination of cysts, can be induced in certain strains of *Acetabularia* under unbalanced conditions.

†*Note added in press:* H.-U. Koop (*Protoplasma*, in press 1975b) has confirmed Woodcock and Miller's electron microscope studies (1973a and b) using Feulgen staining. He found that the first mitotic division of the secondary nuclei occurred while the cyst wall was still forming; the nuclear membrane appeared to remain intact. The second division was synchronous within any one cyst and between neighbouring cysts. After four days, a cyst contained an average of 16 nuclei; after the dormant period, several hundred. Using his method of inducing fairly synchronous gamete release under controlled conditions (1975a) he was able to show that more mitotic divisions occur during a short time period 16–24 hours after the beginning of induction, just prior to gamete formation.

gametes are released per cyst of *A. mediterranea* (Schweiger *et al.* 1974), which implies at least ten mitotic divisions between cyst formation and gamete release. There have apparently been no attempts to study DNA synthesis in the cyst.

It is not at all clear what happens in the primary nucleus between the time when the gamete nuclei fuse in the zygote (Crawley 1966) and the time when the primary nucleus starts to break down into secondary nuclei. The primary nucleus can be seen to increase in diameter from 3 to 100 μm during vegetative growth, i.e. to exhibit more than a thousand-fold increase in volume. It does not stain with the Feulgen reagent once the cell is a few millimetres long (Puiseux-Dao 1962; de Vitry 1965), so it has not been possible to determine the amount of DNA at various stages of growth. No chromosomes have been seen by any staining method. The only attempt to study DNA replication was made by de Vitry (1965), who supplied cells with tritiated thymidine and measured its incorporation. She found that a large fraction of the thymidine label ended up in a RNase-sensitive fraction, indicating either conversion of the labelled compound or impurity of the ribonuclease, but she also detected some DNase-sensitive incorporation in the middle of the vegetative phase.

It is tempting to assume that an increase in DNA content, resulting from polyploidization (endomitosis), parallels the increase in nuclear volume during vegetative growth. However, a number of treatments which diminish metabolic rate or inhibit RNA or protein synthesis cause a marked decrease in nuclear volume (for references, see Puiseux-Dao 1970, Hämmerling 1963). This suggests that a large part of the nuclear volume is attributable to newly synthesized RNA. The changing morphology of the nucleolus has also been used as an argument in favour of polyploidization. The nucleolus is small and round in the zygote and young germling, but becomes large and sausage-shaped in the main stage of vegetative growth (Schulze 1939; Puiseux-Dao 1962). This suggests fusion of multiple nucleoli following endomitosis (Puiseux-Dao 1966); but it could also indicate the presence of multiple nucleolus-organizer regions in a diploid chromosome complement.

Trendelenburg *et al.* (1974) isolated the nucleoli of *A. mediterranea* by micromanipulation and observed that they seem to be aggregates of spherical subunits about ten μm in diameter, which fell apart on isolation. The isolated nucleoli were spread out for electron microscopy by the method of Miller and Beatty (1969), which allows the visualization of RNA molecules still attached to the DNA from which they are being transcribed. Typical brush-like regions with small RNA molecules at one end and long RNA molecules at the other were obtained from *Acetabularia* nucleoli. From the size of these transcriptional units and the number of nucleolar subunits per nucleolus (about 100), Trendelenburg *et al.* calculated that a nucleolus contains about 13,000 rRNA cistrons. They interpreted their results as indicating rDNA amplification, i.e. preferential replication of the DNA coding for ribosomal

RNA. In view of the large amount of ribosomal synthesis required by a cell which grows to 3 cm in length and differentiates a cap 1 cm in diameter before cell division, this is a very plausible hypothesis. On the other hand, polyploidization of the entire genome during vegetative growth might be necessary to supply all types of RNA for such a large cell.

Several weeks after cap formation is completed, the primary nucleus breaks down, budding off large numbers of small secondary nuclei, which are carried up to the cap rays by cytoplasmic streaming (Schulze 1939). The ultrastructural changes of the nuclei during this process have been described by Boulouckhère (1970) and Woodcock and Miller (1973b). Neither ultrastructural nor light-microscopic techniques showed chromosome divisions in the primary nucleus before or during breakdown (Schulze 1939; Puiseux-Dao 1962). However, the latter authors both reported division figures in the migrating secondary nuclei. They concluded that these were mitotic figures; but the chromosomes of *Acetabularia* are small (Puiseux-Dao 1966), so some of the secondary nuclei could be undergoing meiosis while migrating to the cap. On the other hand, meiosis could occur within the primary nucleus, before or during its breakdown (see footnote †, p. 241).

The large number of secondary nuclei produced in a short period of time suggests prior replication of the chromosome complement. Working backwards from the number of cysts formed (with one secondary nucleus each), it is possible to calculate the approximate number of chromosome replications which have occurred. For *A. mediterranea*, Loni and Bonotto (1971) reported a range of 2,287–15,774 and Schweiger *et al.* (1974) reported *ca.* 12,000 cysts per cap. The number of cysts formed seems to depend on the size of the cap and the state of the plant, but these figures nevertheless show that the original diploid zygote nucleus, at some time in its growth, must have undergone 12 or 13 mitoses and one reduction division.

However, there is no particular reason why 12–13 rounds of DNA synthesis should not occur in a limited time period just before nuclear breakdown. Vegetative cells of another green alga, *Chlamydomonas reinhardtii*, can replicate their haploid DNA complement twice in a four-hour period (Kates *et al.* 1968). Although the diploid genome size of *Acetabularia* may be larger and the metabolic rate lower, there is a lag of several weeks between the time the cap reaches its maximum size and the time the primary nucleus starts to break down. Hämmerling (1953) showed that the initiation of nuclear breakdown was dependent on the presence of a mature cap. Breakdown could be delayed indefinitely by repeated amputations of the cap. Perhaps a message transmitted from the mature cap to the primary nucleus triggers the rounds of DNA synthesis needed before the nucleus breaks down.

In summary, the existence of haploid cysts makes it theoretically possible to do genetic studies on *Acetabularia*. There is an urgent need for more work on the nuclear events in the life-cycle. It would be very profitable to study DNA synthesis using modern autoradiographic techniques. This might give some indication of when chromosomal replications occur, and might help to define the meiotic period in the cell cycle.

3 WHY HAVE NO MUTANTS OF *ACETABULARIA* BEEN REPORTED?

No mutants of any species of *Acetabularia* have been isolated. All the 'genetics' has been done with interspecific grafts. The lack of mutants, particularly morphological ones, has probably been a major factor in the slow progress of research on this fascinating organism, which would seem at first glance to be an obvious choice as a model system for the study of morphogenesis at the biochemical level.

A thoughtful inspection of the life-cycle (Fig. 12.1) provides some insight into the difficulties facing a hopeful geneticist. The long life-cycle time, six to nine months under optimum conditions, is enough to discourage even an experienced investigator. The second obstacle is the fact that during most of its life-cycle the cell is at best diploid, and, at worst, extensively polyploid. The haploid cysts are simply a resistant resting stage. The haploid gametes die within 48 hours if they do not fuse. In addition, *Acetabularia* has never been reported to grow heterotrophically, although in sea-water medium cells can survive a month or more in the dark. This would make it difficult to isolate nutritional or photosynthetic mutants.

It is evident that *Acetabularia* is not the organism of choice for studying genetics *per se*. However, if mutants could be isolated, they would provide excellent tools for studying morphogenesis or chloroplast autonomy. The latter is of particular importance because of the ability of this organism to photosynthesize, grow, and differentiate at a normal rate for a period of weeks after removal of the nucleus (Hämmerling 1963; Shephard 1965b, 1970).

Some types of mutants it would be useful to have are:

(1) those having a clear abnormality in some stage of development.

(2) those affecting the process of photosynthesis.

(3) those having resistance or sensitivity to organelle-specific inhibitors such as chloramphenicol, streptomycin, spectinomycin, lincomycin or rifamycin. This type has been particularly useful in studies of chloroplast autonomy in *Chlamydomonas* (Harris *et al*. Chapter 6).

Mutations which affect morphogenesis are at once the most logical ones to look for, considering the experimental history of *Acetabularia*, and the most difficult to pin down. In order to be expressed in the first generation, they would have to be dominant. They would also have to permit normal cyst formation. Aside from the logistic difficulties of screening thousands of cells, there is the problem that most of the developmental characteristics such as cap shape, number of rays, etc., are very variable in expression even in normal cultures (Loni & Bonotto 1971; Janowski & Bonotto 1969; Bonotto & Janowski 1969). There is also considerable variation within any culture in growth rate, percentage of cells forming caps, and percentage of cells forming abnormal caps (Bonotto 1970; Bonotto & Janowski 1969; Bonotto *et al.* 1971a; Green unpublished). Most of the cells forming abnormal caps will regenerate a perfectly normal one if the first cap is amputated before the nucleus starts to divide. Attempts to reduce variation in the population by cloning from one cap have not produced any increase in uniformity after two or three generations (Green unpublished). This may be because our *A. mediterranea* has been in cultivation in Brachet's laboratory for over 15 years, and a lot of inadvertent selection has already been done.

In an attempt to isolate developmental mutants (Green unpublished), cysts of *A. mediterranea* were treated with various concentrations of N-methyl nitrosoguanidine. A small number of the cysts treated with 50 μg/ml survived to release gametes, and the progeny were observed at various stages. Many appeared to be normal. Some grew to the length of a few millimetres and then stopped, i.e. they acted like enucleated basal stalk fragments. Most of the cells which survived to the 10-mm stage formed normal caps. Out of 37 cells with abnormal caps which were decapitated and allowed to regenerate, 24 regenerated normal caps, four regenerated abnormal caps, and the rest didn't regenerate or died. A second decapitation of the four plants with abnormal caps killed three of the cells, and the remaining one produced a normal cap. Similar results were obtained with plants with abnormal caps from control cultures.

Only a few thousand cells were screened in this experiment, a number which is probably totally inadequate. A pre-selection technique cannot be applied to very young cells when one seeks morphological mutants. Raising to maturity the tens of thousands of cells needed would present logistical difficulties. However, there is no reason why a laboratory equipped to work on this scale could not undertake to isolate this type of morphological mutant.

The problem of photosynthetic mutants is a difficult one. *Acetabularia* is generally assumed to be a photoautotroph, although it has never been thoroughly tested for ability to grow heterotrophically in the dark or mixotrophically in the light because of the usual presence of bacterial contaminants. Now that several laboratories have been successful in raising axenic cultures by the methods of Gibor and Izawa (1963) and Shephard (1970), the question should be reinvestigated.

A hunt for mutants resistant to organelle-specific inhibitors might prove rewarding. It should be possible to screen for resistance at the zygote or small-germling stage, when 10^4 cells can be screened in a few test tubes. One should keep in mind, of course, that a mature *Acetabularia* cell has 10^6-10^7 chloroplasts (Shephard 1965); the mutagenic agent should be applied at a stage where there are only one or two chloroplasts, i.e. to the gamete or the

M

zygote. Of course, any resistant mutant could be altered either in its permeability or in a specific chloroplast or mitochondrial process. Such a strain might still be useful for workers studying the electrophysiology of giant cells, as well as being useful as a genetic marker.

4 GENE EXPRESSION: CLASSICAL EXPERIMENTS

Acetabularia was extremely important in early work showing the importance of the nucleus in the control of development. This important role was due to its unique ability to continue growth after removal of the nucleus, and to tolerate micrurgy leading to both intra- and inter-specific grafts. Most of the pioneering work by J. Hämmerling and his collaborators has been extensively reviewed (Hämmerling 1953; Werz 1965; Puiseux-Dao 1970). In brief, Hämmerling's group established that:

(1) 'Morphogenetic substances' involved in cap formation are produced by the nucleus and stored in the cytoplasm, explaining the ability of cells, enucleated after they have reached about one-third of their final length, to continue growing and in some cases to differentiate a cap.

(2) There is an apico-basal gradient of these substances, as shown by the greater cap-forming ability of apical sections as compared to mid-sections.

(3) The morphogenetic substances are continuously produced by the nucleus, as shown by the ability of a nucleated portion of the cell to regenerate several times, and to form caps after repeated amputations.

(4) Some of the morphogenetic substances determine the species-specific form of the cap. Interspecific grafts first form caps intermediate between the two species. If the intermediate cap is removed, the cell then forms a cap with the characteristics of the nuclear donor. This shows that the postulated morphogenetic substances, continuously produced by the nucleus, are used up during cap formation and that there is an interaction between those already stored in the cytoplasm and new ones elaborated by the grafted nucleus.

(5) There is a morphogenetic substance, evidently not species-specific, which is necessary for triggering cap formation.

Students brought up on the Central Dogma of DNA → messenger RNA → protein should not forget the importance of Hämmerling's early work in establishing the dominant role of the nucleus in the control of morphogenesis. As pointed out by J. Brachet (1957), Hämmerling 'had concluded, as early as 1934, that the morphogenetic capability of an anucleate part is determined by the amount of nucleus-dependent morphogenetic substances in it'. Hämmerling's indirect evidence for 'morphogenetic substances' carrying information from nucleus to cytoplasm pre-dated the discovery of messenger RNA by at least 25 years.

5 GENE EXPRESSION: MODERN EXPERIMENTS AND CONCEPTS

The modern interpretation of Hämmerling's discoveries is that his postulated morphogenetic substances are long-lived messenger RNAs, similar to the globin mRNA of erythrocytes. The parallels between an *Acetabularia* cell and an (anucleate) erythrocyte were pointed out by Brachet (1957) before the existence of mRNAs was more than a vague hypothesis. However, the erythrocyte is a specialized cell making large quantities of one protein, so it is not completely analogous to the *Acetabularia* system. An *Acetabularia* cell would need mRNAs for a fairly large number of proteins which continue to be synthesized in anucleate cells (see below), including those involved in the differentiation of the cap. The lack of a predominant mRNA is probably one reason why no mRNAs have yet been isolated from this organism.

There is a more practical reason why no mRNAs have been isolated. *Acetabularia* contains several very active ribonucleases (Schweiger 1966). It has been necessary to resort to such measures as homogenizing the cells at liquid-N_2 temperature before attempting to extract the RNA (Dillard 1970). There has been considerable controversy over the various RNAs isolated from *Acetabularia*, even the chloroplast ribosomal RNAs (Woodcock & Bogorad 1970a; Janowski *et al.* 1969; Farber 1969). One of the problems with *Acetabularia* is that most of the cell volume is occupied by a vacuole, the contents of which have a very low pH and a very high Mg^{++} concentration (Janowski 1966). The cytoplasm consists largely of chloroplasts. The isolation of cytoplasmic 80S ribosomes from *A. mediterranea* and *A. major* has only recently been reported (Kloppstech & Schweiger 1973a). The yields from normal cells were very low although yields from regenerating nucleated cells were somewhat higher (Kloppstech & Schweiger 1973b). Using *A. major*, which has a lower ribonuclease content than the other species (Schweiger *et al.* 1974), they showed that the 80S particles contained 18S and 26·5S RNA, i.e. they were of the correct size to be from eukaryotic cytoplasmic ribosomes. If polyribosomes could be isolated it might be possible to isolate mRNA from them, but even detergent extraction, which should have released membrane-bound polysomes, did not give satisfactory results (Kloppstech & Schweiger 1973b).

In a very recent report, Kloppstech (1975) claims to have detected RNA-containing polyadenylate tails in both nucleate and anucleate *A. mediterranea*. Since mRNA in many eukaryotic systems has a poly-(A) tail at the 3'-end (Brawerman 1974), this is presumptive evidence for mRNA. This material has been shown to increase in amount per cell with age, and to increase in regenerating nucleated cells. However, it has so far been characterized only by its ability to bind to H^3-poly(U) on a filter. One must therefore conclude that there is still no direct biochemical evidence that the 'morphogenetic substances' are RNAs.

Some indirect evidence has been obtained from studies of the effect of ultraviolet irradiation on morphogenetic processes, taking advantage of the fact that nucleic acids (both RNA and DNA) have an absorption maximum at 258 nm, with the extinction coefficient falling off to zero at 310 nm. Six (1956) showed that the growth and development of enucleated stalk fragments were strongly inhibited when these were irradiated with monochromatic light of 254 nm, were less inhibited by light of 281 nm, and were hardly inhibited at all by light of 297 nm. (Doses of equal total energy were given in each case.) It could be argued that the light of 254 nm was inactivating ribosomes or some other component of the cytoplasm necessary for growth, and producing an indirect effect on the expression of the morphogenetic substances. In order to test this more directly, Werz (1965) grafted *A. mediterranea* rhizoids (nucleated) on to stalks of *A. crenulata* which had been treated with various doses of 254 nm radiation. The normal result of a graft between these two species is the production of an intermediate-type cap, resulting from the simultaneous presence of morphogenetic substances stored in the stalk and those newly synthesized by the new nucleus. Werz found that increasing doses of 254 nm light resulted in an increasing percentage of *mediterranea*-type caps, indicating the destruction of the stored *crenulata*-type substances.

A number of enzymes have been shown to increase in activity in enucleated cells for periods of up to four weeks after enucleation (Keck and Clauss 1958; Triplett *et al.* 1965; Clauss 1959; Schweiger 1966; Zetsche 1966, 1968; Schweiger *et al.* 1967a; Reuter & Schweiger 1969). However, only two of these enzymes, malic and lactic dehydrogenase, have been shown to be coded for by the nucleus (Schweiger *et al.* 1967a; Reuter & Schweiger 1969). Two other enzymes, UDP-glucose-4-epimerase and UDPG-pyrophosphorylase, increase markedly in activity just before cap formation in both nucleate and anucleate cells (Zetsche 1966, 1968), and this increase is inhibited by puromycin and cycloheximide. Since cap formation is known to be under the control of the nucleus, presumably these enzymes are coded for by the nucleus and their synthesis is regulated at the translational level. What is really needed is an experiment to determine not only whether an enzyme which is definitely coded for by the nucleus increases in activity in an anucleate cell, but also whether new enzyme molecules are synthesized. The best evidence comes from experimental comparisons of the buoyant densities of malate dehydrogenase extracted from cells grown in media prepared with D_2O and H_2O respectively. In preparations from both nucleate and anucleate cells the D_2O caused a density shift of one fraction, indicating that new molecules had been synthesized.

Much of the evidence for long-lived mRNAs is based on the action of inhibitors on nucleate and anucleate cells. This evidence has been reviewed in detail elsewhere (Puiseux-Dao 1970; Brachet 1967; Schweiger 1969) and only critical points will be dealt with here. Readers interested in pursuing the matter should keep in mind that it was not until late in the 1960s that it was realized that most of the RNA and protein synthesis in anucleate cells occurs in the chloroplast (Schweiger *et al.* 1967b). It should further be remembered that many antibiotics which affect protein synthesis have an equally strong effect on the transport of nucleic acid and protein precursors into the cell (Bonotto *et al.* 1969).

Brachet *et al.* (1964) showed that puromycin, a protein-synthesis inhibitor, inhibits growth and cap formation in both nucleate and anucleate cells. This implies that *de novo* protein synthesis is needed for cap formation, and that the necessary messenger RNAs must have been stored in the cytoplasm before enucleation. This appears to rule out the possibility that proteins are presynthesized and stored in the cytoplasm in an inactive state, since in that case protein synthesis would not be necessary for cap formation. However, as the same group of workers showed a few years later (Bonotto *et al.* 1969), puromycin at the level used (20 μg/ml) inhibits both amino-acid uptake and uridine uptake, and could thereby exert its effect on a variety of other processes.

Actinomycin D, an inhibitor of RNA synthesis, inhibits morphogenesis of the cap in both intact and anucleate cells (Brachet *et al.* 1964; Zetsche 1964). This observation has been used to support the hypothesis that the morphogenetic substances are messenger RNAs. However, it has recently been shown that this drug preferentially inhibits ribosomal RNA synthesis in animal systems (see Brawerman 1974), and if this were true in the *Acetabularia* system it could mean simply that translation was being inhibited. Oddly enough, cap formation is much less inhibited in anucleate cells than in nucleate cells. If actinomycin acts on mRNA synthesis, treated nucleate and anucleate cells of the same age should have the same amount of mRNA stored in their stalks and should not be dependent on new synthesis. If actinomycin acts to decrease the number of ribosomes synthesized, it should not have any effect on anucleate cells and should affect nucleate cells only if ribosome turnover is significant. The experimental results imply that there is a turnover of messenger or ribosomal RNA in nucleate cells, but not in anucleate cells.

Rifampicin, an inhibitor of prokaryotic-type DNA-dependent RNA polymerase (Hartmann *et al.* 1967), has been shown to inhibit chloroplast RNA synthesis but not nuclear RNA synthesis in *Acetabularia* at concentrations up to 10 μg/ml (Brändle & Zetsche 1971). Cap formation and increase in UDPG-pyrophosphorylase activity are only slightly inhibited, whereas chlorophyll synthesis is more strongly inhibited. This indicates that UDPG-pyrophosphorylase is coded for by the nucleus, and suggests that the slight inhibition observed is a secondary effect due to inhibition of photosynthesis. On the other hand, Puiseux-Dao *et al.* (1972) found that rifampicin concentrations as low as 5 μg/ml inhibit cap formation in nucleate cells, but not in anucleate cells, which were inhibited only by concentrations exceeding 20 μg/ml.

Judging by the time taken for cap formation, it is possible that the two groups of workers used cells of slightly different ages, but this should not have affected the results to such an extent. Thanks to the careful work of Brändle and Zetsche (1971), we know that, at concentrations up to 10 µg/ml, rifampicin is a specific inhibitor of chloroplast-RNA synthesis, and it would be worthwhile reinvestigating this discrepancy.

A possible explanation of the problems of long-lived messenger RNAs has been suggested by K. M. Downey (personal communication), who discovered an RNA-dependent RNA polymerase in (anucleate) reticulocytes (Downey *et al.* 1973). The action of this enzyme is inhibited by certain rifamycins but not by others, and it is not inhibited by actinomycin D. If such an enzyme were present in *Acetabularia*, the messenger RNAs could be continually synthesized in the cytoplasm, which would explain their apparent longevity. A similar proposal, on the basis of indirect evidence, was made by Bonotto *et al.* (1971b). Turnover of mRNA in nucleate cells could be due to a specific ribonuclease which decreases in activity once the cell is enucleated. These hypotheses may be difficult to test because of interference from the ribonucleases mentioned above, but they should be considered seriously.

6 HOW AUTONOMOUS ARE *ACETABULARIA* CHLOROPLASTS?

The fact that *Acetabularia* can grow and differentiate in the absence of a nucleus suggests that its chloroplasts may have more genetic autonomy than those of other organisms. Isolated chloroplasts can incorporate RNA precursors into high molecular-weight RNA (Schweiger & Berger 1964; Berger 1967). They can also incorporate amino acids into proteins (Goffeau & Brachet 1965; Goffeau 1969; Ceron & Johnson 1971; Apel & Schweiger 1972), and some of these have the same electrophoretic mobility as proteins synthesized in intact cells. There is indirect evidence that chloroplasts in anucleate cells can synthesize DNA, both from autoradiographic studies of ^{3}H-thymidine incorporation (Shephard 1965b; de Vitry 1965) and from the observed increase in the DNA of anucleate fragments over a period of several weeks (Heilporn-Pohl & Brachet 1966).

It therefore appears that *Acetabularia* chloroplasts can carry out all the necessary steps for replication and utilization of genetic information. However, the degree of their autonomy can be large only if there is enough DNA in the chloroplasts to code for all their components and the enzymes to synthesize them. The first estimates of DNA content suggested there was only 10^{-16} g per chloroplast, or 6×10^7 daltons (Gibor & Izawa 1963). This is about the size of a bacteriophage genome, and is not nearly enough for any sort of autonomy. However, chloroplasts lysed by osmotic shock and spread by the Kleinschmidt technique (1968) release relatively large masses of DNA (Green & Burton 1970; Woodcock & Bogorad 1970b). Although the DNA released in this way is too tangled to be measured, the amount released from a single plastid appears to be about the same as that released from a typical bacterium (MacHattie *et al.* 1965).

Analytical determinations of the amount of DNA per chloroplast, using a diphenylamine assay, confirmed this impression (Green unpublished). *A. mediterranea* contains $2 \cdot 3 \times 10^{-15}$ g or $1 \cdot 4 \times 10^{9}$ daltons per chloroplast, while *A. cliftonii* contains $7 \cdot 9 \times 10^{-15}$ g or $4 \cdot 7 \times 10^{9}$ daltons per chloroplast. These amounts correspond to 700 and 2450 μm of DNA, respectively. Unfortunately, it has been impossible to isolate intact chloroplast 'chromosomes', even by lysing chloroplasts with detergent and spreading the lysate immediately for electron microscopy. The molecules measured had lengths up to 200 μm, although most of them were below 80 μm, probably fragmented as a result of shear (Green unpublished).

In order to determine the unique genome size, or actual information content of the chloroplast DNA, its kinetic complexity was determined by the renaturation kinetics technique of Wetmur and Davidson (1968). Chloroplast DNA of *A. cliftonii* was found to have a kinetic complexity of approximately $1 \cdot 5 \times 10^{9}$ daltons (Padmanabhan & Green unpublished). This means that the average chloroplast of *A. cliftonii* contains three copies of a very large genome. In a control experiment, we confirmed that *Chlamydomonas reinhardtii* chloroplast DNA has a genome size of $0 \cdot 2 \times 10^{9}$ daltons (Bastia *et al.* 1971; Wells & Sager 1971), i.e. it is only one-seventh the size of the *A. cliftonii* chloroplast genome. Renaturation experiments on the chloroplast DNA of *A. mediterranea* indicated a genome size of about $1 \cdot 1 \times 10^{9}$ daltons, or about one genome per chloroplast (Muir 1974).

All this would indicate that the *Acetabularia* chloroplast has the potential of being completely autonomous, since its genome size is as large as that of many kinds of bacteria. It is tempting to speculate that it is an evolutionary relic from the days when chloroplasts were originating by progressive evolution of an endosymbiotic prokaryote (e.g. Margulis 1970). As with all nice theories, however, there are some awkward data which are very hard to explain.

In spite of the apparent potential of the *Acetabularia* chloroplast genome, it is now clear that some chloroplast proteins are either coded for directly by the nucleus or are so tightly under nuclear control that they appear to be coded for by the nucleus and synthesized in the cytoplasm. Apel and Schweiger (1972) showed that three out of 11 chloroplast-membrane protein fractions were species-specific, and changed after nuclear transplantation to resemble those of the nucleus-donor species. Kloppstech and Schweiger (1973c) reported that several proteins of the larger subunit of the chloroplast ribosome were also species-specific. These investigators allowed graft-hybrid cells to form cysts and then tested the progeny. No specific biochemical features of the cytoplasmic parent remained. The inter-specific nuclear transplantation method provides valuable information when there is a detectable difference in electrophoretic mobility between the proteins of the two species, but not with most of the ribosomal proteins, where they are apparently identical. (Only four or five of these proteins were different, and one of those gave questionable results.)

Another approach to the problem of chloroplast autonomy involves the use of inhibitors which are specific for either organellar (prokaryotic-type) protein synthesis or cytoplasmic protein synthesis. Unfortunately, such inhibitors are not always as specific as they are supposed to be, and results obtained with them should be interpreted with caution, particularly if there is any possibility of interaction between the two systems in the intact cell. For example, Apel and Schweiger (1973) showed that chloramphenicol inhibits no more than 60% of amino-acid incorporation into isolated intact chloroplasts, whereas cycloheximide (supposedly specific for 80s cytoplasmic ribosomes) likewise inhibits up to 60%. Both inhibitors at high concentrations are necessary to achieve 90% inhibition.

The chloroplasts in their experiments appeared clean, and there was no evidence of significant contamination by endoplasmic reticulum. Two of the three protein fractions of chloroplast membranes could be labelled *in vitro*, indicating synthesis by the chloroplast protein-synthesizing apparatus. However, the synthesis of one of these fractions was strongly inhibited by as little as 1 μg/ml cycloheximide. The synthesis of both fractions was inhibited by chloramphenicol, although the cycloheximide-sensitive fraction was less inhibited than the other fraction. These workers found a 26s RNA in the chloroplasts, indicating the presence of 80s ribosomes. They interpret their results to mean that chloroplasts contain two classes of ribosomes, both 70s and 80s, with different antibiotic sensitivities. If chloroplasts in general contain two classes of ribosomes, this raises serious questions about the validity of many of the inhibitor studies done to date.

It appears that *Acetabularia* chloroplasts, although they seem to have the potential for complete autonomy, are at least partly dependent on the nuclear genome. Perhaps there is nuclear repression of certain genes carried by the chloroplast, and these are only activated, if at all, after enucleation. Perhaps the chloroplast and cytoplasmic translation mechanisms are interdependent, as proposed by Ellis *et al.* (1973). By studying protein synthesis in isolated chloroplasts, under conditions where completed protein chains are made, and by careful identification of the products of such syntheses, we may hope to obtain a clearer picture of what the chloroplast genome is coding for in *Acetabularia*.

ACKNOWLEDGEMENTS

I should like to thank Dr Klaus Apel and Dr E. Ann Frazier for critical and helpful reading of the manuscript, and Dr H.-U. Koop for sending me his manuscript prior to publication.

7 REFERENCES

APEL K. & SCHWEIGER H.G. (1972) Nuclear dependency of chloroplast proteins in *Acetabularia*. *Eur. J. Bioch.* **25**, 229–38.

APEL K. & SCHWEIGER H.G. (1973) Sites of synthesis of chloroplast membrane proteins. Evidence for three types of ribosomes engaged in chloroplast protein synthesis. *Eur. J. Bioch.* **38**, 373–83.

BASTIA C., CHIANG K.-S., SWIFT H. & SIERSMA P. (1971) Heterogeneity, complexity, and repetition of the chloroplast DNA of *Chlamydomonas reinhardtii*. *Proc. Nat. Acad. Sci.* (*U.S.*) **68**, 1157–61.

BERGER S. (1967) RNA synthesis in *Acetabularia*. II. RNA synthesis in isolated chloroplasts. *Protoplasma*, **64**, 13–25.

BOLOUCKHÈRE M. (1970) Ultrastructure of *Acetabularia mediterranea* in the course of formation of secondary nuclei. In *Biology of Acetabularia*, eds. Brachet J. & Bonotto S., pp. 145–76. Academic Press, New York.

BONOTTO S. (1970) Morphogenèse d'*Acetabularia mediterranea* dans la mer et en laboratoire. *Bull. Soc. Roy. Bot. Belg.* **103**, 213–23.

BONOTTO S. & JANOWSKI M. (1969) Quelques observations sur la forme du chapeau d'*Acetabularia mediterranea*. *Bull. Soc. Roy. Bot. Belg.* **102**, 257–65.

BONOTTO S., GOFFEAU A., JANOWSKI M., VANDEN DRIESSCHE T. & BRACHET J. (1969) Effects of various inhibitors of protein synthesis on *Acetabularia mediterranea*. *Biochim. Biophys. Acta*, **174**, 704–12.

BONOTTO S., KIRCHMANN R., PUISEUX-DAO S. & VALET G. (1971a) Chapeaux en forme de cloche et relations nucléocytoplasmiques chez l'*Acetabularia mediterranea*. *C.R. Acad. Sci.* **272**, 545–8.

BONOTTO S., PUISSEUX-DAO S., KIRCHMANN R. & BRACHET J. (1971b) Faits et hypothèses sur le contrôle de l'alternance morphogenetique: croissance végétatif-differenciation de l'appareil reproducteur chez les *Acetabularia mediterranea Lam.*, *A. crenulata Lam.* et *Halicoryne spicata* (Kutzing) Solms-Laubach. *C.R. Acad. Sci.* **272**, 392–5.

BRACHET J. (1957) *Biochemical Cytology*, p. 303. Academic Press, New York.

BRACHET J. (1965) Acetabularia. *Endeavor*, **24**, 155–61.

BRACHET J. (1967) Protein synthesis in the absence of the nucleus. *Nature, Lond.* **213**, 650–5.

BRACHET J., DENIS H. & DE VITRY F. (1964) The effects of actinomycin D and puromycin on morphogenesis in amphibian eggs and *Acetabularia mediterranea*. *Dev. Biol.* **9**. 398–434.

BRÄNDLE E. & ZETSCHE K. (1971) Die Wirkung von Rifampicin auf die RNA-und Proteinsynthese sowie die Morphogenese und den Chlorophyll-gehalt kernhältiger und kernloser *Acetabularia*-Zellen. *Planta*, **99**, 46–55.

BRAWERMAN G. (1974) Eukaryotic messenger RNA. *Ann. Rev. Bioch.* **43**, 621–42.

CERON G. & JOHNSON E.M. (1971) Control of protein synthesis during the development of *Acetabularia*. *J. Embryol. Exp. Morph.* **26**, 323–38.

CLAUSS H. (1959) Das Verhalten der Phosphorylase in kernhältigen und kernlosen Teilen von *Acetabularia mediterranea*. *Planta*, **52**, 534–43.

CRAWLEY J.C.W. (1966) Some observations on the fine structure of the gametes and zygotes of *Acetabularia*. *Planta*, **69**, 365–76.

DAO S. (1957) La gamétogenèse chez l'*Acetabularia mediterranea* Lam. *C.R. Acad. Sci.* **243**, 1552–4.

DAO S. (1958) Recherches caryologiques chez le *Neomeris annulata*. *Rev. Algol.* **3**, 192–201.

DILLARD W.L. (1970) RNA synthesis in *Acetabularia*. In *Biology of Acetabularia*, eds. Brachet J. & Bonotto S., pp. 13–15. Academic Press, New York.

DOWNEY K.M., BYRNES J.J., JURMARK B.S. & SO A.G. (1973) Reticulocyte RNA-dependent RNA polymerase. *Proc. Nat. Acad. Sci.* (*U.S.*) **70**, 3400–4.

ELLIS R.J., BLAIR G.E. & HARTLEY M.R. (1973) The nature and function of chloroplast protein synthesis. *Biochem. Soc. Symp.* **38**, 137–62.

FARBER F.E. (1969) Studies on RNA metabolism in *Acetabularia mediterranea*. I. The isolation of RNA and labelling studies of RNA on whole plants and plant fragments, *Biochim. Biophys. Acta*, **174**, 1–11.

GIBOR A. & IZAWA M. (1963) The DNA content of the chloroplasts of *Acetabularia*. *Proc Nat. Acad. Sci.* (*U.S.*) **50**, 1164–9.

GOFFEAU A. (1969) Incorporation of amino acids into the soluble and membrane-bound protein of chloroplasts isolated from enucleated *Acetabularia*. *Biochim. Biophys. Acta*, **174**, 340–50.

GOFFEAU A. & BRACHET J. (1965) Deoxyribonucleic acid-dependent incorporation of amino acids in the proteins of the chloroplasts isolated from anucleate *Acetabularia* fragments. *Biochim. Biophys. Acta*, **95**, 302–13.

GREEN B.R. (1973) Evidence for the occurrence of meiosis before cyst formation in *Acetabularia mediterranea*. *Phycologia*, **12**, 233–5.

GREEN B.R. & BURTON H. (1970) *Acetabularia* chloroplast DNA: Electron microscopic visualization. *Science*, **168**, 981–2.

HÄMMERLING J. (1934) Über die Geschlechtsverhältnisse von *Acetabularia mediterranea* und *Acetabularia Wettsteinii*. *Arch. Protistenk.* **83**, 57–97.

HÄMMERLING J. (1944) Zur Lebenweisse, Fortpflanzung und Entwicklung verschiedener Dasycladaceae. *Arch. Protistenk.* **97**, 7–56.

HÄMMERLING J. (1953) Nucleocytoplasmic relationships in the development of *Acetabularia*. *Int. Rev. Cytol.* **2**, 475–98.

HÄMMERLING J. (1963) Nucleocytoplasmic interactions in *Acetabularia* and other cells. *Ann. Rev. Plant Physiol.* **14**, 65–92.

HÄMMERLING J. (1964) Gibt es bie Dasycladaceen Zoosporen? *Ann. Biol.* **3**, 33–6.

HARTMANN G., HONIKEL K.O., KNÜSEL F., NÜESCH J. (1967) The specific inhibition of the DNA-directed RNA synthesis by rifampicin. *Biochim. Biophys. Acta*, **145**, 843–4.

HEILPORN-POHL V. & BRACHET J. (1966) Net DNA synthesis in anucleate fragments of *Acetabularia mediterranea*. *Biochim. Biophys. Acta*, **119**, 429–31.

JANOWSKI M. (1966) Detection of ribosomes and polysomes in *Acetabularia mediterranea*. *Life Sciences*, **5**, 2113–6.

JANOWSKI M. & BONOTTO S. (1969) Les loges du chapeau d'*Acetabularia mediterranea*. *Bull. Soc. Roy. Belg.* **102**, 267–76.

JANOWSKI M., BONOTTO S. & BOLOUKHÈRE M. (1969) Ribosomes of *Acetabularia*. *Biochim. Biophys. Acta*, **174**, 525–35.

KATES J.R., CHIANG K.S. & JONES R.F. (1968) Studies on DNA replication during synchronized vegetative growth and gametic differentiation in *Chlamydomonas reinhardtii*. *Exp. Cell Res.* **49**, 121–35.

KECK K. & CLAUSS H. (1958) Nuclear control of enzyme synthesis in *Acetabularia*. *Bot. Gaz.* **120**, 43–9.

KLEINSCHMIDT A.K. (1968) Monolayer techniques in electron microscopy of nucleic acid molecules. In *Methods in Enzymology*, *Vol.* 12*B*, eds. Grossman L. & Moldave K., pp. 361–77. Academic Press, New York.

KLOPPSTECH K. (1975) Poly-A Containing RNA in *Acetabularia*. *Protoplasma*, **83**, 177 (abst.).

KLOPPSTECH K. & SCHWEIGER H.G. (1973a) 80S ribosomes from *Acetabularia*. *Biochim. Biophys. Acta*, **324**, 365–74.

KLOPPSTECH K. & SCHWEIGER H.G. (1973b) Synthesis of chloroplast and cytosol ribosomes in regenerating *Acetabularia* cells. *Differentiation*, **1**, 331–8.

KLOPPSTECH K. & SCHWEIGER H.G. (1973c) Nuclear genome codes for chloroplast ribosomal proteins in *Acetabularia*. II. Nuclear transplantation experiments. *Exp. Cell Res.* **80**, 69–78.

KOOP H.-U. (1975a) Germination of cysts in *Acetabularia mediterranea*. *Protoplasma*, **84**, 137–46.

KOOP H.-U. (1975b) Über den Ort der Meiose bei *Acetabularia mediterranea*. *Protoplasma*, in press.

LONI M.C. & BONOTTO S. (1971) Statistical studies on cap morphology in normal and branched *Acetabularia mediterranea*. *Arch. Biol.* **82**, 225–44.

MACHATTIE L.A., BERNS K.I. & THOMAS C.A. (1965) Electron microscopy of DNA from *Hemophilus influenzae*. *J. Mol. Biol.* **11**, 648–9.

MARGULIS L. (1970) *Origin of Eukaryotic Cells*. Yale University Press, New Haven.

MILLER O.L. & BEATTY B.R. (1969) Visualization of nucleolar genes. *Science*, **164**, 955–7.

MUIR B.L. (1974) Studies on *Acetabularia* Chloroplast DNA. M.Sc. Thesis, University of British Columbia.

PUISEUX-DAO S. (1962) Recherches biologiques et physiologiques sur quelques Dasycladacées. *Rév. Gen. Bot.* **819**, 409–503.

PUISEUX-DAO S. (1966) Siphonales and Siphonacladales. In *Chromosomes of the Algae.*, ed. Godward M.B.E., pp. 52–77. Arnold, London.

PUISEUX-DAO S. (1970) *Acetabularia and Cell Biology*. Springer, New York.

PUISEUX-DAO S., AKSIYOTE-BEN BASET J. & BONOTTO S. (1972) Effets biologiques de la rifampicine chez l'*Acetabularia mediterranea*. *C.R. Acad. Sci.* **274**, 1678–81.

REUTER W. & SCHWEIGER H.G. (1969) Kernkontrollierte Lactadehydrogenase in *Acetabularia*. *Protoplasma*, **68**, 357–68.

RICHTER G. (1962) Nuclear-cytoplasmic Interactions. In *Physiology and Biochemistry of Algae*, ed. Lewin R.A., pp. 633–52. Academic Press, New York.

SCHULZE K.L. (1939) Cytologische Untersuchungen an *Acetabularia mediterranea* und *Acetabularia wettsteinii*. *Arch. Protistenk.* **92**, 179–225.

SCHWEIGER H.G. (1966) Ribonuclease-Aktivität in *Acetabularia*. *Planta*, **68**, 247–55.

SCHWEIGER H.G. (1969) Cell Biology of *Acetabularia*. *Curr. Topics Microb. Immunol.* **50**, 1–36.

SCHWEIGER H.G. & BERGER S. (1964) DNA-dependent RNA synthesis in chloroplasts of *Acetabularia*. *Biochim. Biophys. Acta*, **87**, 533–5.

SCHWEIGER H.G., MASTER R.S.P. & WERZ G. (1967a) Nuclear control of a cytoplasmic enzyme in *Acetabularia*. *Nature, Lond.* **216**, 554–7.

SCHWEIGER H.G., DILLARD W.L., GIBOR A. & BERGER S. (1967b) RNA synthesis in *Acetabularia*. I. RNA synthesis in enucleated cells. *Protoplasma*, **64**, 1–12.

SCHWEIGER H.G., APEL K. & KLOPPSTECH K. (1972) Source of genetic information of chloroplast proteins in *Acetabularia*. *Adv. Biosciences*, **8**, 249–62.

SCHWEIGER H.G., BERGER S., KLOPPSTECH K., APEL K. & SCHWEIGER M. (1974) Some fine structural and biochemical features of *Acetabularia major* (Chlorophyta, Dasycladaceae) grown in the laboratory. *Phycologia*, **13**, 11–20.

SHEPHARD D.S. (1965a) Chloroplast multiplication and growth in the unicellular alga *Acetabularia mediterranea*. *Exp. Cell. Res.* **37**, 93–110.

SHEPHARD D.S. (1965b) An autoradiographic comparison of the effects of enucleation and actinomycin D on the incorporation of nucleic acid and protein precursors by *Acetabularia* chloroplasts. *Biochim. Biophys. Acta*, **108**, 635–43.

SHEPHARD D.S. (1970) Axenic culture of *Acetabularia* in synthetic media. In *Methods in Cell Physiology*, Vol. 4, pp. 49–69, ed. Prescott D. Academic Press, New York.

SIX E. (1956) Die Wirkung von Strahlen auf *Acetabularia*. I. Die Wirkung von ultravioletten Strahlen auf kernlose Teile von *Acetabularia mediterranea*. *Z. Naturf.* **11b**, 463–70.

TRENDELENBURG M.F., SPRING H., SCHEER U., FRANKE W.W. (1974) Morphology of nucleolar cistrons in a plant cell, *Acetabularia mediterranea*. *Proc. Nat. Acad. Sci.* (*U.S.*) **71**, 3626–30.

TRIPLETT E.L., STEENS-LEIVENS A. & BALTUS E. (1965) Rates of synthesis of acid phosphatases in nucleate and enucleate *Acetabularia* fragments. *Exp. Cell Res.* **38**, 366–78.

VALET G. (1968) Contribution à l'étude des Dasycladales. 1. Morphogenèse. *Nova Hedwigia*, **16**, 22–84.

VALET G. (1969) Contribution à l'étude des Dasycladales. 2. Cytologie et Reproduction. 3. Révision systématique. *Nova Hedwigia*, **17**, 551–644.

DE VITRY F. (1965) Etude du métabolisme des acides nucléiques chez *Acetabularia mediterranea*. I. Incorporation de précurseurs de DNA, de RNA et de protéines chez *Acetabularia mediterranea*. *Bull. Soc. Chim. Biol.* **47**, 1325–51.

WELLS R. & SAGER R. (1971) Denaturation and renaturation kinetics of chloroplast DNA from *Chlamydomonas reinhardi*. *J. Mol. Biol.* **58**, 611–22.

WERZ G. (1965) Determination and realization of morphogenesis in *Acetabularia*. *Brookhaven Symp. Biol.* **18**, 185–203.

WETMUR J.G. & DAVIDSON N. (1968) Kinetics of renaturation of DNA. *J. Mol. Biol.* **31**, 349–70.

WOODCOCK C.F.L. & BOGORAD L. (1970a) On the extraction and characterization of ribosomal RNA from *Acetabularia*. *Biochim. Biophys. Acta*, **224**, 639–43.

WOODCOCK C.F.L. & BOGORAD L. (1970b) Evidence for variation in the quantity of DNA among plastids of *Acetabularia*. *J. Cell Biol.* **44**, 361–75.

WOODCOCK C.F.L. & MILLER G.L. (1973a) Ultrastructural features of the life cycle of *Acetabularia mediterranea*. I. Gametogenesis. *Protoplasma*, **77**, 313–29.

WOODCOCK C.F.L. & MILLER G.L. (1973b) Ultrastructural features of the life cycle of *Acetabularia mediterranea*. II. Events associated with the division of the primary nucleus and the formation of cysts. *Protoplasma*, **77**, 331–41.

ZETSCHE K. (1964) Hemmung der Synthese morphogenetischer Substanzen im Zellkern von *Acetabularia mediterranea* durch Actinomycin D. *Z. Naturf.* **19b**, 751–9.

ZETSCHE K. (1966) Regulation of UDPglucose-4-epimerase synthesis in nucleate and anucleate *Acetabularia*. *Biochim. Biophys. Acta*, **124**, 332–8.

ZETSCHE K. (1968) Regulation der UDPG-Pyrophosphorylase-Aktivität in *Acetabularia*. I. Morphogenese und UDPG-Pyrophosphorylase-Synthese in kernhaltigen und kernlosen Zellen. *Z. Naturf.* **23b**, 369–76.

CHAPTER 13

INHERITANCE AND SYNTHESIS OF CHLOROPLASTS AND MITOCHONDRIA OF *EUGLENA GRACILIS*

GREGORY W. SCHMIDT AND HARVARD LYMAN

Department of Cellular and Comparative Biology,
The State University of New York at Stony Brook,
Stony Brook, New York, U.S.A.

1 Introduction 257

2 The *Euglena* mitochondria 259

3 Chloroplast DNA 262

4 Role of the plastid genome in chloroplast synthesis 267

5 The genetic control of light-mediated chloroplast development 286

6 Autonomy and origin of *Euglena* chloroplasts 289

7 References 290

1 INTRODUCTION

The order Euglenales comprises a few dozen genera and many hundreds of species. Nevertheless, almost all experimental studies of these organisms have been carried out with a single strain, generally designated as *Euglena gracilis* var. *bacillaris*, which has the almost unique ability to survive, and to grow heterotrophically for an indefinitive number of generations, even after the chloroplasts have been eliminated by natural or artificial means. (This phenomenon—induced apochlorosis—has been exhaustively discussed elsewhere, e.g. by Pringsheim, 1963.) To the best of our knowledge, all of the research work reported in this chapter refers solely to this strain.

Since no sexual reproduction is known in *Euglena gracilis*, no formal genetic analysis is possible. Nevertheless, mutant strains are isolated with relative ease (Lewin 1960; Schiff *et al.* 1971) and have been most useful in the study of photosynthesis (Russell & Lyman 1968; Russell *et al.* 1969; Schwelitz

The experimental work reported here was supported by National Science Foundation Grant BO 38262 to H. Lyman.

et al. 1972), phototaxis and motility (Diehn 1973), cell cycles and rhythms (Jarrett & Edmunds 1970, Edmunds *et al.* 1971); respiration (Danforth 1968; Raison & Smillie 1969; Sharpless & Butow 1970) and intermediary metabolism (Barras & Stone 1968; Erwin 1968, Smillie 1968). Metabolic activities have been studied independently of photosynthesis in strains of *Euglena* lacking chloroplasts. Studies of chloroplast heredity and synthesis have been particularly aided by the use of mutant strains (Schiff 1973; Marćenko 1973; Schmidt & Lyman 1973).

Most mutants of *Euglena* are probably non-nuclear. X-ray and ultraviolet killing curves indicate that the nucleus is octaploid (Schiff & Epstein 1965; Hill *et al.* 1966) making the expression of recessive nuclear mutations unlikely (Schiff *et al.* 1971). Thus the chloroplast and mitochondrial genomes become candidates for consideration as sites of readily detectable mutation. *Euglena* is apparently an obligate aerobe, unable to grow by fermentative processes (Smillie *et al.* 1963); therefore all mutations affecting mitochondrial function should be lethal, at least in non-photosynthetizing cells. It is possible that phosphorylation and photoreduction in the chloroplasts might support the growth of a strain possessing defective mitochondria, but such a mutant has not been reported. There are, perhaps, 200 to 500 mitochondria per cell (Edelman *et al.* 1966), so there is only a low probability for the segregating out of mutant organelles. It appears likely, therefore, that most of the mutants described represent mutations of the chloroplast genome, although it is important to examine the evidence for this closely.

Euglena gracilis has about ten chloroplasts (or ten proplastids when grown in the dark). A number of ultraviolet-sensitive entities within the cytoplasm affect chloroplast and/or proplastid replication (Lyman *et al.* 1961; Gibor & Granick 1962). It is generally assumed that these are identical with plastid DNA, each entity being one plastid genome (Lyman *et al.* 1961). In the earlier studies ultraviolet target analysis indicated that there were 30 plastid genomes per cell (Lyman *et al.* 1961, Hill *et al.* 1966), but recent work indicates that 60 is a more likely number (Uzzo & Lyman 1972). The very high sensitivity of chloroplast replication to various physical and chemical treatments (Schiff *et al.* 1971) suggests the following hypothesis to explain why the bulk of *Euglena* mutants isolated are mutations of the chloroplast genomes. Mutant strains probably represent clones derived from cells in which all but one of the plastid genomes were prevented from replicating by the mutagenic treatment, while a mutation was induced in the surviving genome. The mutated genome then replicated many times until ultimately a cell was produced in which all of the chloroplast genomes carried the mutation. An experimental observation which supports this hypothesis comes from ultraviolet target analysis of chloroplast replication (Lyman *et al.* 1961; Hill *et al.* 1966). Target theory applied to this system indicates the existence of a number of genetically identical, ultraviolet-sensitive entities, any one of which can mediate chloroplast replication. Furthermore, experiments with nalidixic acid, a specific inhibitor of chloroplast replication in *Euglena* (Lyman 1967; Ebringer 1970), indicates that the cell is soon able to regenerate chloroplast DNA lost through mutagenic treatment. Cells exposed to nalidixic acid for six hours show a loss of over 50% of the chloroplast-associated DNA, and yet, when the drug is removed and the cells are plated, every clone is found to consist of cells with the normal complement of chloroplasts (Lyman *et al.* 1975). Since many of the agents that specifically affect the chloroplasts of *Euglena* are mutagens (Schiff *et al.* 1971), we conclude that the chloroplast genome in this organism is especially sensitive to mutagenesis. At present, mutations of the mitochondrial genome cannot be studied in this way.

In this chapter we shall first review the characteristics of the *Euglena* mitochondrial genome and gene products. This will be followed by a discussion of the chloroplast genome and a summary of its role in the replication and synthesis of the chloroplast and proplastid.

2   THE *EUGLENA* MITOCHONDRIA

Mutant strains of *Euglena* which lack chloroplasts have been very useful in the characterization of mitochondrial nucleic acids because the possibility of contamination of preparations with chloroplast nucleic acids is eliminated. Some physical properties of *Euglena* mitochondrial DNA are summarized in Table 13.1.

A number of methods have been employed to determine the size of the mitochondrial genome. Measurements in electron micrographs of the contour length of linear DNA molecules from *Euglena* mitochondria range from 0·6 to 19 μm with a mean distribution of 0·9 to 1·7 μm (Schori *et al.* 1970; Manning *et al.* 1971; Nass & Ben-Shaul 1972; Nass *et al.* 1974). A few circular forms of mitochondrial DNA with a contour length of 0·95 to 1·01 μm have been reported (Schori *et al.* 1970; Nass *et al.* 1974). Assuming a molar density of $2·07 \times 10^6$ daltons/μm (Lang 1970), the mean molecular weight of *Euglena* mitochondrial DNA determined by this method would be 1·86 to $3·52 \times 10^6$ daltons. These estimates of the genome size compare well with molecular weight determinations of 3·5 to $3·7 \times 10^6$ daltons obtained by sucrose density-gradient analysis (Ray & Hanawalt 1965; Talen, Sanders & Flavell 1974; Crouse 1974). Manning *et al.* (1971) suggest the 19 μm molecules (39×10^6 daltons) might be tandem repeats of shorter DNA sequences. However, the kinetic complexity determined by renaturation rates of *Euglena* mitochondrial DNA reported by Talen *et al.* (1974) indicates weights as high as 36 to 45×10^6 daltons, and Crouse, Vandrey and Stutz (1974b) obtained an even higher value of 100×10^6. Assuming that there is only one ribosomal RNA cistron per DNA molecule with a molecular weight of $1·6 \times 10^6$, and that it represents 3·7% of the genome, Crouse *et al.* (1974b) calculate a molecular weight of 43×10^6 daltons for *Euglena* mitochondrial DNA. 40×10^6 daltons approximates the molecular weight of mitochondrial DNA in yeast, higher plants and *Tetrahymena* (Kroon 1971; Borst 1972). On the other hand, 2 to 4×10^6 daltons is smaller than the mitochondrial genome size of any animal tissue (Kroon 1971) and is, in fact, nearly the lowest value reported for any mitochondrial DNA.

Table 13.1. Characteristics of mitochondrial DNA (mtDNA) of *Euglena*

Buoyant density (gm/cm³)	
1·691	Edelman Epstein & Schiff 1966[a]
	Uzzo & Lyman 1972
	Nass Schori Ben-Shaul & Edelman 1974
1·690	Richards & Ryan 1974[b]
1·689	Manning *et al.* 1971[c]
1·688	Kraweic & Eisenstadt 1970[d]
A+T* (mole %)	
68	Edelman Epstein & Schiff 1965
69	Richards & Ryan 1974
	Crouse Vandrey & Stutz 1974b
70	Manning *et al* 1971
71	Krawiec & Eisenstadt 1970
75	Crouse 1974
Daltons/genome	
$2 \cdot 1 \times 10^6$	Nass Schori Ben-Shaul & Edelman 1974[e]
$3 \cdot 0 \times 10^6$	Ray & Hanawalt 1965
$3 \cdot 5 \times 10^6$	Crouse 1974
$1 \cdot 6 - 4 \cdot 5 \times 10^6$	Schori Ben-Shaul & Edelman 1970
$40 \cdot 0 \times 10^6$	Talen Sanders & Flavell 1974[f]
	Crouse Vandrey & Stutz 1974b[g]
Genomes/mitochondrion	
11	[h]
215	[i]

[a] Wild type *bacillaris* strain and a streptomycin-bleached strain, $W_{10}SmL$.
[b] Wild type 'Z' and a white mutant, W_3BUL.
[c] Wild type 'Z' and two white mutants W_3BUL and ZsmL.
[d] Two white strains, HB_3 and SB_3.
[e] Based on 1·0 μm contour length measurement.
[f] Based on kinetic complexity measurement, assuming that kinetic complexity equals molecular weight of DNA.
[g] Based on 3·7% hybridization of mtrRNA with mtDNA, and assuming one cistron of $1 \cdot 6 \times 10^6$ daltons for 23S and 16S rRNA.
[h] Based on a molecular weight of 4×10^7 daltons and the amount of mtDNA per organelle calculated by Schiff, 1970.
[i] Based on a molecular weight of $2 \cdot 1 \times 10^6$ and the amount of mtDNA per organelle calculated by Schiff, 1970.

* Adenine+thymine (molar percentage in DNA).

Because the coding capacity of mitochondrial DNA is important if its genetic role in *Euglena* is to be assessed, resolution of the discrepant molecular weights should be rigorously pursued. A mitochondrial genome of 3 or 40×10^6 daltons could code for 10 or 133 proteins of 15,000 molecular weight, respectively. No apparent explanation for the variance in genome size is seen in the published reports. All workers employ DNA isolation techniques which include precautions against endogenous nuclease activity and shearing, although in no case is the possibility of such degradation unequivocally precluded.

The mitochondria in *Euglena*, estimated at 200 to 500 per cell (Edelman, Epstein & Schiff 1966), have been observed to undergo extensive structural modulation. At interphase a giant network or reticulum of a single mitochondrion is formed which later fragments to form many smaller organelles during cell division (Konitz 1965; Calvayrac *et al.* 1972; Osafune 1973). The formation of giant mitochondria has been associated also with changes in the metabolic state of the organelles (Brandes *et al.* 1964; Leedale & Buetow 1970). The number of mitochondria seen in sections may thus be deceptive, because what appear to be 'individual' organelles might actually be sections of meshes or arms of a reticulum which would tend to overestimate the number of genomes/organelles as given in Table 13.1.

Mitochondrial DNA replication in *Euglena*, as determined by density-transfer experiments, appears to be dispersive rather than semi-conservative (Richards & Ryan 1974). The absence of heavy strands of mitochondrial DNA in these studies may be due to rapid metabolic turnover during the cell cycle. However, although in *Euglena* mitochondrial and chloroplast DNA turnover rates are approximately the same, with half-lives of $1 \cdot 8$ and $1 \cdot 6$ generations respectively (Manning & Richards 1972a; Richards & Ryan 1974), chloroplast DNA replication is clearly semi-conservative (Manning & Richards 1972). It would seem reasonable to suggest, therefore, that mitochondrial DNA replication in *Euglena* appears dispersive because of extensive recombination between template and replicate strands of DNA. Recombination of mitochondrial genomes has been well characterized in other organisms (Coen *et al.* 1970; Gillham 1974). In *Euglena* the association of mitochondrial DNA with membranes has been noted (Schori *et al.* 1970; Nass *et al.* 1974), and Nass *et al.* (1974) have suggested that enzymes active in nucleic acid metabolism may be associated with certain membrane sites.

The sensitivity of mitochondrial DNA synthesis to a variety of inhibitors is shown in Table 13.3. Cycloheximide appears to be more effective than either chloramphenicol or ethidium bromide, suggesting that the components required for mitochondrial DNA replication may be synthesized on cytoplasmic ribosomes. Nass and Ben-Shaul (1973) have demonstrated that ethidium bromide reversibly affects mitochondrial structure and cell division in *Euglena*. Unlike mitochondria of other organisms (Slonimski *et al.* 1968; Goldring *et al.* 1970; Nass 1970; Mahler & Perlman 1972), those of *Euglena* appear to be refractory to mutagenesis by ethidium bromide.

Unfortunately, there have been only a few studies of the mitochondrial gene products in *Euglena*. The isolation of RNA from mitochondria has yielded predominantly ribosomal species. Although Avadhani and Buetow (1972a) were able to characterize 21.3S and 16.3S mitochondrial RNAs, others (Kraweic & Eisenstadt 1970; Crouse *et al* 1974a) could detect only 14S and 11S ribosomal RNA species in *Euglena* mitochondria. The smaller RNAs have base compositions quite similar to the larger species (Kraweic & Eisenstadt 1970; Avadhani & Buetow 1972b) and are, therefore, presumably degradation products. Ribosomal RNA and ribosome monomers from mitochondria have sedimentation coefficients nearly identical to those from chloroplasts in *Euglena*, but the base compositions of the organellar ribosomal RNAs differ considerably (Avadhani & Buetow 1972b). Significant hybridization of ribosomal RNA from mitochondria occurs only with DNA of mitochondrial origin (Crouse *et al.* 1974a). Although Crouse *et al.* (1974a) have concluded that ribosomal RNA cistrons represent 3·6 to 3·8% of the *Euglena* mitochondrial genome, they used in their hybridization studies the small, probably degraded RNA. In fact, the ribosomal RNA genes of *Euglena* may be considerably larger or more numerous than their experiments would indicate.

The ability of purified mitochondria and isolated mitochondrial polysomes to carry out protein synthesis *in vitro* has been studied (Avadhani *et al.* 1971; Kislev & Eisenstadt 1972; Avadhani & Buetow 1972a, b). *Euglena* mitochondria have been shown to contain unique tRNAs and tRNA synthetases (Kislev *et al.* 1972; Kislev & Eisenstadt 1972), and may therefore be capable of at least some autonomous protein synthesis. The incorporation of amino acids into protein is characteristically susceptible to inhibition by antibiotics (Table 13.4). The mitochondrial protein-synthesizing system is like that in prokaryotes because chloramphenicol and streptomycin, but not cycloheximide, strongly inhibit its activity. However, chloramphenicol and streptomycin have little effect on the rates of growth and respiration in whole cells (Smillie *et al.* 1963; Stewart & Gregory 1969; Ben-Shaul & Ophir 1970). It is probable that the lack of effect of these inhibitors on *Euglena* mitochondria *in vivo* may be due to an impermeability of the mitochondrial membranes to these antibiotics, as has been demonstrated for other organisms (Kroon & Devries 1970; Ibrahim *et al.* 1974).

3   CHLOROPLAST DNA

The characteristics of plastid-associated DNA in *Euglena gracilis* are summarized in Table 13.2.

Table 13.2. Characteristics of chloroplast-associated DNA of *Euglena*.

Buoyant density (gm/cm³)

1·684	Brawerman & Eisenstadt 1964
1·685	Manning & Richards 1972b
1·686	Schiff 1973

A+T* (mole %)

70–79	Schiff 1973
75	Brawerman & Eisenstadt 1964
76·3	Ray & Hanawalt 1964

DNA/plastid (10^{-14} gm)

0·12	Edelman Cowan Epstein & Schiff 1964
0·9–1·1	Brawerman & Eisenstadt 1964
1·2–1·5	Scott 1973

Daltons/genome ($\times 10^7$)

2·0–4·0	Ray & Hanawalt 1964[a]
7·5	[b]
8·3	Manning et al. 1971[c]
9·0	Manning et al. 1971[d]
9·2	Manning & Richards 1972[e]
15·0	Schiff 1973[f]
18·0	Stutz 1970[g]

Genomes/plastid

Chloroplast

3	Schiff 1973
6	Uzzo & Lyman 1972
9	Manning et al 1971[h]
72	Manning et al 1971[i]

Proplastid

3	Uzzo & Lyman 1972
3	Schiff 1973

rRNA Cistrons/genome

1	Rawson & Haselkorn 1973
2	Stutz Crouse & Graf 1974
3	Stutz & Vandrey 1971
3–6	Scott 1973

---

[a] Determined by sucrose density-gradient sedimentation analysis.

[b] Calculated from the data of Uzzo & Lyman (1972) indicating 6 genomes per chloroplast and the molecular weight given in Schiff (1973).

[c] Determined from contour length measurements of covalently-closed circular DNA molecules.

[d] Calculated from the kinetic complexity determination of Stutz (1970), correcting for the low G+C‡ content of *Euglena* chloroplast DNA.

[e] Determined by sucrose density-gradient sedimentation analysis and contour length measurements.

[f] Calculated from a molecular weight of $4·5 \times 10$ daltons of DNA per chloroplast and 3 genomes per chloroplast.

[g] Determined by kinetic complexity analysis.

[h] Calculated from $1·2 \times 10^{-15}$ gm DNA per chloroplast.

[i] Calculated from $1·0 \times 10^{-14}$ gm DNA per chloroplast.

* Adenine and thymine (molar percentage in DNA)

The synthesis of chloroplast DNA in synchronous cultures of *Euglena* occurs prior to and following cell division (Cook 1966). Chloroplasts are seen to divide just before cytokinesis (Cook 1966). The turnover of chloroplast DNA has been estimated by Manning and Richards (1972) to be 1·5 times as fast as that of nuclear DNA. Chloroplast replication is sensitive to ultraviolet irradiation (Lyman *et al.* 1961), nalidixic acid (Lyman 1967), streptomycin (Provasoli *et al.* 1948), growth at 32 to 34°C (Pringsheim & Pringsheim 1952; Uzzo & Lyman 1969), and other treatments (Mego 1968; Ebringer 1972). In strains bleached with these agents chloroplasts and plastid-associated DNA cannot be detected (Leff *et al.* 1963; Edelman *et al.* 1965; Ray & Hanawalt 1965; Uzzo & Lyman 1972; Lyman *et al.* 1975). Certain membrane-limited structures have been reported in some of these strains (Siegesmund *et al.* 1962; Moriber *et al.* 1963; Kivic & Vesk 1974), but they may not be derived from plastids.

Uzzo and Lyman (1972) observed that dark-grown *Euglena* cells have less plastid-associated DNA than light-grown cells. They found that when dark-grown cells are illuminated, chloroplasts develop and there is a concomitant synthesis of plastid-associated DNA. The amount of chloroplast DNA in fully green cells was observed to be about double that in dark-grown cells. Previously, it had been noted that the ultraviolet target number in both green and dark-grown cells was about 30 (Lyman *et al.* 1961). From the difference in DNA content one might have anticipated that green cells would have a higher target number than dark-grown cells. Uzzo and Lyman (1972) discovered that in non-dividing green cells incubated at 32°C the ultraviolet target number rises to 60; when such cells are returned to 26°C, the target number drops again to 30. They therefore proposed that there are six copies of the genome in each chloroplast and three in each proplastid; chloroplast DNA exists in two different states or conditions; three of the genomes can participate in chloroplast replication at 26°C, as defined by ultraviolet target analysis, while the other three cannot; in light-grown cells at 26°C the three genomes mediating chloroplast replication are tightly bound in some fashion, possibly to the chloroplast membrane. They also suggested that at 32°C the affinity of the binding site is loosened so that all the genomes may compete for association with it. It is interesting to note that if one uses a value of six for the number of genomes and $4·5 \times 10^8$ daltons as the molecular weight of the total DNA in a single chloroplast (Schiff 1973), the molecular weight of one genome can be estimated to be $7·5 \times 10^7$. This compares well with the value of $8·3 \times 10^7$ calculated from the contour length measurement of covalently-closed circular chloroplast DNA molecules (Manning *et al.* 1971).

The replication of proplastids and chloroplasts of *Euglena* is temperature-sensitive (Pringsheim & Pringsheim 1952; Uzzo & Lyman 1969). Uzzo and Lyman (1972) showed that the effect of temperature is solely on replication; elevated temperatures have no effect on the synthesis of any chloroplast component or plastid-associated DNA. Chloroplast DNA is lost by dilution from cells dividing at 'bleaching' temperatures, due to the inability of the entire plastid to replicate. The actual site(s) of temperature-sensitivity is unknown. However, dark-grown cells were shown to be more sensitive to 'heat-bleaching' than are fully green cells. From the kinetics of temperature-induced inhibition of plastid replication Uzzo and Lyman (1972) calculated that there are two theoretical temperature-sensitive sites per chloroplast but only one per proplastid. They speculated that these sites function in the binding of plastid DNA.

In addition to light-dependent changes of DNA during chloroplast development, some modulations of novel components of chloroplast DNA during the cell cycle have been observed. The main band of chloroplast DNA in *Euglena* has a mean buoyant density of $1\cdot685$ gm/cm^3 in neutral cesium chloride (Table 13·2). An additional DNA from purified chloroplasts, with a buoyant density of $1\cdot701$ gm/cm^3, was first identified by Stutz and Vandrey (1971). They demonstrated, on the basis of DNA/RNA hybridization, that this heavy component is enriched in chloroplast ribosomal RNA cistrons. Subsequently, Manning and Richards (1972b) noted that the $1\cdot701$ gm/cm^3 chloroplast DNA is more prominent in preparations of chloroplasts from cells dividing exponentially than in non-dividing, stationary-phase cells. Similarly, Scott (1973) was not able to detect the heavy component in his studies with non-dividing *Euglena*. By mechanically shearing chloroplast DNA from stationary-phase cells Rawson and Haselkorn (1973) were able to generate a derivative with a buoyant density of $1\cdot700$ gm/cm^3. They postulated that the heavy DNA component observed by Stutz and Vandrey (1971) and Manning and Richards (1972b) might be an early product of chloroplast DNA replication. Rawson and Haselkorn (1973) proposed that the DNA enriched in ribosomal RNA cistrons might be generated because these genes are located near the origin of DNA replication. Greater amounts of the heavy species of chloroplast DNA would therefore be expected in chloroplasts of dividing cells. Vandrey and Stutz (1973) have, in fact, determined the existence of a minor component of chloroplast DNA with a buoyant density of $1\cdot700$ gm/cm^3 in preparations of chloroplasts from exponentially dividing cells. On the other hand, the preferential replication of 'covalently-closed circular chloroplast DNA' species during the transition of cells from exponential to stationary growth stages has been reported by Mielenz and Hershberger (1974). They observe that in dividing cells there is only one circular chloroplast DNA species, with a buoyant density of $1\cdot686$ gm/cm^3. They have also reported that when cell division is arrested, several circular DNA species with buoyant densities of $1\cdot700$, $1\cdot699$, $1\cdot692$ and $1\cdot690$ gm/cm^3 are synthesized. They assume these to be of chloroplast origin, although they were not able to account for mitochondrial DNA in their study.

It is important to note that Nass and Ben-Shaul (1972) have characterized a different covalently-closed circular DNA with a contour length of 3·13 μm and a buoyant density of 1·701 gm/cm³ in *Euglena*. This DNA, found in dark-grown, green and aplastidic cells, was observed most distinctly in dividing cells, but was not evidently associated with chloroplasts or mitochondria.

Small amounts of circular DNA molecules from chloroplasts with a contour length of 3·6 μm have been noted by Nass and Ben-Shaul (1972) and Manning *et al.* (1971), in addition to the bulk of covalently-closed circular chloroplast DNA which has a length of 44 to 45 μm. Manning *et al.* (1971) related the 3·6 μm DNA molecules to minor chloroplast DNA species with buoyant densities of 1·690 and 1·699 gm/cm³, which evidently comprise less than 20% of the total DNA in *Euglena* chloroplasts.

Other chloroplast DNA components which represent species enriched in ribosomal RNA cistrons have been characterized. 1 to 2·9% of the main band of chloroplast DNA (1·684 to 1·686 gm/cm³) has been calculated to be complementary to ribosomal RNA in most RNA/DNA hybridization studies (Stutz & Rawson 1970; Stutz & Vandrey 1971; Vandrey & Stutz 1973; Rawson & Haselkorn 1973; Crouse *et al.* 1974a) although a value of 6·2% was reported by Scott (1973). Scott (1973) found that the greatest amount of hybridization of ribosomal RNA was with a 1·696 gm/cm³ fraction of chloroplast DNA from non-dividing cells. Vandrey and Stutz (1973) isolated a 1·692 gm/cm³ species of chloroplast DNA from exponentially-dividing cells which they found to comprise 25 to 30% of the total chloroplast DNA. Brawerman and Eisenstadt (1964) also noted this species in their DNA preparation from chloroplasts, but they attributed it to contamination from mitochondria. 6·9% of this DNA species was determined to be complementary to chloroplast ribosomal RNA (Vandrey & Stutz 1973; Crouse *et al.* 1974a). The 1·692 gm/cm³ component of chloroplast DNA was shown to give rise to a 1·701 gm/cm³ species upon mechanical shearing (Vandrey & Stutz 1973). Crouse *et al.* (1974a) have demonstrated that the 1·692 gm/cm³ component of chloroplast DNA is distinct from the mitochondrial DNA of *Euglena* which has a similar buoyant density (Table 13.1). Vandrey and Stutz (1973) observed that the amount of the 1·692 gm/cm³ DNA associated with chloroplasts fluctuates with the cell cycle and that it may replicate at a different rate from the main DNA species at certain stages. Although Stutz *et al.* (1974) did not find that chloroplast ribosomal genes are selectively amplified during the light-dependent development of chloroplasts from proplastids, Scott (1973) has concluded that there are no chloroplast ribosomal RNA cistrons in dark-grown *Euglena*. Scott (1973) did not attempt RNA/DNA hybridization with purified plastid DNA and for that reason was probably unable to demonstrate significant levels of annealment of ribosomal RNA. Hirvonen and Price (1971) have determined that proplastids of *Euglena* do in fact contain ribosomes.

The molecular nature of the many subspecies of plastid DNA and their role in chloroplast synthesis is not clear. It is apparent that further characterization of these DNAs must be accomplished through the use of methods of purification which insure against degradation and/or shearing. Circularity, buoyant density, contour length measurement and the capacity of any of these DNA species to hybridize with ribosomal RNA should be determined as matters of standard procedure. The existence of putative novel plastid-associated DNA in *Euglena* should be routinely confirmed through the use of aplastidic strains. Finally, there is further need to correlate the presence of the minor plastid DNAs with various stages of cell growth, in dark-grown cells and in cells grown in the light with or without an assimilable substrate.

4   ROLE OF THE PLASTID GENOME IN CHLOROPLAST SYNTHESIS

It is of current interest to define the role of chloroplast and nuclear genes in the inheritance and control of synthesis of the chloroplast. *Euglena* has been useful in this endeavour because work with mutant strains has shown that the plastid DNA codes for structural, functional and control molecules of the plastid (Schiff 1973; Schmidt & Lyman 1973). Various workers have estimated that the plastid DNA of *Euglena* can code for 50 to 300 different proteins (Manning *et al.* 1971; Schiff 1973). The estimates vary, depending upon the size of the genome, the number of ribosomal-RNA cistrons (Table 13.2) and the average size of the proposed proteins. The immediate products of the plastid genome are the plastid RNAs and therefore, before proceeding with a description of plastid protein synthesis, we shall examine the characteristics and synthesis of plastid RNAs in *Euglena*.

The study of RNA synthesis in *Euglena* has been facilitated by studies of the light-dependent development of chloroplasts. A number of distinct RNA species are preferentially synthesized during this process. The synthesis of chloroplast ribosomal RNA is especially dramatic during the development of functional chloroplasts from proplastids. Proplastids contain fully-formed ribosomes which can be visualized by electron microscopy (Schiff 1970). The subunits of proplastid 70S ribosomes have been isolated and shown to be identical to those of mature chloroplasts (Hirvonen & Price 1971). However, the amount of ribosomes in proplastids is quite diminutive, as exemplified by several reports of failure to detect either ribosomal particles or their RNAs in dark-grown cells (Brown *et al.* 1970, Brown & Haselkorn 1970; Scott *et al.* 1971). After two to three hours of illumination of dark-grown cells rapid synthesis of chloroplast ribosomal RNA begins (Heizmann 1970; Brown & Haselkorn 1971). Within the first 16 hours of development, the specific activity of chloroplast ribosomal RNA is five times higher than that of the cytoplasmic ribosomes when cells are incubated with a radioactively labelled RNA precursor (Scott *et al.* 1971). Zeldin and Schiff (1967) demonstrated that synthesis of chloroplast and cytoplasmic ribosomal RNA is stimulated by light when they compared the incorporation of $^{32}PO_4$ into RNA in dark-grown and in greening *Euglena*. Their results complemented earlier reports of light-stimulation of total RNA synthesis in whole cells (Brawerman & Chargaff 1959a, b; Brawerman *et al.* 1962; Smillie *et al.* 1963). Brawerman and Chargaff (1959a, b) observed that during chloroplast development the increase of RNA, mostly ribosomal, occurs before chloroplast proteins and chlorophyll are actively synthesized. Heizmann (1970) and Munns *et al.* (1972) have shown that the most rapid rate of light-dependent synthesis of chloroplast RNA occurs during the first eight hours of development.

Mutant stains of *Euglena* have been employed to determine the nature of the light-mediated control of synthesis of RNA. Zeldin and Schiff (1968) reported that, even in some aplastidic strains, light-stimulated RNA synthesis persists. Cohen and Schiff (1973) concluded, however, that there is no change in the net amount of cytoplasmic ribosomal RNA during chloroplast development. Although they described a light-induced increase in the rate of $^{32}PO_4$-labelling of cytoplasmic ribosomal RNA in a chloroplast-less mutant, they interpreted their results in terms of a light-stimulated increase in the rate of turnover of cytoplasmic ribosomes. These findings, together with an observed induction of polysome formation by light in a strain of *Euglena* which lacks chloroplasts (Freyssinet & Schiff 1974), are strong evidence that products of the chloroplast genome are not involved in the regulation of cytoplasmic ribosomal RNA metabolism during the illumination of dark-grown cells. Cohen and Schiff (1973) reported that the regulation of cytoplasmic and chloroplast-ribosomal-RNA synthesis differ because, after light-induction, labelling of chloroplast rRNA with $^{32}PO_4$ continues when cells are returned to darkness while cytoplasmic rRNA is labelled at progressively decreased rates. Heizmann (1970) also noted that chloroplast-ribosomal-RNA synthesis is not dependent upon continual illumination once it is light-induced.

The transcription of ribosomal RNA genes in isolated chloroplasts of *Euglena* has been demonstrated (Carritt & Eisenstadt 1973a, b; Harris *et al.* 1973). As shown in Table 13.3, this synthesis is sensitive to actinomycin but is not consistently inhibited by rifampicin. Although Brown *et al.* (1970) concluded that chloroplast-ribosomal-RNA synthesis in whole cells during chloroplast development is inhibited by rifampicin if the cells are rendered permeable to the drug by dimethylsulphoxide, Heizmann (1974a) was not able to confirm their observations.

Chloroplasts of *Euglena* contain unique transfer RNAs for isoleucine, leucine, glutamic acid, aspartic acid, phenylalanine, proline and valine. These are recognized by chloroplast-specific aminoacyl-synthetases but not by the corresponding cytoplasmic enzymes (Kislev *et al.* 1972). Kislev *et al.* (1972) also demonstrated that transfer RNAs for arginine, glutamine, histidine,

Table 13.3. Sensitivity of nucleic acid synthesis in *Euglena* to inhibitors

	Time* (hrs)	Per cent of control	Reference
1. Chloroplast DNA synthesis			
Chloramphenicol			
1mg/ml	70	73	Richards Ryan & Manning 1971
Cycloheximide			
15µg/ml	70	46	Richards Ryan & Manning 1971
Nalidixic acid			
50µg/ml	24	0	Lyman Jupp & Larrinua 1975
2. DNA synthesis in isolated chloroplasts			
Actinomycin D			
1µg/ml		88–89	Scott Shah & Smillie 1968
10µg/ml		57–61	Scott Shah & Smillie 1968
3. Mitochondrial DNA synthesis			
Chloramphenicol			
1mg/ml	70	73–74	Richards Ryan & Manning 1971
Cycloheximide			
15µg/ml	70	44–82	Richards Ryan & Manning 1971
Ethidium bromide			
1µg/ml	72	106	Nass & Ben-Shaul 1973
5µg/ml	72	73	Nass & Ben-Shaul 1973

	Time* (hrs)	Per cent of control	Reference
4. Nuclear DNA synthesis			
Chloramphenicol			
1mg/ml	70	73–85	Richards Ryan & Manning 1971
Cycloheximide			
15µg/ml	70	17–29	Richards Ryan & Manning 1971
5. Chloroplast 16S-ribosomal RNA synthesis			
Rifampicin+1% dimethyl sulfoxide			
10µg/ml	9	82	Heizmann 1974a
20µg/ml	12	0	Brown Bastia & Haselkorn 1970
6. RNA synthesis by isolated chloroplasts			
Actinomycin D			
10µg/ml		21	Carritt & Eisenstadt 1973a
15µg/ml		27	Harris Preston & Eisenstadt 1973
20µg/ml		3	Shah & Lyman 1966
Rifampicin			
5µg/ml		36	Carritt & Eisenstadt 1973a; Harris, Preston & Eisenstadt 1973
100µg/ml		30–98	Carritt & Eisenstadt 1973a

* Duration of treatment of cells undergoing cell division or chloroplast development with inhibitors.

lysine and serine, isolated from chloroplasts, are preferentially aminoacylated by chloroplast synthetases but that they also show some activity with cytoplasmic enzymes. A third group of chloroplast transfer RNAs, including those for alanine, glycine, methionine, threonine and tryptophan, were found to be more active with cytoplasmic synthetases than with the homologous aminoacylating enzymes isolated from chloroplasts.

During chloroplast development, some of the chloroplast transfer RNAs which are synthesized appear to be products of chloroplast gene transcription. During incubation in the light, cells synthesize the tRNAs specific for amino acids, phenylalanine, glutamate, methionine and isoleucine, none of which are detectable in dark-grown cells (Barnett et al. 1969; Reger et al. 1970; Goins et al. 1973). Goins et al. (1973) showed that the synthesis of isoleucyl-tRNA exhibits about an eight-hour lag following exposure of cells to light. Information on the kinetics of synthesis of other light-induced transfer RNAs is lacking.

The effect of light on RNA synthesis in *Euglena* has been examined in some detail by Nigon et al. (1967), who were able to distinguish two classes of polysomes produced during the illumination of dark-grown cells. Photoactive polysomes were characterized by their rapid degradation upon return of briefly illuminated cells to the dark. Photoinduced polysomes were defined as those light-induced species which continue to be formed for some time in cells subsequently incubated in the dark (Heizmann et al. 1972). Verdier et al. (1973) demonstrated that the polysomal RNAs could be distinguished on the basis of association with polyadenylic acid sequences. Poly(A)-containing polysomal RNAs are synthesized only under conditions of continuous light and are not formed when cells given brief illumination are incubated in the dark. Verdier et al. (1973) concluded that the polyribosomal species containing poly(A) belong to the photoactive type defined by Nigon et al. (1967) and Heizmann et al. (1972). Since prokaryotic messenger RNAs do not contain poly(A) sequences (Perry et al. 1972), they suggested that the photoinduced polysomes may contain messenger RNA transcribed from chloroplast genes and that the photoactive polysomes originate in the nucleus. This hypothesis is supported by the observation of Heizmann (1970, 1974b) and Cohen and Schiff (1973) that chloroplast-ribosomal-RNA synthesis continues in cells incubated in the dark following brief light exposure.

Although polysomes have been isolated from chloroplasts of *Euglena* (Harris & Eisenstadt 1971; Avadhani & Buetow 1972b), *Chlamydomonas* (Hoober & Blobel 1969; Chua et al. 1973) and pea seedlings (Philippovich et al. 1973), whether these contain poly(A) has not been established. [A poly(A) polymerase has been isolated from wheat chloroplasts (Burkard & Keller 1974). The presence of poly(A) polymerase in mitochondria of some mammalian cells (Jacob & Schindler 1972) may be correlated with the presence of polyadenylated RNA transcribed from the organellar genome (Hirsch & Penman 1974; Ojala & Attardi 1974). However, Avadhani et al.

(1973) have determined that at least some of the poly(A)-RNA in mammalian mitochondria is of nuclear origin. Groot *et al.* (1974) have concluded that yeast mitochondria contain no polyadenylated RNA. They propose that lower eukaryotic mitochondria have retained more of the characteristics of prokaryotic genetic systems than have mammalian mitochondria through evolution and, consequently, have not adopted the functional requirements fulfilled by poly(A)-RNA in higher forms. Chloroplasts have a greater degree of genetic autonomy than mitochondria in terms of genetic complexity. An extension of the hypothesis of Groot *et al.* (1974), therefore, would predict the absence of polyadenylated messenger RNA in chloroplasts.] It also remains equivocal whether in *Euglena* the poly(A)-RNA functions in chloroplast as well as cytoplasmic protein synthesis, and whether this RNA originates genetically from the nucleus or from the chloroplasts.

Because the synthesis of chloroplast RNA is environmentally regulated, one could postulate that the activity of plastid-associated RNA polymerase would likewise be subject to light-mediated control. Munns *et al.* (1971) reported that a DNA-dependent RNA polymerase normally present in isolated chloroplasts is not found when dark-grown cells are maintained in an inorganic medium in the dark. There is, however, plastid-RNA polymerase in cells grown in the dark under normal organotrophic conditions. RNA polymerase activity in non-dividing dark-grown cells exposed to light increases after a lag period of six to eight hours, reaching maximum levels after 24 hours of light treatment. Because cycloheximide inhibits the synthesis of plastid-RNA polymerase but has little effect on the differentiation of proplastids to chloroplasts, Munns *et al.* (1971) concluded that the enzyme functions in the synthesis of RNA required for chloroplast replication but is not essential for chloroplast development.

Cells of *Euglena gracilis*, grown organotrophically for many generations in the dark, possess about ten proplastids. These contain only small amounts of most soluble enzymes involved in photosynthetic carbon reduction and electron-transport components, no chlorophyll and few internal membranes. Upon illumination, a dramatic differentiation of the proplastids into chloroplasts ensues. Table 13.5 illustrates the synthesis of chloroplast components during light-induction; the levels found in dark-grown cells are compared to those of fully mature chloroplasts and to cases where a plastid constituent is measured in aplastidic strains. The work with chloroplast-less mutants implies that among constituents coded for by the chloroplast genome are at least one aminoacyl-tRNA synthetase, a number of transfer RNA species, cytochrome c_{552} and ribulose 1,5-diphosphate carboxylase.

An indirect approach to the location of genetic information for many chloroplast components is through the use of protein-synthesis inhibitors specific for either chloroplast or cytoplasmic ribosomes (Table 13.4). Inhibition of the synthesis of proteins by antibiotics which preferentially block protein synthesis in chloroplasts may suggest that the site of synthesis (and,

Table 13.4. Sensitivity of protein synthesis in *Euglena* chloroplasts, mitochondria and cytoplasm to inhibitors

1. Chloroplast protein synthesis

A. Isolated chloroplasts

	Per cent of control	Reference
Actinomycin D		
25µg/ml	88	Harris Preston & Eisenstadt 1973
27µg/ml	88	Eisenstadt & Brawerman 1963
50µg/ml	70	Gnaman & Kahn 1967
Chloramphenicol		
5µg/ml	54	Reger Smillie & Fuller 1972
20µg/ml	22	Reger Smillie & Fuller 1972
40µg/ml	13	Reger Smillie & Fuller 1972
60µg/ml	99	Eisendstadt & Brawerman 1964
100µg/ml	11	Gnaman & Kahn 1967
100µg/ml	22	Harris Preston & Eisenstadt 1973
120µg/ml	84	Eisenstadt & Brawerman 1964
200µg/ml	8	Reger Smillie & Fuller 1972
210µg/ml	55	Eisendstadt & Brawerman 1964
Cycloheximide		
5µg/ml	89	Reger Smillie & Fuller 1972
100µg/ml	83	Harris Preston & Eisenstadt 1973
200µg/ml	89	Reger Smillie & Fuller 1972
Oligomycin		
8µM	98	Harris Preston & Eisenstadt 1973
Penicillin G		
50µg/ml	98	Gnaman & Kahn 1967
Rifampicin		
25µg/ml	85	Harris Preston & Eisenstadt 1973

	Per cent of control	Reference
Spectinomycin		
72µg/ml	98	Harris Preston & Eisenstadt 1973
Streptomycin		
50µg/ml	69	Gnaman & Kahn 1967
200µg/ml	40	Gnaman & Kahn 1967
250µg/ml	9	Gnaman & Kahn 1967
Tetracycline		
200µg/ml	32	Harris Preston & Eisenstadt 1973

B. Isolated proplastids

	Per cent of control	Reference
Chloramphenicol		
5µg/ml	89	Reger Smillie & Fuller 1972
20µg/ml	55	Reger Smillie & Fuller 1972
40µg/ml	41	Reger Smillie & Fuller 1972
200µg/ml	24	Reger Smillie & Fuller 1972
Cycloheximide		
5µg/ml	100	Reger Smillie & Fuller 1972
200µg/ml	96	Reger Smillie & Fuller 1972

C. Chloroplast ribosomes

	Per cent of control	Reference
Chloramphenicol		
60µg/ml	99	Eisenstadt & Brawerman 1964
120µg/ml	74–88	Eisenstadt & Brawerman 1964
210µg/ml	61–87	Eisenstadt & Brawerman 1964
360µg/ml	59–74	Eisenstadt & Brawerman 1964

1. Chloroplast protein synthesis (*continued*)
  D. Chloroplast polyribosomes
  Chloramphenicol
    250μg/ml 48 Avadhani & Buetow 1972b
  Cycloheximide
    250μg/ml 93 Avadhani & Buetow 1972b

2. Mitochondrial protein synthesis
  A. Isolated mitochondria
  Chloramphenicol
    400μg/ml 15 Kislev & Eisenstadt 1972
    500μg/ml 12 Avadhani Lynch & Buetow 1971
  Cycloheximide
    200μg/ml 95 Kislev & Eisenstadt 1972
    500μg/ml 89 Avadhani Lynch & Buetow 1971

  B. Mitochondrial ribosomes
  Cycloheximide
    250μg/ml 90 Avadhani & Buetow 1972a
  Streptomycin
    250μg/ml 18 Avadhani & Buetow 1972a

  C. Mitochondrial polyribosomes
  Chloramphenicol
    250μg/ml 47 Avadhani & Buetow 1972b
  Cycloheximide
    250μg/ml 90 Avadhani & Buetow 1972a, b
  Streptomycin
    250μg/ml 35 Avadhani & Buetow 1972a

3. Cytoplasmic protein synthesis
  A. Cytoplasmic ribosomes
  Chloramphenicol
    60μg/ml 99 Eisenstadt & Brawerman 1964
    120μg/ml 103 Eisenstadt & Brawerman 1964
    210μg/ml 94 Eisenstadt & Brawerman 1964
    360μg/ml 94 Eisenstadt & Brawerman 1964
    500μg/ml 100 Avadhani Lynch & Buetow 1971
  Cycloheximide
    500μg/ml 17 Avadhani Lynch & Buetow 1971

  B. Cytoplasmic polyribosomes
  Chloramphenicol
    250μg/ml 91 Avadhani & Buetow 1972b
  Cycloheximide
    250μg/ml 22 Avadhani & Buetow 1972b

Table 13.5. Comparison of chloroplast components in dark-grown, light-grown and chloroplast-less *Euglena*

	Dark-grown	Chloroplast-less (light-grown)	Light-grown	Reference
1. Aminoacyl-tRNA synthetase				
(10^{-3} μM/hr/mg protein)				
Alanine	2·8		6·4	Parthier Krauspe & Samtleben 1972
Arginine	1·6		11·2	Parthier Krauspe & Samtleben 1972
Aspartate	5·2		50·0	Parthier Krauspe & Samtleben 1972
Glycine	3·6		8·0	Parthier Krauspe & Samtleben 1972
Isoleucine	3·2		8·2	Parthier Krauspe & Samtleben 1972
	0	0 (W3BUL)	Inducible	Reger Fairfield Epler & Barnett 1970
Leucine	14·2		158·0	Parthier Krauspe & Samtleben 1972
Methionine	3·0		34·2	Parthier Krauspe & Samtleben 1972
Phenylalanine	10·0		83·0	Parthier Krauspe & Samtleben 1972
	0·027	present (W3BUL)	0·108	Hecker Egan, Reynolds, Nix, Schiff & Barnett 1974 Reger Fairfield Epler & Barnett 1970
Proline	6·6		6·4	Parthier Krauspe & Samtleben 1972
Serine	6·0		9·6	Parthier Krauspe & Samtleben 1972
Threonine	102·4		113·0	Parthier Krauspe & Samtleben 1972
Tyrosine	1·8		8·8	Parthier Krauspe & Samtleben 1972
Valine	10·2		53·2	Parthier Krauspe & Samtleben 1972
	0·012	present (W3BUL)	0·084	Hecker Egan Reynolds Nix Schiff & Barnett 1974
2. Aminolaevulinate dehydratase				
(10^{-3} μM/hr/10^6 cells)	4·81	18·0 (W34ZUD)	22·9	Hovenkamp-Obbema Moorman & Stegwee 1974
	8·0–18·0		38·0	Salvador Nigon Richard & Nicolas 1972
3. CO₂ Fixation				
(μM/hr/mg protein)	0	0 (W3BUL)	3·4	Latzko & Gibbs 1969
(10^{-6} μM/hr/cell)	0		0·28–1·41	Bovarnick Schiff Freedman & Egan 1974
(10^{-6} μM/hr/10^{-6} μg chlorophyll)			0·126	Bovarnick Zeldin & Schiff 1969

4. Cytochrome 552 (10⁻¹¹μM/cell)	0 (W3BUL)		2·79	Bovarnick Schiff Freedman & Egan 1974
			3·6	Smillie 1963
	0 (W3 & W8)		4·2	Perini Schiff & Kamen 1964
			4·64	Smillie Graham Dwyer Grieve & Tobin 1967
5. Cytochrome 561 (10⁻¹¹μM/cell)			4·4	Smillie Graham Dwyer Grieve & Tobin 1967
6. DNA (molecules per cell)		30	60	Uzzo & Lyman 1972
			30	Schiff 1973
7. DNAase, alkaline (10⁻³μM/hr/mg protein)	20·4 (W3BUL)	22·9	138·5	Egan & Carrell 1972
8. Ferredoxin-NADP reductase (μM/hr/10⁹ cells)		8·0	863·0	Bishop Bain & Smillie 1973
9. Fructose 1,6-diphosphatase, alkaline (μM/hr/mg protein) (μM/hr/10⁹ cells)		1·2 206·0	2·0 1212·0	Latzko & Gibbs 1969 Smillie 1963
10. Fructose 1,6-diphosphate aldolase, Class I (μM/hr/mg protein)	0·41 (W14ZNaIL)	0·49	4·68	Latzko & Gibbs 1969 Schmidt & Lyman 1974
11. Glyceraldehyde 3-phosphate dehydrogenase, NADP-dependent (μm/hr/mg protein)	4·01 (W3BUL) 9·8 (W14ZNaIL)	4·9 4·3–6·2	71·0 31–40	Latzko & Gibbs 1969 Bovarnick Schiff Freedman & Egan 1974 Schmidt & Lyman 1974

Table 13.5—*continued.* Comparison of chloroplast components in dark-grown, light-grown and chloroplast-less *Euglena*

	Dark-grown	Chloroplast-less (light-grown)	Light-grown	Reference
12. Glycerate 3-phosphate kinase (μM/hr/mg protein)	25·4	23·3 (W14ZNaIL)	163	Latzko & Gibbs 1969 Schmidt & Lyman 1974
13. Glycollate-DPIP oxidoreductase (μM/hr/mg protein)	0·2		2·05	Codd & Merrett 1970
14. O_2 evolution (μM/hr/mg protein) (μL/hr/10^6 cells) (μM/hr/10^6 cells)	0 4·4 0		3·4 27·9 120·2	Latzko & Gibbs 1969 Smillie 1963 Gnaman & Kahn 1967
15. Phosphoglycollate phosphatase (μM/hr/mg protein)	28		85	Codd & Merrett 1970
16. Photophosphorylation (μM/hr/mg protein)	0		3·42	Gnaman & Kahn 1967
17. Photosystem I (μM/hr/mg chlorophyll)	0		61	Bishop *et al.* 1973
18. Photosystem II (μM/hr/mg chlorophyll)	0		220	Bishop *et al.* 1973
19. Photosystem I+II (μM/hr/mg protein)	0		173	Bishop *et al.* 1973
20. Ribose 5-phosphate isomerase (μM/hr/mg protein)	17·5		22·7	Latzko & Gibbs 1969

21. Ribulose 5-phosphate kinase				
(μM/hr/mg protein)	5·2		23·4	Latzko & Gibbs 1969
N (μM/hr/10⁹ cells)	0·87	0·15 (W34ZUD)	8·2	Salvador et al. 1972
22. Ribulose 1,5-diphosphate carboxylase				
(μM/hr/mg protein)	0·07	0 (W3BUL)	1·9	Latzko & Gibbs 1969
	0·16	0 (W14ZNaIL)	4·14	Bovarnick et al. 1974
				Schmidt & Lyman 1974
23. Ribosomal RNA				
(10⁻⁹ mg/cell)	0·5		2·5	Heizmann 1974b
24. RNA polymerase				
(CPM*/10⁶ cells)	6·9		16·0	Smillie et al. 1971
25. Lamellar structural protein, determined by				
the procedure of Criddle (1966)				
(mg/10⁹ cells)	1·6		4·6	Smillie et al. 1968
26. Transfer RNA				
Glutamyl II	+	+(W3BUL)	++	Barnett et al. 1969
Glutamyl III	0	0 (W3BUL)	+	Barnett et al. 1969
Isoleucyl II	0	0 (W3BUL)	+	Barnett et al. 1969
				Reger et al. 1970
Isoleucyl III, IV, V	0	0 (W8BHL)	+	Goins et al. 1973
Methionyl III	0		+	Goins et al. 1973
Phenylalanyl I	0	0 (W3BUL)	+	Barnett et al. 1969
				Reger et al. 1970
27. Triosephosphate isomerase				
(μM/hr/mg protein)	485		525	Latzko & Gibbs 1969
28. Xylulose 5-phosphate epimerase				
(μM/hr/mg protein)	10·5		73	Latzko & Gibbs 1969

* Incorporation of tritiated UTP per 10⁶ cells.

perhaps, the probable origin of genetic information) is with the chloroplast. The results of such studies are tabulated in Table 13.6. The origin of genetic information for chloroplast proteins would be more definitely resolved if inhibitors were found which specifically prevent RNA synthesis in chloroplasts or in the nucleus. As shown in Table 13.3, rifampicin is, unfortunately, not a suitable agent for inhibition of chloroplast-RNA synthesis in *Euglena*. A questionable feature in the use of protein-synthesis inhibitors in this endeavour is the possibility that messenger RNA of nuclear origin is translated within organelles (Avadhani *et al.* 1973) or that messenger RNA of chloroplast origin is translated on cytoplasmic ribosomes (Yu & Stewart 1974). Other problems arising from the assumption that protein-synthesis inhibitors can be used for locating the respective genes for chloroplast components in *Euglena* are examined below.

In addition to the synthesis of ribosomal and transfer RNAs, other constituents of the translational machinery of *Euglena* chloroplasts are made during chloroplast development. Among these are the transfer-RNA aminoacyl-synthetases, which have been isolated from chloroplasts and characterized by Reger *et al.* (1970), Kislev *et al.* (1972) and Hecker *et al.* (1974). Parthier, Krauspe and Samtleben (1972) examined the light-induced chloroplast-synthetase activities in crude whole-cell extracts of *Euglena* by employing transfer RNAs from the blue-green alga *Anacystis nidulans* in their assays. (The similarity of *Euglena* chloroplast-transfer RNAs to those of prokaryotes had been demonstrated by Fairfield and Barnett in 1971.) Parthier *et al.* (1972) determined that upon illumination of dark-grown cells aminoacyl-synthetases are synthesized with no apparent lag and therefore represent the earliest proteins made during chloroplast development. The synthesis of many of these enzymes has been shown to be inhibited by chloramphenicol, but also by cycloheximide if high concentrations are used (Table 13.6). If the increase of the chloroplast aminoacyl-transfer RNA synthetases is required for protein synthesis during chloroplast development, then either cycloheximide or chloramphenicol would be expected to affect the synthesis of proteins on chloroplast ribosomes.

The capacity of isolated plastids to synthesize proteins during development was examined by Reger *et al.* (1972), who found that the rate of incorporation of leucine by proplastids is only 5 to 10% of that of fully-developed chloroplasts. The protein-synthesizing capacity of developing chloroplasts was shown not to change for approximately 12 hours during illumination of dark-grown cells; during the subsequent 12 hours the rate of leucine incorporation by plastids increases to nearly that of mature chloroplasts. Cycloheximide was shown not to affect development of the capacity of plastids to synthesize proteins. Therefore, if chloroplast ribosomal proteins, possible initiation factors and other proteins necessary for chloroplast protein synthesis are generated in the cytoplasm, their synthesis does not appear to be required for chloroplast development. One must conclude that cycloheximide inhibition of protein synthesis is specific for cytoplasmic ribosomes in *Euglena* (Table 13.4) and that this agent does not directly or indirectly inhibit protein synthesis by chloroplast ribosomes.

It is apparent from Table 13.6 that the effects of inhibitors of protein synthesis in *Euglena* are very dependent upon the concentration and duration of treatment with antibiotics. Discrepant reports of inhibitor effects can be largely explained on this basis, and Table 13.6 illustrates the need for standard conditions in studies of this kind.

If transport of messenger RNA of nuclear origin into chloroplasts does not occur, Kirk (1972) has argued that identification of a protein synthesized by isolated plastids suggests that it may have been coded for by plastid DNA. Characterization of the proteins synthesized by isolated *Euglena* chloroplasts has been carried out by Harris *et al.* (1973), who showed that plastids *in vitro* can actively incorporate radioactive amino acids into several protein species and that the proteins synthesized in isolated chloroplasts are electrophoretically identical to those labelled by ^{14}C-amino acids in cells treated with cycloheximide *in vivo*. The predominant proteins generated by isolated chloroplasts were fractions corresponding to ribulose 1,5-diphosphate carboxylase and lamellar structural proteins. Several electrophoretic bands and gelfiltration peaks indicated to Harris *et al.* (1973) that various soluble proteins were also synthesized by isolated plastids, but they were not further identified. In these experiments they found no evidence for synthesis of chloroplast ribosomal proteins, cytochrome c_{552} or ferredoxin, although some unidentified cytochromes were labelled.

Although the studies of Harris *et al.* (1973) substantiate the validity of conclusions drawn from the use of inhibitors of protein synthesis *in vivo*, there is some evidence that in *Euglena* antibiotics may interfere with nuclearchloroplast interactions which regulate chloroplast synthesis. Wolfovitch and Perl (1972) and Perl (1972) observed that low concentrations of cycloheximide can stimulate the development of the ability of *Euglena* chloroplasts to fix CO_2 and synthesize chlorophyll, and can thereby counteract the normally inhibitory affects of chloramphenicol on this process. Since high concentrations of cycloheximide are strongly inhibitory (Table 13.6), they concluded that the proteins involved in chlorophyll synthesis and CO_2 fixation are synthesized on cytoplasmic ribosomes. Wolfovitch and Perl postulated that the regulation of synthesis of these proteins is effected by a repressor which also is synthesized on cytoplasmic ribosomes. In their scheme, low concentrations of cycloheximide inhibit the synthesis of the repressor and therefore stimulate the synthesis of chloroplast-associated proteins. They furthermore proposed that chloroplast ribosomes are involved in the synthesis of a *de*repressor which is not made during chloramphenicol treatment of cells undergoing chloroplast development. In the absence of such a derepressor, they explain, chloroplast development is inhibited to the same extent as in cells treated with high concentrations of cycloheximide. A similar repressor-

Table 13.6. Effects of inhibitors on the synthesis of components during *Euglena* chloroplast development.

Time* (hrs)	Per cent of control	Reference	Time* (hrs)	Per cent of control	Reference
1. Chlorophyll synthesis					
Actinomycin D (µg/ml)					
10	32	33–44	Pogo & Pogo 1964		
20	32	32	Pogo & Pogo 1964		
20	36–38	16	Shah & Lyman 1966		
20	48	83	Bovarnick Zeldin & Schiff 1969		
25	48	54	Bovarnick Zeldin & Schiff 1969		
35	48	25	Bovarnick Zeldin & Schiff 1969		
45	48	18	Bovarnick Zeldin & Schiff 1969		
Chloramphenicol (mg/ml)			1·0	48	15	Bishop Bain & Smillie 1973
2·0	12	23	Kirk 1968	1·0	48	22	Bishop & Smillie 1970
0·65	24	29	Perl 1972	1·0	48	6	Smillie Evans & Lyman 1963
2·0	24	5	Parthier 1973	1·0	48	16	Smillie Graham Dwyer Grieve & Tobin 1967
0·06	48	72	Smillie Evans & Lyman 1963	1·3	48	0	Perl 1972
0·125	48	50	Smillie Evans & Lyman 1963	3·9	48	17	Stewart & Gregory 1969
0·16	48	82	Perl 1972	2·0	72	49	Bovarnick Schiff Freedman & Egan 1974
0·25	48	38	Bishop & Smillie 1970	**Cycloheximide (µg/ml)**			
0·25	48	29	Smillie Evans & Lyman 1963	0·003	48	133	Perl 1972
0·32	48	59	Perl 1972	0·028	48	166	Perl 1972
0·50	48	43	Ben-Shaul & Markus 1969	0·28	48	367	Perl 1972
0·50	48	42	Hovenkamp-Oobema Moorman & Stegwee 1974	1·0	48	317	Parthier 1973
				2·0	48	255	Parthier 1973
				2·0	48	199	Reger Smillie & Fuller 1972
				2·5	48	107	Bishop Bain & Smillie 1973
0·50	48	16	Smillie Evans & Lyman 1963	2·5	48	135	Bishop & Smillie 1970
0·65	48	29	Perl 1972	2·81	48	387	Perl 1972
0·75	48	22	Reger Smillie & Fuller 1972	3·0	48	72	Ben-Shaul & Ophir 1970
				4·0	48	100	Parthier 1973
				5·0	48	56	Ben-Shaul & Ophir 1970
				5·0	48	50	Bishop Bain & Smillie 1973
				5·0	48	72	Hovenkamp-Oobema Moorman & Stegwee 1974
				6·0	48	21	Parthier 1973

1. Chlorophyll synthesis (*continued*)

Cycloheximide (μg/ml)

8·0	48	15	Ben-Shaul & Ophir 1970
10·0	48	42	Ben-Shaul & Ophir 1970
10·0	48	5	Bishop & Smillie 1970
15·0	48	0	Ben-Shaul & Ophir 1970
15·0	48	14	Smillie Graham Dwyer Grieve & Tobin 1967
20·0	48	0	Ben-Shaul & Ophir 1970
28·1	48	67	Perl 1972
281·0	48	0	Perl 1972

Mitomycin C (μg/ml)

30·0	36	188	Shah & Lyman 1966

Myxin (μg/ml)

8·0	48	15	McCalla & Baerg 1969

Rifampicin (μg/ml)

250·0	48	100	Theiss-Seuberling 1973

Streptomycin (mg/ml)

1·0	6	43	Kirk 1962
2·0	6	60	Kirk 1962
0·5	12	105	Bovarnick Chang Schiff & Schwartzbach 1974
0·15	24	54	Gnaman & Kahn 1967
0·5	48	16	Ben-Shaul Silman & Ophir 1972
0·5	49	23	Hecker Egan Reynolds Nix Schiff & Barnett 1974
0·5	72–96	8	Bovarnick & Schiff (Schiff 1970)

2. Carotenoid synthesis

Actinomycin D (μg/ml)

0·13	4	100	Kirk & Allen 1965
0·32	4	80	Kirk & Allen 1965
1·0	4	80	Kirk & Allen 1965
10·0	4	78	Kirk & Allen 1965

Chloramphenicol (mg/ml)

2·0	12	79	Kirk 1968
1·0	48	46	Hovenkamp-Obbema 1974

Streptomycin (mg/ml)

0·5	42–54	47	Bovarnick Chang Schiff & Schwartzbach 1974
0·5	69–72	39	Bovarnick Chang Schiff & Schwartzbach 1974

3. Aminolaevulinate dehydratase

Chloramphenicol (mg/ml)

0·5	48	98	Hovenkamp-Obbema Moorman & Stegwee 1974

Cycloheximide (μg/ml)

5·0	48	26	Hovenkamp-Obbema Moorman & Stegwee 1974

4. Cytochrome 552

Chloramphenicol (mg/ml)

1·0	72	17	Smillie Graham Dwyer Grieve & Tobin 1967

Table 13.6—*continued.* Effects of inhibitors on the synthesis of components during *Euglena* chloroplast development.

	Time* (hrs)	Per cent of control	Reference
4. Cytochrome 552 (continued)			
Cycloheximide (μg/ml)			
15·0	72	25	Smillie Graham Dwyer Grieve & Tobin 1967
Streptomycin (mg/ml)			
0·5	12	88	Bovarnick Chang Schiff & Schwartzbach 1974
0·5	72–96	13	Bovarnick & Schiff (Schiff 1970)
0·5	68–117	14	Bovarnick Schiff Freedman & Egan 1974
5. Cytochrome 561			
Chloramphenicol (mg/ml)			
1·0	72	2	Smillie Graham Dwyer Grieve & Tobin 1967
Cycloheximide (μg/ml)			
15·0	72	30	Smillie Graham Dwyer Grieve & Tobin 1967
6. Ferredoxin-NADP reductase			
Chloramphenicol (mg/ml)			
1·0	48	48	Bishop Bain & Smillie 1973
1·0	72	19	Smillie Graham Dwyer Grieve & Tobin 1967
Cycloheximide (μg/ml)			
2·5	48	73	Bishop Bain & Smillie 1973
15·0	72	36	Smillie Graham Dwyer Grieve & Tobin 1967
7. Fraction I protein			
Chloramphenicol (mg/ml)			
1·0	72	5	Smillie Graham Dwyer Grieve & Tobin 1967
Cycloheximide (μg/ml)			
15·0	72	90	Smillie Graham Dwyer Grieve & Tobin 1967
8. Fructose 1,6-diphosphate aldolase Class I			
Chloramphenicol (mg/ml)			
0·5	48	0	Mo Harris & Gracy 1973
1·5	66	29	Smillie & Scott 1969
Cycloheximide (μg/ml)			
3·0	66	90	Smillie & Scott 1969
9. Glyceraldehyde 3-phosphate dehydrogenase, NADP-dependent			
Actinomycin D (μg/ml)			
134·0	74	86	Theiss-Seuberling 1973

9. Glyceraldehyde 3-phosphate dehydrogenase (continued)

Chloramphenicol
(mg/ml)

	48		
1·0	48	24	Hovenkamp-Obbema & Stegwee 1974
1·0	72	0	Smillie Graham Dwyer Grieve & Tobin 1967
2·0	72	49–92	Bovarnick Schiff Freedman & Egan 1974 Bovarnick & Schiff (Schiff 1970)
15·0	36–48	9–11	Bovarnick Schiff Freedman & Egan 1974
15·0	72	118	Smillie Graham Dwyer Grieve & Tobin 1967
15·0	72–96	5	Bovarnick Schiff Freedman & Egan 1974

Rifampicin
(µg/ml)

	48		
250·0	48	95	Theiss-Seuberling 1973

Streptomycin
(mg/ml)

0·5	36–48	87	Bovarnick Schiff Freedman & Egan 1974
0·5	72	52	Bovarnick & Schiff (Schiff 1970)
0·5	72–96	59	Bovarnick Schiff Freedman & Egan 1974

10. Glycollate-DCPIP oxidoreductase

Chloramphenicol
(mg/ml)

1·0	48	60	Codd & Merrett 1970

Cycloheximide
(µg/ml)

	48		
15·0	48	0	Codd & Merrett 1970

11. Phosphoglycollate phosphatase

Chloramphenicol
(mg/ml)

1·0	48	70	Codd & Merrett 1970

Cycloheximide
(µg/ml)

15·0	48	0	Codd & Merrett 1970

12. Photosystems I+II

Chloramphenicol
(mg/ml)

1·0	48	4	Bishop Bain & Smillie 1973

Cycloheximide
(µg/ml)

2·5	48	57	Bishop Bain & Smillie 1973

13. Photosystem I

Chloramphenicol
(mg/ml)

1·0	48	0	Bishop Bain & Smillie 1973

Cycloheximide
(µg/ml)

2·5	48	38	Bishop Bain & Smillie 1973

14. Photosystem II

Chloramphenicol
(mg/ml)

1·0	48	14	Bishop Bain & Smillie 1973

Table 13.6—*continued*. Effects of inhibitors on the synthesis of components during *Euglena* chloroplast development.

	Time* (hrs)	Per cent of control	Reference
14. Photosystem II (*continued*)			
Cycloheximide (µg/ml)			
2·5	48	48	Bishop Bain & Smillie 1973
15. RNA polymerase			
Chloramphenicol (mg/ml)			
1·0	24	0	Smillie Munns Graham Scott & Grieve 1971
1·0	48	0	Smillie Munns Graham Scott & Grieve 1971
16. Ribulose 1,5-diphosphate carboxylase			
Chloramphenicol (mg/ml)			
2·0	24	3	Parthier 1973
1·0	48	15	Hovenkamp-Obbema & Stegwee 1974
1·0	72	5	Smillie Graham Dwyer Grieve & Tobin 1967
2·0	72	15–38	Bovarnick Schiff Freedman & Egan 1974 Bovarnick & Schiff (Schiff 1970)
Cycloheximide (µg/ml)			
4·0	24	130	Parthier 1973
6·0	24	55	Parthier 1973
15·0	72	216	Smillie Graham Dwyer Grieve & Tobin 1967
Streptomycin (mg/ml)			
0·5	55–115	3	Bovarnick Schiff Freedman & Egan 1974
0·5	72	0·5	Bovarnick & Schiff (Schiff 1970)
17. Structural protein (Criddle 1966)			
Chloramphenicol (mg/ml)			
1·0		93	Smillie et al. 1968
18. Triosephosphate isomerase (Type A)			
Chloramphenicol (mg/ml)			
0·5	48	0	Mo et al. 1973
19. Aminoacyl-tRNA synthetases			
Chloramphenicol (mg/ml)			
Ala 1·0	48	78	Parthier et al. 1972
Arg 2·0	24	97	Parthier 1973
1·0	48	55	Parthier et al. 1972
Asp 2·0	24	71	Parthier 1973
1·0	48	79	Parthier et al. 1972
Gly 1·0	48	73	Parthier et al. 1972
Ile 2·0	24	93	Parthier 1973
1·0	48	63	Parthier et al. 1972
Leu 2·0	24	60	Parthier 1973
1·0	48	48	Parthier et al. 1972

19. Aminoacyl-tRNA synthetases (continued)

Chloramphenicol (mg/ml)

Lys	2·0	24	90	Parthier 1973
	1·0	48	59	Parthier et al. 1972
Met	1·0	48	52	Parthier et al. 1972
Phe	2·0	24	62	Parthier 1973
	1·0	48	61	Parthier et al. 1972
Pro	1·0	48	97	Parthier et al. 1972
Ser	2·0	24	105	Parthier 1973
	1·0	48	75	Parthier et al. 1972
Thr	2·0	24	105	Parthier 1973
	1·0	48	99	Parthier et al. 1972
Tyr	2·0	24	51	Parthier 1973
	1·0	48	53	Parthier et al. 1972
Val	2·0	24	66	Parthier 1973
	1·0	48	64	Parthier et al. 1972
Lys	6·0	48	49	Parthier 1973
	6·0	24	21	Parthier 1973
	4·0	48	163	Parthier et al. 1972
Met	4·0	48	171	Parthier et al. 1972
Phe	6·0	24	40	Parthier 1973
	15·0	36	1	Hecker Egan Reynolds Nix Schiff & Barnett 1974
	4·0	48	192	Parthier et al. 1972
Pro	4·0	48	91	Parthier et al. 1972
Ser	6·0	24	62	Parthier 1973
	4·0	48	128	Parthier et al. 1972
Thr	6·0	24	93	Parthier 1973
	4·0	48	131	Parthier et al. 1972
Tyr	6·0	24	23	Parthier 1973
	4·0	48	284	Parthier et al. 1972
Val	6·0	24	34	Parthier 1973
	15·0	36	0	Hecker Egan Reynolds Nix Schiff & Barnett 1974

Cycloheximide (µg/ml)

Ala	4·0	48	203	Parthier et al. 1972
Arg	6·0	24	60	Parthier 1973
	4·0	48	171	Parthier et al. 1972
Asp	6·0	24	26	Parthier 1973
	4·0	48	155	Parthier et al. 1972
Gly	4·0	48	105	Parthier et al. 1972
Ile	6·0	24	77	Parthier 1973
	4·0	48	144	Parthier et al. 1972
Leu	6·0	24	32	Parthier 1973
	1·0	48	119	Parthier 1973
	2·0	48	116	Parthier 1973
	3·0	48	107	Parthier 1973
	4·0	48	89	Parthier 1973
	4·0	48	152	Parthier et al. 1972

Streptomycin (mg/ml)

	4·0	48	42	Parthier et al. 1972
Phe	0·5	48	134	Hecker et al. 1974
Val	0·5	48	108	Hecker et al. 1974

20. Protein-synthesizing capacity

Chloramphenicol (mg/ml)

	0·75	48	6	Reger et al. 1972

Cycloheximide (ug/ml)

	2·0	48	86	Reger et al. 1972

Cycloheximide (µg/ml)

	2·0	48	86	Reger et al. 1972

* Length of period of illumination and inhibitor treatment of cells undergoing chloroplast development.

derepressor interaction was suggested by Smillie *et al.* (1968), who described experiments in which dark-grown *Euglena* were incubated in chloramphenicol, washed, and then illuminated in media containing cycloheximide at concentrations which normally prevent chlorophyll synthesis. The chloramphenicol treatment apparently reversed the normal cycloheximide inhibition of chlorophyll and electron-transport protein synthesis; how this is effected is not clear. Evans *et al.* (1967) showed that at low concentrations cycloheximide reduces the inhibitory affect of chloramphenicol when both antibiotics are present during light-induction of chlorophyll synthesis. Further work to establish the significance of these observations in terms of the control of gene expression is anticipated.

5 THE GENETIC CONTROL OF LIGHT-MEDIATED CHLOROPLAST DEVELOPMENT

Chloroplast development in *Euglena* is ultimately regulated by one or more light reactions. The light control must involve a mechanism for the co-ordinated expression of both nuclear and chloroplast genes coding for proteins synthesized on both cytoplasmic and chloroplast ribosomes. The genetic control of the light-induced synthesis of the chloroplast has been the subject of only a few studies in *Euglena*.

Photosynthesis itself seems not to be involved in chloroplast synthesis, even in the later stages of development when photosynthetic capacity has been established. This was demonstrated through the use of a mutant (P_4ZUL) blocked in photosystem II (Russell & Lyman 1968; Russell *et al.* 1969; Schwelitz *et al.* 1972a, b). This mutant can synthesize a fully green chloroplast with normal amounts of enzymes involved in carbon reduction (Schmidt & Lyman 1972). Wild-type dark-grown *Euglena* illuminated in the presence of 3-(3,4-dichlorophenyl)-1, 1-dimethylurea, an inhibitor of Photosystem II activity, likewise produce essentially normal chloroplasts (Schiff *et al.* 1967).

The control of chlorophyll synthesis by light during chloroplast development was first studied by Wolken *et al.* (1955), who determined that in dividing cells the action spectrum for the synthesis of chlorophyll is similar to the absorption spectrum of protochlorophyll or chlorophyll. Nishimura and Huizige (1959) obtained a similar action spectrum for chlorophyll synthesis in non-dividing cells and determined that the earliest observable photochemical action is the conversion of protochlorophyll(ide) to chlorophyll. It was suggested that chloroplast development was induced by this photoconversion (Schiff & Epstein 1965), but it was shown later that the control of chlorophyll synthesis is subject to two light controls. Holowinsky and Schiff (1970) performed experiments in which dark-grown cells were illuminated for 90 minutes, returned to the dark for 12 hours, and then exposed to continuous light. This treatment resulted in an elimination of the 12-hour lag in the synthesis of chlorophyll, normally observed when dark-grown cells are continuously illuminated. This potentiation of chlorophyll synthesis was found to be dependent on the wavelength and intensity of the pre-illumination light. Both red and blue light were effective in this process, indicating that it involves the conversion of protochlorophyll(ide) to chlorophyll. However, at saturating intensities, blue light was shown to be considerably more effective than red light. Holowinsky and Schiff (1970) concluded that there are two light receptors involved in the process of potentiation, one absorbing blue light and one absorbing red *and* blue light. These receptors were proposed to act synergistically in the blue region of the spectrum and therefore to have enhanced effects relative to red light in which only one of the receptors is active. The potentiation of chlorophyll synthesis by red light was shown not to be reversible by subsequent exposure of the cells to far-red light. This observation, confirmed by Boutin and Klein (1972), indicates that potentiation in *Euglena* does not involve phytochrome as it does in higher plants (Mitrakos 1961; Price & Klein 1961; Virgin 1961; Augustinussen & Madsen 1965).

The events provoked by potentiation are not clear. Klein *et al.* (1972) demonstrated that potentiation does not induce any structural changes in the chloroplasts either during or following pre-illumination in *Euglena*. Lazzarini and Woodruff (1964) observed that the lag in synthesis of NADP-transhydrogenase during chloroplast development can be eliminated by pre-illumination of cells with red light. Schwartzbach and Schiff (1971) reported that potentiation of chlorophyll synthesis is inhibited by both chloramphenicol and cycloheximide. Holowinsky and Schiff (1970) postulated that potentiation unblocks protein synthesis at either a transcriptional or a translational level. The induction of chloroplast development, which normally is carried out during the 12-hour lag in cells not pre-illuminated, is therefore initiated by potentiation. The stimulated synthesis of chloroplast ribosomal RNA by a brief exposure of dark-grown *Euglena* to light, as discussed by Cohen and Schiff (1973) and Heizmann (1974a, b), may be considered a product of potentiation. However, since pre-illumination does not alter the capacity of chloroplasts to synthesize proteins (Reger *et al.* 1972), potentiation of chloroplast development must involve more than simple augmentation of the chloroplast-translational mechanisms.

Schmidt and Lyman (1972, 1974) have extended their studies on the photocontrol of chloroplast development in terms of genetic control. They found that two light-dependent controls regulate the synthesis of NADP-dependent glyceraldehyde 3-phosphate dehydrogenase, Class 1 fructose 1,6-diphosphate aldolase, phosphoglycerate kinase and ribulose 1,5-diphosphate carboxylase during chloroplast development in wild-type *Euglena* (Table 13.7). Synthesis of these enzymes is nearly twice as effective in blue light as in red light at saturating intensities. In accord with the conclusions of Holowinsky and

Table 13.7. Evidence for spectral requirements of chloroplast enzyme synthesis in wild type and regulatory mutants of *Euglena* (Schmidt & Lyman 1974).

(Cells were illuminated for 48 hours in light of equal energy (6000 ergs/cm².sec); for red and blue light this had been determined to be above saturating intensity for light-mediated enzyme synthesis. Colored glass filters G-772-6300 and G-774-4450, manufactured by the Oriel Optic Corp., were used to provide spectral outputs of light with peaks at 675 nm and 450 nm respectively. Specific activities are given in µM/mg protein.hr.)

Cell type	Light colour	NADP-glyceraldehyde 3-phosphate dehydrogenase		Phosphoglycerate kinase		Class I fructose 1,6-diphosphate aldolase		Ribulose 1,5-diphosphate carboxylase	
		Activity	Stimulation %	Activity	Stimulation %	Activity	Stimulation %	Activity	Stimulation %
Wild	Blue	21·9	200	76·0	230	1·3	465	0·278	1012
	Red	15·2	108	56·0	144	0·86	274	0·211	744
	Dark	7·3	—	23·0	—	0·23	—	0·025	—
Y₉ZNalL	Blue	11·9	75	50·5	146	0·66	175	0·066	136
	Red	7·0	3	22·1	8	0·29	21	0·027	0
	Dark	6·8	—	20·5	—	0·24	—	0·028	—
Y₁₁P₂₇DL	Blue	14·3	64	40·0	111	0·86	169	0·170	608
	Red	15·0	72	42·0	121	0·94	194	0·200	733
	Dark	8·7	—	19·0	—	0·32	—	0·024	—
W₁₄ZNalL	Blue	10·9	11	25·7	10	0·41	114	0	0
	Red	9·8	0	25·1	8	0·30	0	0	0
	Dark	9·8	0	23·3	0	0·36	0	0	0

Schiff (1970), they suggested that there is a synergistic action of two photo-controls in blue light, whereas only one is active in red light. The hypothesis that two chromophores control the synthesis of these enzymes is strongly supported by the characterization of mutants lacking one or the other photoresponse. Mutant Y_9ZNalL was shown to make no chlorophyll or protochlorophyll and, as shown in Table 13.7, in this mutant synthesis of all four chloroplast enzymes is induced by blue but not red light. In a chlorophyll-deficient mutant, $Y_{11}P_{27}DL$, synthesis of these four enzymes is seen to be equally effective in blue and red light. Because Y_9ZNalL lacks chlorophyll and protochlorophyll, Schmidt and Lyman (1974) concluded that the chromophore for the red-blue photocontrol in wild-type and $Y_{11}P_{27}DL$ is one of these pigments. An action spectrum for light-mediated synthesis of NADP-dependent glyceraldehyde 3-phosphate dehydrogenase has been reported which confirms the hypothesis that protochlorophyll is one of the chromophores for the control of chloroplast development (Egan *et al.* 1974).

A third mutant characterized by Schmidt and Lyman (1974) lacks chloroplasts, chloroplast DNA and ribulose 1,5-diphosphate carboxylase. $W_{14}ZNalL$ is shown in Table 13.7 to contain the same levels of NADP-dependent glyceraldehyde 3-phosphate dehydrogenase, phosphoglycerate kinase and Class I fructose 1,6-diphosphate aldolase in the light as dark-grown wild-type cells. The presence of these enzymes in an aplastidic strain provides evidence that they are coded for by the nuclear genome and synthesized on cytoplasmic ribosomes. There is no light-induced synthesis of these chloroplast enzymes in this mutant. Bovarnick *et al.* (1974) have also shown that NADP-dependent glyceraldehyde 3-phosphate dehydrogenase activity in a light-grown *Euglena* mutant lacking chloroplast DNA (W_3BUL) is the same as in dark-grown wild-type cells.

Schmidt and Lyman (1974) concluded that mutations of the chloroplast genome can block the activity of either or both photocontrol systems for enzyme synthesis and that chloroplast gene products are required for the function of the photoregulatory mechanisms that control chloroplast development. Other photocontrols independent of the chloroplast genome also apparently exist in *Euglena*. In some aplastidic strains light-induced carotenoid synthesis (Dolphin 1970; Gross & Stroz 1974), RNA metabolism (Cohen & Schiff 1973) and cytoplasmic polysome formation (Freyssinet & Schiff 1974) have been observed, although it is not yet possible to assess whether these phenomena are related to chloroplast development.

6   AUTONOMY AND ORIGIN OF
EUGLENA CHLOROPLASTS

Most of the evidence indicates a complex of interactions between the chloroplast and its cytoplasmic milieu in *Euglena*. The chloroplasts and the proplastids are only semiautonomous, in that they require products of the nuclear genome for their synthesis. On the other hand, chloroplast replication is probably wholly controlled by the plastids. The chloroplasts of *Euglena* appear to make regulatory molecules that control the synthesis of their constituents on both plastid and cytoplasmic ribosomes. Possibly both of the photosystems regulating chloroplast development may reside in the plastid, although their effects are manifest outside it.

If chloroplasts originated as prokaryotic symbionts in some primitive eukaryotic host, as some believe (Margulis 1970), they appear to have lost most of their synthetic ability but to have retained regulatory mechanisms which have been evolved to function as well outside the organelle in the host cytoplasm.

In conclusion, we should like to project that the future of genetics in *Euglena* is promising, even in the absence of conventional means of gene mapping in this organism. Exciting developments in the methodologies of parasexual hybrid formation in other organisms (Bonnett & Eriksson 1974; Chaleff & Carlson 1974; Hess *et al.* 1974) and mapping of DNA with molecular techniques (Brown & Vinograd 1974; Casey *et al.* 1974; Lee & Sinsheimer 1974, Robberson *et al.* 1974) will, without doubt, be eventually applied to the greater understanding of gene action in *Euglena*.

7 REFERENCES

AUGUSTINUSSEN E. & MADSEN A. (1965) Regeneration of protochlorophyll in etiolated barley seedling following different light treatments. *Physiol. Plant.* **18**, 828–37.

AVADHANI N.G., BATTULA N. & RUTMAN R.J. (1973) Messenger RNA metabolism in mammalian mitochondria. Origins of ethidium bromide resistant poly(adenylic acid) containing RNA in Ehrlich ascite mitochondria. *Biochem.* **12**, 4122–8.

AVADHANI N.G. & BUETOW D.E. (1972a) Protein synthesis with isolated mitochondrial polysomes. *Biochem. Biophys. Res. Comm.* **46**, 773–8.

AVADHANI N.G. & BUETOW D.E. (1972b) Isolation of active polyribosomes from the cytoplasm, mitochondria, and chloroplasts of *Euglena gracilis. Biochem J.* **128**, 353–65.

AVADHANI N.G., LYNCH M.J. & BUETOW D.E. (1971) Protein synthesis on polysomes in mitochondria isolated from *Euglena gracilis. Exp. Cell. Res.* **69**, 226–8.

BARNETT W.E., PENNINGTON C.J. & FAIRFIELD S.A. (1969) Induction of *Euglena* transfer RNA's by light. *Proc. Nat. Acad. Sci. U.S.A.* **63**, 1261–8.

BARRAS D.R. & STONE B.A. (1968) Carbohydrate composition and metabolism in *Euglena.* In *The Biology of Euglena.* Vol. II, ed. Buetow D.E., pp. 149–86. Academic Press, New York & London.

BEN-SHAUL Y. & MARKUS Y. (1969) Effects of chloramphenicol on growth, size distribution, chlorophyll synthesis and ultrastructure of *Euglena. J. Cell Sci.* **4**, 627–44.

BEN-SHAUL Y. & OPHIR I. (1970) Effects of streptomycin on plastids in dividing *Euglena.* Planta, **91**, 195–203.

BEN-SHAUL Y., SILMAN R. & OPHIR I. (1972) Effects of streptomycin on the ultrastructure of plastids in *Euglena. Physiol. Veg.* **10**, 255–68.

BISHOP D.G., BAIN J.M. & SMILLIE R.M. (1973) The effect of antibiotics on the ultrastructure and photochemical activity of a developing chloroplast. *J. Exp. Bot.* **24**, 361–76.

BISHOP D.G. & SMILLIE R.M. (1970) The effect of chloramphenicol and cycloheximide on lipid synthesis during chloroplast development in *Euglena gracilis. Arch. Biochem. Biophys.* **137,** 179–89.

BONNETT H.T. & ERIKSSON T. (1974) Transfer of algal chloroplasts into protoplasts of higher plants. *Planta,* **120,** 71–80.

BORST P. (1970) Mitochondrial DNA: structure, information content, replication and transcription. *Soc. Exptl. Biol. Symp.* **24,** 201–26.

BOUTIN M.E. & KLEIN R.M. (1972) Absence of phytochrome participation in chlorophyll synthesis in *Euglena. Plant Physiol.* **49,** 656–7.

BOVARNICK J.G., CHANG S.-W., SCHIFF J.A. & SCHWARTZBACH S.D. (1974) Events surrounding the early development of *Euglena* chloroplasts: Experiments with streptomycin in non-dividing cells. *J. Gen. Microbiol.* **83,** 51–62.

BOVARNICK J.G., SCHIFF J.A., FREEDMAN Z. & EGAN J.M. Jr. (1974) Events surrounding the early development of *Euglena* chloroplasts: Cellular origins of chloroplast enzymes in *Euglena. J. Gen. Microbiol.* **83,** 63–71.

BOVARNICK J.G., ZELDIN M.H. & SCHIFF J.A. (1969) Differential effects of actinomycin D on cell division and light-induced chloroplast development in *Euglena. Devel. Biol.* **19,** 321–40.

BRANDES D., BUETOW D.E., BERTINI E. & MALKOFF D.B. (1964) Role of lysosomes in cellular lytic processes. 1. Effect of carbon starvation in *Euglena gracilis. Exptl. Mol. Pathol.* **3,** 583–609.

BRAWERMAN G. & CHARGAFF E. (1959a) Factors involved in the development of chloroplasts in *Euglena gracilis.* Changes in protein and RNA during the formation of chloroplasts in *Euglena gracilis. Biochim. Biophys. Acta.* **31,** 164–71.

BRAWERMAN G. & CHARGAFF E. (1959b) Relation of ribonucleic acid to the photosynthetic apparatus in *Euglena gracilis. Biochim. Biophys. Acta,* **31,** 172–7.

BRAWERMAN G. & EISENSTADT J.M. (1964) Deoxyribonucleic acid from the chloroplasts of *Euglena gracilis. Biochim. Biophys. Acta,* **91,** 477–85.

BRAWERMAN G., POGO A.O. & CHARGAFF E. (1962) Induced formation of RNAs and plastid protein in *Euglena gracilis* under the influence of light. *Biochim. Biophys. Acta,* **55,** 326–34.

BROWN R.D., BASTIA D. & HASELKORN R. (1970) Effect of rifampicin on transcription in chloroplasts of *Euglena.* In *RNA Polymerase and Transcription, Lepetit Colloquium on Medicine and Biology,* ed. Silvestri L., pp. 309–27. North-Holland, Amsterdam.

BROWN R.D. & HASELKORN R. (1971) Chloroplast RNA populations in dark-grown, light-grown and greening *Euglena gracilis. Proc. Nat. Acad. Sci. U.S.A.* **68,** 2536–9.

BROWN W.M. & VINOGRAD J. (1974) Restriction endonuclease cleavage maps of animal mitochondrial DNAs. *Proc. Nat. Acad. Sci. U.S.A.* **71,** 4617–21.

BURKARK G. & KELLER E.B. (1974) Poly(A)polymerase and poly(G)polymerase in wheat chloroplasts. *Proc. Nat. Acad. Sci. U.S.A.* **71,** 389–93.

CALVAYRAC R., BUTOW R.A. & LEFORT-TRAN M. (1972) Cyclic replication of DNA and changes in mitochondrial morphology during the cell cycle of *Euglena gracilis. Exp. Cell Res.* **71,** 422–32.

CARRITT B. & EISENSTADT J.M. (1973a) RNA synthesis in isolated chloroplasts: characterization of the newly synthesized RNA. *FEBS Letters,* **36,** 116–20.

CARRITT B. & EISENSTADT J.M. (1973b) Synthesis *in vitro* of high-molecular-weight RNA by isolated *Euglena* chloroplasts. *Eur. J. Biochem.* **36,** 482–8.

CASEY J., GORDON P. & RABINOWITZ M. (1974) Characterization of mitochondrial DNA from Grande and Petite yeasts by renaturation and denaturation analysis and by tRNA hybridization; Evidence for internal repetition or heterogeneity in mitochondrial DNA populations. *Biochem.* **13,** 1059–66.

CHALEFF R.S. & CARLSON P.S. (1974) Somatic cell genetics of higher plants. *Ann. Rev. Genetics* **8,** 267–78.

CHUA N.-H., BLOBEL G., SIEKEVITZ P. & PALADE G.E. (1973) Attachment of chloroplast polysomes to thylakoid membranes in *Chlamydomonas reinhardtii*. *Proc. Nat. Acad. Sci. U.S.A.*, **70**, 1554–8.

CODD G.A. & MERRETT M.J. (1970) Enzymes of the glycollate pathway in relation to greening in *Euglena gracilis*. *Planta*, **95**, 127–32.

COEN D., DEUTSCH J., NETTER P., PETROCHILLO E. & SLONIMSKI P.P. (1970) Mitochondrial genetics. I. Methodology and phenomenology. *Soc. Exptl. Biol. Symp.* **24**, 449–496.

COHEN D. & SCHIFF J.A. (1973) Photoregulation of formation and turnover of chloroplast rRNA (ChlrRNA) and cytoplasmic rRNA (CytrRNA) during chloroplast development in *Euglena gracilis* var. *bacillaris* Pringsheim. *Biophys. Soc. Abstr.* p. 111a.

COOK J.R. (1966) The synthesis of cytoplasmic DNA in synchronized *Euglena*. *J. Cell Biol.* **29**, 369–73.

CRIDDLE R.S. (1966) Protein and lipoprotein organization in the chloroplast. In *Biochemistry of Chloroplasts*. **I**, 203., ed. Goodwin T.W. Academic Press, New York & London.

CROUSE E.J. (1974) Isolation and characterization of *Euglena gracilis* mitochondrial DNA. Doctoral Thesis, Northwestern University.

CROUSE E.J., VANDREY J.P. & STUTZ E. (1974a) Hybridization studies with RNA and DNA isolated from *Euglena gracilis* chloroplasts and mitochondria. *FEBS Letters*, **42**, 262–6.

CROUSE E.J., VANDREY J.P. & STUTZ E. (1974b) Comparative analyses of chloroplast and mitochondrial DNAs from *Euglena gracilis*. 3rd Internat. Congress Photosynth., ed. Avron M.

DANFORTH W.F. (1968) Respiration. In *The Biology of Euglena Vol. II*, ed. Buetow D.E., pp. 55–72. Academic Press, New York & London.

DIEHN B. (1973) Phototaxis and sensory transduction in *Euglena*. *Science*, **181**, 1009–1015.

DOLPHIN W.D. (1970) Photoinduced carotenogenesis in chlorotic *Euglena gracilis*. *Plant Physiol.* **46**, 685–91.

EBRINGER L. (1970) The action of nalidixic acid on *Euglena* plastids. *J. Gen. Microbiol.* **61**, 141–4.

EBRINGER L. (1972) Are plastids derived from prokaryotic microorganisms? Action of antibiotics on chloroplasts of *Euglena gracilis*. *J. Gen. Microbiol.* **71**, 35–52.

EDELMAN M., COWAN C.A., EPSTEIN H.T. & SCHIFF J.A. (1964) Studies of chloroplast development in *Euglena gracilis*. VIII. Chloroplast-associated DNA. *Proc. Nat. Acad. Sci. U.S.A.*, **52**, 1214–9.

EDELMAN M., EPSTEIN H.T. & SCHIFF J.A. (1966) Isolation and characterization of DNA from the mitochondrial fraction of *Euglena*. *J. Mol. Biol.* **17**, 463–9.

EDELMAN M., SCHIFF J.A. & EPSTEIN H.T. (1965) Studies of chloroplast development in *Euglena*. XII. Two types of satellite DNA. *J. Mol. Biol.* **11**, 769–74.

EDMUNDS L.N., CHUANG L., JARRETT R.M. & TERRY O.W. (1971) Long-term persistence of free-running circadian rhythms of cell division in *Euglena*. *J. Interdiscipl. Cycle Res.* **2**, 121–32.

EGAN J.M. JR. & CARRELL E.F. (1972) Studies on chloroplast development and replication in *Euglena*. III. A study of the site of synthesis of alkaline deoxyribonuclease induced during chloroplast development in *Euglena gracilis*. *Plant Physiol.* **50**, 391–5.

EGAN J.M.JR, DORKY D. & SCHIFF J.A. (1974) Action spectra for various processes associated with light-induced chloroplast development in *Euglena gracilis* var. *bacillaris*. *Plant Physiol.* **53** suppl. 4.

EISENSTADT J.M. & BRAWERMAN G. (1963) The incorporation of amino acids into the proteins of chloroplasts and chloroplast ribosomes of *Euglena gracilis*. *Biochim. Biophys. Acta*, **76**, 319–21.

EISENSTADT J.M. & BRAWERMAN G. (1964) The protein synthesizing systems from the cytoplasm and the chloroplasts of *Euglena gracilis*. *J. Mol. Biol.* 10, 392–402.

ERWIN J.A. (1968) Lipid metabolism. In *The Biology of Euglena. Vol. II.*, ed. Buetow D.E., pp. 133–48. Academic Press, New York & London.

EVANS W.R., WALENGA R. & JOHNSON C. (1967) Effect of cycloheximide on chloroplast development in *Euglena*. In *4th Ann. Res. Report of the Charles F. Kettering Research Lab.*, pp. 99–101. Ampersand Press, Yellow Springs, Ohio.

FAIRFIELD S.A. & BARNETT W.E. (1971) On the similarity between the tRNAs of organelles and prokaryotes. *Proc. Nat. Acad. Sci. U.S.A.*, 68, 2972–6.

FREYSSINET G. & SCHIFF J.A. (1974) The chloroplast and cytoplasmic ribosomes of *Euglena*. II. Characterization of ribosomal proteins. *Plant Physiol.* 53, 543–54.

GIBOR A. & GRANICK S. (1962) Ultraviolet sensitive factors in the cytoplasm that affect the differentiation of *Euglena* plastids. *J. Cell Biol.* 15, 599–603.

GILLHAM N.W. (1974) Genetic analysis of the chloroplast and mitochondrial genomes. *Ann. Rev. Genetics*, 8, 347–92.

GNAMAN A. & KAHN J.S. (1967) Biochemical studies on the induction of chloroplast development in *Euglena gracilis*. II. Protein synthesis during induction. *Biochim. Biophys. Acta*, 142, 486–92.

GOINS D.J., REYNOLDS R.J., SCHIFF J.A. & BARNETT W.E. (1973) A cytoplasmic regulatory mutant of *Euglena*: Constitutivity for the light-inducible chloroplast transfer RNAs *Proc. Nat. Acad. Sci. U.S.A.*, 70, 1749–52.

GOLDRING E.S., GROSSMAN L.I., KRUPNICK D., COYER D.R. & MARMUR J. (1970) The petite mutation in yeast: Loss of mitochondrial deoxyribonucleic acid during induction of petites with ethidium bromide. *J. Mol. Biol.* 52, 323–35.

GROOT G.S.P., FLAVELL R.A., VAN OMMEN G.J.B. & GRIVELL L.A. (1974) Yeast mitochondrial RNA does not contain poly(A). *Nature, Lond.* 252, 168–9.

GROSS J.A. & STROZ R.J. (1974) Evidence for phytoene in *Euglena* mutants. *Plant Sci. Letters*, 3, 67–73.

HARRIS E.H. & EISENSTADT J.M. (1971) Initiation of polysome formation in chloroplasts isolated from *Euglena gracilis*. *Biochim. Biophys. Acta*, 232, 167–70.

HARRIS E.H., PRESTON J.F. & EISENSTADT J.M. (1973) Amino acid incorporation and products of protein synthesis in isolated chloroplasts of *Euglena gracilis*. *Biochem.* 12, 1227–34.

HECKER L.I., EGAN J., REYNOLDS R.J., NIX C.E., SCHIFF J.A. & BARNETT W.E. (1974) The sites of transcription for *Euglena* chloroplastic aminoacyl tRNA synthetases. *Proc. Nat. Acad. Sci.* 71, 1910–14.

HEIZMANN P. (1970) Proprietes des ribosomes et des RNA ribosomiques d'*Euglena gracilis*. *Biochim. Biophys. Acta*, 224, 144–54.

HEIZMANN P. (1974a) Maturation of chloroplast rRNA in *Euglena gracilis*. *Biochem. Biophys. Res. Comm.* 56, 112–18.

HEIZMANN P. (1974b) La synthèse des RNA ribosomiques au cours de l'éclairement d'*Euglenes* étiolées. *Biochim. Biophys. Acta*, 353, 301–12.

HEIZMANN P., TRABUCHET G., VERDIER G., FREYSSINET G. & NIGON V. (1972) Influence de l'éclairement sur l'évolution des polysomes dans des cultures d'*Euglena gracilis*. *Biochim. Biophys. Acta*, 277, 149–60.

HESS D., LORZ H. & WEISSERT E.M. (1974) Uptake of bacterial DNA into swelling and germinating pollen grains of *Petunia-hybrida* and *Nicotiana glauca*. *Z. Pflanzen-physiol.* 74, 52–63.

HILL H.Z., SCHIFF J.A. & EPSTEIN H.T. (1966) Studies of chloroplast development in *Euglena*. XIII. Variation of U.V. sensitivity with extent of chloroplast development. *Biophys. J.* 6, 125–33.

HIRSCH M. & PENMAN S. (1974) The messenger-like properties of the poly(A) + RNA in mammalian mitochondria. *Cell*, 3, 335–9.

HIRVONEN A.P. & PRICE C.A. (1971) Chloroplast ribosomes in the proplastids of *Euglena gracilis*. *Biochim. Biophys. Acta*, **232**, 696–704.

HOLOWINSKY A.W. & SCHIFF J.A. (1970) Events surrounding the early development of *Euglena* chloroplasts. I. Induction by preillumination. *Plant Physiol.* **45**, 339–47.

HOOBER J.K. & BLOBEL G. (1969) Characterization of the chloroplastic and cytoplasmic ribosomes of *Chlamydomonas reinhardi*. *J. Mol. Biol.* **41**, 121–38.

HOVENKAMP-OBBEMA R. (1974) Effect of chloramphenicol on development of proplastids in *Euglena gracilis*. 2. Synthesis of carotenoids. *Z. Pflanzenphysiol.* **73**, 439–47.

HOVENKAMP-OBBEMA R., MOORMAN A. & STEGWEE D. (1974) Aminolaevulinate dehydratase in greening cells of *Euglena gracilis*. *Z. Pflanzenphysiol.* **72**, 277–86.

HOVENKAMP-OBBEMA R. & STEGWEE D. (1974) Effect of chloramphenicol on development of proplastids in *Euglena gracilis*. 1. Synthesis of ribulosediphosphate carboxylase, NADP-linked glyceraldehyde 3-phosphate dehydrogenase and aminolaevulinate dehydratase. *Z. Pflanzenphysiol.* **73**, 430–8.

IBRAHIM N.G., BURKE J.P. & BEATTIE D.S. (1974) The sensitivity of rat liver and yeast mitochondrial ribosomes to inhibitors of protein synthesis. *J. Biol. Chem.* **249**, 6808–6811.

JACOB S.T. & SCHINDLER D.G. (1972) Polyriboadenylate polymerase solubilized from rat liver mitochondria. *Biochem. Biophys. Res. Comm.* **48**, 126–34.

JARRETT R.M. & EDMUNDS L.N. JR. (1970) Persisting circadian rhythm of cell division in a photosynthetic mutant of *Euglena*. *Science*, **167**, 1730–3.

KIRK J.T.O. (1962) Effect of streptomycin on greening and biosynthesis in *Euglena gracilis*. *Biochim. Biophys. Acta*, **56**, 139–51.

KIRK J.T.O. (1968) Dependence of chlorophyll synthesis on protein synthesis in *Euglena gracilis*, together with a nomogram for determination of chlorophyll concentration. *Planta*, **78**, 200–7.

KIRK J.T.O. (1972) Genetic control of plastid formation: Recent advances and strategies for the future. *Sub-Cell. Biochem.* **1**, 333–61.

KIRK J.T.O. & ALLEN R.L. (1965) Dependence of chloroplast pigment synthesis on protein synthesis: effect of actidione. *Biochem. Biophys. Res. Comm.* **21**, 523–30.

KISLEV N. & EISENSTADT J.M. (1972) Protein synthesis in mitochondria of *Euglena gracilis*. *Eur. J. Biochem.* **31**, 226–9.

KISLEV N., SELSKY M.I., NORTON C. & EISENSTADT J.M. (1972) TRNA and tRNA aminoacyl synthetases of chloroplasts, mitochondria and cytoplasm from *Euglena gracilis*. *Biochim. Biophys. Acta*, **287**, 256–69.

KIVIC P.A. & VESK M. (1974) An electron microscope search for plastids in bleached *Euglena gracilis* and in *Astasia longa*. *Can. J. Bot.* **52**, 695–9.

KLEIN S., SCHIFF J.A. & HOLOWINSKY A.W. (1972) Events surrounding the early development of *Euglena* chloroplasts. II. Normal development of fine structure and the consequences of preillumination. *Devel. Biol.* **28**, 253–73.

KONITZ W. (1965) Elektronenmikroskopische Untersuchungen an *Euglena gracilis* im tagesperiodischen Licht-Dunkel-Wechsel. *Planta*, **66**, 345–73.

KRAWEIC W. & EISENSTADT J.M. (1970) Ribonucleic acids from the mitochondria of bleached *Euglena gracilis* Z. I. Isolation of mitochondria and extraction of nucleic acids II. Characterization of highly polymeric ribonucleic acids. *Biochim. Biophys. Acta*, **217**, 120–41.

KROON A.M. (1971) Structure and function of mitochondrial nucleic acids. *Chimia*, **25**, 114–21.

KROON A.M. & DE VRIES H. (1970) Antibiotics: a tool in the search for the degree of autonomy of mitochondria in higher animals. *Symp. Soc. Exptl. Biol.* **24**, 181–200.

LANG D. (1970) Molecular weights of coliphages and coliphage DNA III. Contour length and molecular weight of DNA from bacteriophages T_4, T_5 and T_7, and from bovine papilloma virus. *J. Mol. Biol.* **54**, 557–65.

LATZKO E. & GIBBS M. (1969) Enzyme activities of the carbon reduction cycle in some photosynthetic organisms. *Plant. Physiol.* **44**, 295–300.

LAZZARINI R.A. & WOODRUFF M. (1964) The photoinduction of transhydrogenase in *Euglena. Biochim. Biophys. Acta,* **79**, 412–5.

LEE A.S., SINSHEIMER R.L. (1974) A cleavage map of bacteriophage øX174 genome. *Proc. Nat. Acad. Sci. U.S.A.,* **71**, 2882–6.

LEEDALE G.F. & BUETOW D.E. (1970) Observations on the mitochondrial reticulum in living *Euglena gracilis. Cytobiol.* **1**, 195–202.

LEFF J., MANDEL M., EPSTEIN H.T. & SCHIFF J.A. (1963) DNA satellites from cells of green and aplastidic algae. *Biochem. Biophys. Res. Comm.* **13**, 126–30.

LEWIN R.A. (1960) A device for obtaining mutants with impaired motility. *Can. J. Microbiol.* **6**, 21–5.

LYMAN H. (1967) Specific inhibition of chloroplast replication in *Euglena gracilis* by nalidixic acid. *J. Cell Biol.* **35**, 726–30.

LYMAN H., EPSTEIN H.T. & SCHIFF J.A. (1961) Studies of chloroplast development in *Euglena.* I. Inactivation of green colony formation by U.V. light. *Biochim. Biophys Acta,* **50**, 301–9.

LYMAN H., JUPP A.S. & LARRINUA I. (1975) Action of nalidixic acid on chloroplast replication in *Euglena gracilis. Plant Physiol.* **55**, 390–2.

MAHLER H.R. & PERLMAN P.S. (1972) Mitochondrial membranes and mutagenesis by ethidium bromide. *J. Supramolec. Struct.* **1**, 105–24.

MANNING J.E. & RICHARDS O.C. (1972a) Synthesis and turnover of *Euglena gracilis* nuclear and chloroplast deoxyribonucleic acid. *Biochem.* **11**, 2036–43.

MANNING J.E. & RICHARDS O.C. (1972b) Isolation and molecular weight of circular chloroplast DNA from *Euglena gracilis. Biochim. Biophys. Acta,* **259**, 285–96.

MANNING J.E., WOLSTENHOLME D.R., RYAN R.S., HUNTER J.A. & RICHARDS O.C. (1971) Circular chloroplast DNA from *Euglena gracilis. Proc. Nat. Acad. Sci. U.S.A.* **68**, 1169–73.

MARCENKO E. (1973) Plastids of the yellow Y-1 strain of *Euglena gracilis. Protoplasma* **76**, 417–34.

MARGULIS L. (1970) *Origin of Eukaryotic Cells.* Yale Univ. Press, New Haven & London.

McCALLA D.R. & BAERG W. (1969) Action of myxin on the chloroplast system of *Euglena gracilis. J. Protozool.* **16**, 425–8.

MEGO J.L. (1968) Inhibitors of the chloroplast system in *Euglena.* In *The Biology of Euglena. Vol. II,* ed. Buetow D.E., pp. 351–81. Academic Press, New York & London.

MIELENZ J.R. & HERSHBERGER C.L. (1974) Are segments of chloroplast DNA differentially amplified? *Biochem. Biophys. Res. Comm.* **58**, 769–77.

MITRAKOS K. (1961) The participation of the red far-red reaction in chlorophyll metabolism. *Physiol. Plant.* **14**, 497–503.

MO Y., HARRIS B.G. & GRACY R.W. (1973) Triosephosphate isomerases and aldolases from light- and dark-grown *Euglena gracilis. Arch. Biochem. Biophys.* **157**, 580–7.

MORIBER L.G., HERSHENOV B., AARONSON S. & BENSKY B. (1963) Teratological chloroplast structures in *Euglena gracilis* permanently bleached by exogenous physical and chemical agents. *J. Protozool.* **10**, 80–6.

MUNNS R., SCOTT N.S. & SMILLIE R.M. (1971) Changes in RNA polymerase activity during chloroplast development in *Euglena gracilis. Proc. Austral. Biochem. Soc.* **4**, 95.

MUNNS R., SCOTT N.S. & SMILLIE R.M. (1972) RNA synthesis during chloroplast development in *Euglena gracilis. Photochem.* **11**, 45–52.

NASS M.M.K. (1970) Abnormal DNA patterns in animal mitochondria: Ethidium bromide-induced breakdown of closed circular DNA and conditions leading to oligomer accumulation. *Proc. Nat. Acad. Sci. U.S.A.,* **67**, 1926–33.

NASS M.M.K. & BEN-SHAUL Y. (1972) A novel closed circular duplex DNA in bleached mutant and green strains of *Euglena gracilis. Biochim. Biophys. Acta,* **272**, 130–6.

NASS M.M.K. & BEN-SHAUL Y. (1973) Effects of ethidium bromide on growth, chlorophyll synthesis, ultrastructure and mitochondrial DNA in green and bleached mutant of *Euglena gracilis. J. Cell Sci.* **13**, 567–90.

NASS M.M.K., SCHORI L., BEN-SHAUL Y. & EDELMAN M. (1974) Size and configuration of mitochondrial DNA in *Euglena gracilis. Biochim. Biophys. Acta*, **374**, 283–91.

NIGON V., HEIZMANN P., DION M. & PORTIER C. (1967) Étude cinétique des modifications induites par l'éclairement chez des populations d'euglenes étiolées. *Bull. Soc. fr. Physiol. veg.* **13**, 233–49.

NISHIMURA M. & HUIZIGE H. (1959) Studies on the chlorophyll formation in *Euglena gracilis* with special reference to the action spectrum of the process. *J. Biochem.* **46** 225–34.

OJALA D. & ATTARDI G. (1974) Identification of discrete polyadenylate containing RNA components transcribed from HeLa cell mitochondrial DNA. *Proc. Nat. Acad. Sci. U.S.A.*, **71**, 563–7.

OSAFUNE I. (1973) Three-dimensional structures of giant mitochondria, dictyosomes and 'concentric lamellar bodies' formed during the cell cycle of *Euglena gracilis* (Z) in synchronous culture. *J. Electron Microscopy*, **22**, 51–62.

PARTHIER B. (1973) Cytoplasmic site of chloroplast aminoacyl tRNA synthetases in *Euglena gracilis. FEBS Letters*, **38**, 70–4.

PARTHIER B., KRAUSPE R. & SAMTLEBEN S. (1972) Light-stimulated synthesis of aminoacyl tRNA synthetases in greening *Euglena gracilis. Biochim. Biophys. Acta*, **277**, 335–41.

PERINI E., SCHIFF J.A. & KAMEN M.D. (1964) Iron-containing proteins in *Euglena*. II. Functional localization. *Biochim. Biophys. Acta*, **88**, 91–8.

PERL M. (1972) A possible ribosomal-directed regulatory system in *Euglena gracilis*. Chlorophyll synthesis. *Biochem. J.* **130**, 813–8.

PERRY R.P., KELLEY D.E. & LA TORRE J. (1972) Lack of polyadenylic acid sequences in the messenger RNA of *E. coli. Biochem. Biophys. Res. Comm.* **48**, 1593–1600.

PHILIPPOVICH I.I., BEZSMERTNAYA I.N. & OPARIN A.I. (1973) On the localization of polyribosomes in the system of chloroplast lamellae. *Exp. Cell Res.* **79**, 159–68.

POGO B.G.T. & POGO A.O. (1964) DNA dependence of plastid differentiation—Inhibition by actinomycin D. *J. Cell Biol.* **22**, 296–301.

PRICE L. & KLEIN W.H. (1961) Red, far-red response and chlorophyll synthesis. *Plant Physiol.* **36**, 733–5.

PRINGSHEIM E.G. & PRINGSHEIM O. (1952) Experimental elimination of chromatophores and eyespot in *Euglena gracilis. New Phytol.* **51**, 65–76.

PRINGSHEIM E.G. (1963) *Farblose Algen*. Fischer, Stuttgart.

PROVASOLI L., HUTNER S., SCHATZ A. (1948) Streptomycin-induced chlorophyll-less races of *Euglena. Proc. Soc. Exptl. Biol. Med.* **69**, 279–82.

RAISON J.K. & SMILLIE R.M. (1969) Respiratory cytochromes of *Euglena gracilis. Biochim. Biophys. Acta*, **180**, 500–8.

RAWSON J.R. & HASELKORN R. (1973) Chloroplast ribosomal RNA genes in the chloroplast DNA of *Euglena gracilis. J. Mol. Biol.* **77**, 125–32.

RAY D.S. & HANAWALT P.C. (1964) Properties of the satellite DNA associated with the chloroplasts of *Euglena gracilis. J. Mol. Biol.* **9**, 812–24.

RAY D.S. & HANAWALT P.C. (1965) Satellite DNA components in *Euglena gracilis* cells lacking chloroplasts. *J. Mol. Biol.* **11**, 760–8.

REGER B.J., FAIRFIELD S.A., EPLER J.L. & BARNETT W.E. (1970) Identification and origin of some chloroplast aminoacyl-tRNA synthetases and tRNAs. *Proc. Nat. Acad. Sci. U.S.A.* **67**, 1207–13.

REGER B.J., SMILLIE R.M. & FULLER R.C. (1972) Light-stimulated production of a chloroplast-localized system for protein synthesis in *Euglena gracilis. Plant Physiol.* **50**, 24–7.

RICHARDS O.C. & RYAN R.S. (1974) Synthesis and turnover of *Euglena gracilis* mitochondrial DNA. *J. Mol. Biol.* **82**, 57–76.

RICHARDS O.C., RYAN R.S. & MANNING J.E. (1971) Effects of cycloheximide and of chloramphenicol on DNA synthesis in *Euglena gracilis. Biochim. Biophys. Acta,* **23,** 190–201.

ROBBERSON D.L., CLAYTON D.A. & MORROW J.F. (1974) Cleavage of replicating forms of mitochondrial DNA by EcoRI endonuclease. *Proc. Nat. Acad. Sci. U.S.A.* **71,** 4447–51.

RUSSELL G. & LYMAN H. (1968) Isolation of mutants of *Euglena gracilis* with impaired photosynthesis. *Plant Physiol.* **43,** 1284–90.

RUSSELL G.K., LYMAN H. & HEATH R. (1969) Absence of fluorescent quenching in a photosynthetic mutant of *Euglena gracilis. Plant Physiol.* **44,** 929–31.

SALVADOR G., NIGON V., RICHARD F. & NICOLAS P. (1972) Structures et propriétés d'un nouveau mutant blanc d'*Euglena gracilis. Protistologica,* **8,** 533–40.

SCHIFF J.A. (1970) Developmental interactions among cellular compartments in *Euglena. Symp. Soc. Exptl. Biol.* **24,** 277–302.

SCHIFF J.A. (1973) The development, inheritance, and origin of the plastid in *Euglena. Adv. Morphogenesis.* **10,** 265–312.

SCHIFF J.A. & EPSTEIN H.T. (1965) The continuity of the chloroplast in *Euglena.* In *Reproduction: Molecular, Subcellular & Cellular,* pp. 131–89, ed. Locke M. Academic Press, New York.

SCHIFF J.A., LYMAN H. & RUSSELL G.K. (1972) Isolation of mutants fro m *Euglena gracilis* In *Methods in Enzymology Vol.* 23*A,* ed. San Pietro A.A., pp. 143–62. Academic Press, New York.

SCHIFF J.A., ZELDIN M.H. & RUBMAN J. (1967) Chlorophyll formation and photosynthetic competence in *Euglema* during light-induced chloroplast development in the presence of 3,(3,4-dichlorophenyl)1,1-dimethyl urea (DCMU). *Plant Physiol.* **42,** 1716–25.

SCHMIDT G.W. & LYMAN H. (1972) Spectral requirements for chloroplast development in *Euglena gracilis. Plant Physiol.* **49,** s28.

SCHMIDT G.W. & LYMAN H. (1973) Studies on a mutant of *Euglena* with a lesion in the control of chloroplast development. *J. Cell Biol.* **59,** 305a.

SCHMIDT G.W. & LYMAN H. (1974) Photocontrol of chloroplast enzyme synthesis in mutant and wild-type *Euglena gracilis. 3rd Internat. Congress Photosynth., Rehovot, Israel,* ed. Avron M., pp. 1755–64. Elsevier, Excerpta Medica, North-Holland, Amsterdam.

SCHORI L., BEN-SHAUL Y. & EDELMAN M. (1970) The size of mitochondrial DNA in *Euglena gracilis. Israel J. Chem.* **8,** 117.

SCHWARTZBACH S.D. & SCHIFF J.A. (1971) Synthetic events during the lag period of chloroplast development in *Euglena gracilis* var. *bacillaris. Plant Physiol.* **47,** s45.

SCHWELITZ F.D., DILLEY R.A. & CRANE F.L. (1972) Biochemical and biophysical characteristics of a photosynthetic mutant of *Euglena gracilis* blocked in photosystem II. *Plant Physiol.* **50,** 161–5.

SCHWELITZ F.D., DILLEY R.A. & CRANE F.L. (1972) Structural characteristics of a photosynthetic mutant of *Euglena gracilis* blocked in photosystem II. *Plant Physiol.* **50,** 166–70.

SCOTT N.S. (1973) Ribosomal RNA cistrons in *Euglena gracilis. J. Mol. Biol.* **81,** 327–36.

SCOTT N.S., MUNNS R., GRAHAM D. & SMILLIE R.M. (1971) Origin and synthesis of chloroplast ribosomal RNA and photoregulation during chloroplast biogenesis. In *Autonomy and Biogenesis of Mitochondria and Chloroplast,* eds. Boardman N.K., Linnane A.W. & Smillie R.M., pp. 383–92. North-Holland, Amsterdam, London.

SCOTT N.S., SHAH V.C. & SMILLIE R.M. (1968) Synthesis of chloroplast DNA in isolated chloroplasts. *J. Cell Biol.* **38,** 151–7.

SHAH V.C. & LYMAN H. (1966) DNA-dependent RNA synthesis in chloroplasts of *Euglena gracilis. J. Cell Biol.* **29,** 174–6.

SHARPLESS T.K. & BUTOW R.A. (1970) Phosphorylation sites, cytochrome complement an alternate pathways of coupled electron transport in *Euglena gracilis* mitochondria. *J. Biol. Chem.* **245,** 50–7.

SIEGESMUND K.A., ROSEN W.G. & GAWLIK S.R. (1962) Effects of darkness and of strepto-mycin on the fine structure of *Euglena gracilis*. *Amer. J. Bot.* **49**, 137–45.

SLONIMSKI P.P., PERRODIN G. & CROFT J.H. (1968) Ethidium bromide induced mutation of yeast mitochondria: complete transformation of cells into respiratory-deficient non-chromosomal petites. *Biochem. Biophys. Res. Comm.* **30**, 232–9.

SMILLIE R.M. (1963) Formation and function of soluble proteins in chloroplasts. *Can. J. Bot.* **41**, 123–54.

SMILLIE R.M. (1968) Enzymology of *Euglena*. In *The Biology of* Euglena. *Vol. II*, ed. Buetow D.E., pp. 2–54. Academic Press, New York & London.

SMILLIE R.M., EVANS W.R. & LYMAN H. (1963) Metabolic events during the formation of a photosynthetic from a nonphotosynthetic cell. *Brookhaven Symp. Biol.* **16**, 89–108.

SMILLIE R.M., GRAHAM D., DWYER M.R., GRIEVE A. & TOBIN N.F. (1967) Evidence for the synthesis *in vivo* of proteins of the Calvin cycle and of the photosynthetic electron-transfer pathway on chloroplast ribosomes. *Biochem. Biophys. Res. Comm.* **28**, 604–10.

SMILLIE R.M., MUNNS R., GRAHAM D., SCOTT N.S., GRIEVE A.M. (1971) Comparative study of the development of protein synthetic capacity in chloroplasts. *2nd International Congress Photosynt. Res. Stresa, Italy*, pp. 2511–8.

SMILLIE R.M. & SCOTT N.S. (1969) Organelle biosynthesis: The chloroplast. *Progr. Molecular & Subcell. Biol.* **1**, 136–202.

SMILLIE R.M., SCOTT N.S. & GRAHAM D. (1968) Biogenesis of chloroplasts: Roles of chloroplast DNA and chloroplast ribosomes. In *Comparative Biochemistry and Biophysics of Photosynthesis*, eds. Shibata K., Takamiya K., Jagendorf A.T. & Fuller R.C., pp. 332–53. Univ. of Tokyo Press, Tokyo.

STEWART P.R. & GREGORY P. (1969) Effect of antibiotics on chloroplast and mitochondrial development in *Euglena gracilis*. *Microbios*, **3**, 253–66.

STUTZ E. (1970) The kinetic complexity of *Euglena gracilis* chloroplast DNA. *FEBS Letters*, **8**, 25–8.

STUTZ E., CROUSE E.J. & GRAF L. (1974) Does gene amplification occur within the chloro-plast DNA of developing chloroplast of *Euglena gracilis*? *Colloq. Int. C.N.R.S. Cycles Cellulaires et Leur Blockage, Gif-sur-Yvette, France.*

STUTZ E. & RAWSON J.R. (1970) Separation and characterization of *Euglena gracilis* chloroplast single-strand DNA. *Biochim. Biophys. Acta*, **109**, 16–23.

STUTZ E. & VANDREY J.P. (1971) Ribosomal DNA satellite of *Euglena gracilis* chloroplast DNA. *FEBS Letters*, **17**, 277–80.

TALEN J.L., SANDERS J.P.M. & FLAVELL R.A. (1974) Genetic complexity of mitochondrial DNA from *Euglena gracilis*. *Biochim. Biophys. Acta*, **374**, 129–35.

THEISS-SEUBERLING H.-B. (1973) Einfluss von Röntgenstrahlen und Hemmstoffen der Proteinsynthese auf die Synthese von Chlorophyll und NADP-abhängigen Glycerin-aldehyde 3-Phosphat-Dehydrogense in ergrünender *Euglena gracilis*. *Arch. Mikrobiol.* **92**, 331–44.

UZZO A. & LYMAN H. (1969) Light-dependence of temperature bleaching in *Euglena gracilis*. *Biochim. Biophys. Acta*, **180**, 573–5.

UZZO A. & LYMAN H. (1972) The nature of the chloroplast genome of *Euglena gracilis*. *2nd International Congress Photosynth. Res.*, *Stresa, Italy*, pp. 2585–99.

VANDREY J.P. & STUTZ E. (1973) Evidence for a novel DNA component in chloroplasts of *Euglena gracilis*. *FEBS Letters*, **37**, 174–7.

VERDIER G., TRABUCHET G., HEIZMANN P. & NIGON V. (1973) Effect de l'éclairement sur les synthèses de RNA et de séquences polyadényliques dans des cultures d'*Euglena gracilis* étiolées. *Biochim. Biophys. Acta*, **312**, 528–39.

VIRGIN H.I. (1961) Action spectrum for the elimination of the lag phase in chlorophyll formation in previously dark grown leaves of wheat. *Physiol. Plant.* **14**, 439–52.

WOLFOVITCH R. & PERL M. (1972) A possible ribosomal-directed regulatory system in *Euglena gracilis*. Carbon dioxide fixation. *Biochem. J.* **130**, 819–23.

WOLKEN J.J., MELLON A.D. & GREENBLATT C.L. (1955) Environmental factors affecting growth and chlorophyll synthesis in *Euglena*. 1. Physical and chemical. II. The effectiveness of the spectrum for chlorophyll synthesis. *J. Protozool.* **2**, 89.

YU R.S.T. & STEWART P.R. (1974) Comparative studies on mitochondrial development in yeasts. III. Growth-phasic affects of antibiotics on mitochondrial differentiation in *Candida*. *Cytobios*, **9**, 175–92.

ZELDIN M.H. & SCHIFF J.A. (1967) RNA metabolism during light-induced chloroplast development in *Euglena*. *Plant Physiol.* **42**, 922–32.

ZELDIN M.H. & SCHIFF J.A. (1968) A comparison of light-dependent RNA metabolism in wild-type *Euglena* with that of mutants impaired for chloroplast development. *Planta*, **81**, 1–15.

PUBLICATIONS BY A. PASCHER ON THE GENETICS OF ALGAE

TRANSLATIONS AND COMMENTARIES BY R. E. REICHLE
Department of Biology, California State University,
Sacramento, California 95819, U.S.A.

1 **Introductory remarks** 300

2 PASCHER A. (1918) Ueber die Beziehung der Reduktionsteilung zur Mendelschen Spaltung. *Ber. Deut. Botan. Gesellschaft*, **36,** 163–8. (English translation) 203

3 PASCHER A. (1918) *Oedogonium,* ein geeignetes Objekt für Kreuzungsversuche an einkernigen, haploiden Organismen. *Ber. Deut. Botan. Gesellschaft*, **36,** 168–72. (English translation) 306

1 INTRODUCTORY REMARKS

R. E. REICHLE

In 1916 Pascher published a paper entitled 'Crosses of unicellular, haploid organisms: *Chlamydomonas*' (*Berichte der Deutschen Botanischen Gessellschaft* **34,** 228–43). He stated that this was a preliminary paper, reporting the results of his experiments involving crosses of different strains (species) of *Chlamydomonas*. He described his experimental work in great detail, as was customary at that time, presented a brief discussion of his observations, and promised to discuss the further significance of these results in a later paper. In brief, he reported that gametes of two different haploid types, which he could distinguish morphologically, would unite to form zygotes intermediate in appearance between those of the two parental strains. When such hybrid zygotes ultimately germinated, four haploid cells emerged. Generally two corresponded morphologically to one parental strain and two to the other, but in a few cases he found that four recombinant types of gametes had been formed. He pointed out that such haploid algal cells correspond genetically to the haploid pollen and egg cells of 'higher' diploid plants. The fusion of such gametes leads to the formation of a diploid zygote. However, whereas in higher plants this zygote normally develops into the sporophyte generation, this does not happen in *Chlamydomonas*, where the zygote undergoes meiosis (reduction division) to form four haploid cells. In this respect the algal zygote corresponds also to the spore mother cell of a higher plant. By examining the products of meiosis in *Chlamydomonas* one could genetically characterize all four products of the reduction division, which is not normally possible in higher plants. No specific mention of Mendel's laws was made in this article.

In 1918 Pascher published two more papers on this subject. The first is the discussion of his 1916 *Chlamydomonas* paper which he had promised, while the second points out that *Oedogonium*, too, could be a very useful object for genetic experimentation. The significance of the latter paper, which also deals with *Chlamydomonas*, is that Pascher realized some of the advantages of haploid organisms for genetic experimentation. He had apparently received considerable response to his 1916 paper, had probably discussed the subject with other people and had looked into the early literature in greater detail. From his discussion it is obvious that he then realized how much easier it is to analyse the results of a cross when one has all four products of a single meiotic event (a tetrad). (It is unfortunate that the importance of his discovery was not more widely realized, although as early as the 1930s tetrad analyses were being studied in the fungus *Neurospora*.)

During the following 25 years, Pascher went on to become the world's leading taxonomist of freshwater algae, especially microscopic forms. Regrettably, for reasons we can only conjecture now, he did not follow up his experimental studies, though he clearly realized their importance. Even more regrettably, in 1943 he committed suicide, one of the many tragic losses during the Second World War.

In the 1920s and 1930s, other German scientists, notably Kniep and Hartmann, studied sexuality in algae. A student of theirs, Moewus, published, ca. 1940, a number of fantastic accounts of Mendelian inheritance in *Chlamydomonas* and other algae—accounts which we now believe to have been largely if not entirely fabricated (see Gowans, Chapter 7 in this volume). It was only after 1950, when studies on the genetics of *Chlamydomonas* species were published by Ebersold, Gowans, Lewin, Sager and their colleagues, that algal genetics can be said to have begun in earnest.

In view of their historic interest, I have prepared the following English translations of Pascher's 1918 papers. Hopefully they will have more impact now than they did then!

2 THE RELATIONSHIP OF THE REDUCTION DIVISION TO THE MENDELIAN SEGREGATION

A. PASCHER

(Received 21 March 1918)

In attempting to explain Mendel's rules of segregation the assumption is made that the alleles which are combined in the diploid heterozygote generation are separated from each other during the reduction division. This results in two types of sexual cells, half of which carry one allele while the other half carry the other allele. This assumption is the fundamental hypothesis on which are based the Mendalian rules—with the exception of the rule of dominance, with which we are not concerned here.

Since we are dealing with a combined pair of alleles, the assumption is made that two kinds of sexual cells are formed. This assumption can be retrospectively verified (proven *indirectly*) by mathematically analysing the ratios of the offspring among the hybrids, and by the fact that the characters under investigation segregate independently. The correctness of this assumption has a very high degree of probability because we are able to explain clearly most of the pertinent experimental results.

No *direct* observations have been made to show that the Mendelian segregation is actually due to the reduction division. The haploid sexual cells (egg cells or pollen grains) of diploids are morphologically alike. The genetic traits which they carry are not expressed morphologically until they reach the diploid phase. Moreover, the formation of the sexual cells is very drawn out, and the haploid generation which they represent is rather obscure.

These conditions do not apply to haploid organisms. Here the organism spends most of its life in the haploid, often single-celled and uninucleate state. The genetic characteristics of such organisms are readily apparent in the haploid phase. Gametes formed by normal vegetative division (mitosis) from the original haploid cell have the same ploidy, and yield upon fusion with another haploid cell the only cell of the diploid phase of the life cycle, the diploid zygote. Four new haploid individuals which arise from the zygote as a consequence of the reduction division initiate the new haploid phase of the life cycle. (In some instances there may be a secondary suppression (of some of the segregants) and some nuclei—up to three—may be eliminated).

In these haploid organisms the diploid phase of the life cycle is limited to the single cell of the zygote, which corresponds to any single diploid cell of a vegetative diploid organism, e.g. a seed plant. Since the reduction division in such a zygote leads to the formation of haploid cells, the zygote is also homologous to the spore mother cell of a diploid organism. A single haploid cell, and gametes which are of equal ploidy, are homologous to haploid sexual cells or their division products, formed by the diploid organism through tetrad formation.

All cells, including those of haploid organisms which differ by one or several characteristics, express these characters in this haploid phase, which directly corresponds to the sexual cells formed by the reduction division of spore mother cells. If gametes of two haploid organisms, which differ in one or several characters, fuse, they form a heterozygote which corresponds to the amphimictically formed hybrid arising from two diploids, e.g. two seed plants. More specifically it resembles an egg cell of a diploid organism which has been fertilized by a nucleus of another strain. At the same time this cell is homologous to any spore mother cell of an amphimictic hybrid in which the reduction division takes place. Of course, in the diploid organism the reduction division is delayed, in comparison with that of the haploid organism, because the vegetative phase of the life cycle is shifted to the diploid phase. In a haploid organism the four haploid cells which emerge from such a heterozygote after the reduction division are therefore homologous to sexual cells arising from the spore mother cells by reduction division in amphimictic hybrid diploids.

The following hypothesis is given to explain Mendel's rule of segregation: During the formation of sexual cells in hybrid, diploid organisms the following events take place. The alleles from every allele pair (in the hybrid) are distributed to the sexual cells in such a way that one-half of the sex cells receive half of the alleles while the other half of the sex cells get the other half. This necessarily means that the four reduced haploid cells arising from the heterozygote of a haploid organism must show (segregation): two carry one allele, while the other two carry the other allele of the original allele pair. Alleles are expressed morphologically and physiologically in the haploid phase of the life cycle. It should therefore be possible to observe the segregation of the markers in the four cells which arise from a heterozygote directly, not merely by a retrospective deduction as is done for the haploid cells of diploid organisms which carry only the segregated determinants (alleles) but not their phenotypic expression.

If this proof can be accepted, then the hypothesis which forms the basis of Mendel's rule of segregation is proved correct. This has already been shown to be the case by Hartmann (1912), who used haploid parthenogenic animals in his experiments (*Zool. Jahrb. Suppl. Festschrift Sprengel.*).

The experimental proof for this has now also been presented (for a plant). I refer to my paper, published in this journal (1916, vol. XXXIV, p. 228), dealing with *Chlamydomonas*, though at that time only the experimental results were reported.

At that time two species of *Chlamydomonas* were crossed. (*Chlamydomonas* is a typical haploid organism. Gametes arise by vegetative division (mitosis) of individual cells which have the same ploidy.) The following characteristics differentiated the two species used from each other.

	Chl.-I	Chl.-II
Shape	Egg-shaped or pear-shaped with a narrower anterior end (i.e. pyriform)	Almost spherical
Cell wall	Thin; without anterior papilla	Thick; papilla present
Chromatophore (plastid)	Lateral	Basal
Eyespot	Linear	Dot-shaped

The (diploid) heterozygotes were intermediate in their appearance between the two homozygotes. Like the latter they formed four new individuals (gametes) after the reduction divisions. Thirteen of these heterozygotes germinated, and in five of these the germination and release of the four new individuals were observed.

The cultures of cells resulting from these heterozygotes can be divided into two groups. One group formed exact replicas of the parental strains. Another group formed only cells of intermediate types (Mischformen), or formed, in addition, new recombinants (Kombinations-Mischformen).

One of the cultures arising from a germinating heterozygote contained four new recombinants:

Cell shape		Cell wall	Papilla	Plastid	Eyespot
Gamete a.	piriform	thin	absent	lateral	linear
Gamete b.	piriform	thin	absent	basal	dot
Gamete c.	spherical-ellipsoid	thick	present	lateral	linear
Gamete d.	spherical	thick	present	basal	dot

Without going into the question of how these recombinations can be explained cytologically, one should point out that every allele of an allele pair occurs in two combinations: two cell types are piriform; two are spherical to ellipsoidal; two have a thin wall without a papilla, two a thick wall with a papilla; two have a basal plastid with a dot-shaped eyespot, two a lateral plastid with a linear eyespot. Every one of the allele pairs united in the heterozygote was later separated by the reduction division and distributed among the four reduced cells in such a way that two of the cells received one allele while the other two got the other. The same holds if we assume, as is probably true, that some of the alleles which belong together internally, e.g. the cell wall and the papilla, or the plastid and the eyespot, are always associated in a particular way.

For our discussion here it is of no importance that, after the reduction division, individual markers were not assorted among the zoospores in the same combinations as in the parental strains. Rather they are exchanged,* leading to hybrid forms.

The offspring of five other cultures originating from these heterozygotes were just like their parents in appearance, morphologically as well as physiologically. In these cultures the ratio of the offspring was not 1:1, as expected, because of the different durations of the generation time (time required to replicate) of the two strains; but, if one takes this into consideration, it becomes very probable that the cultures originated from two zoospores of each of the pure (parental) type strains which were produced by the heterozygote.

This could also be confirmed by direct observation. On four separate occasions I was able to observe that four zoospores were released from a germinating heterozygote. Two of the released zoospores corresponded to one parental strain while the other two were like cells of the other.†

It therefore can be seen that in the reduced zoospores there was a complete separation of characters, quite homologous to the formation of sexual cells during the reduction division in diploid organisms. Moreover, in two pairs of individuals the characters which underwent segregation during the reduction division in the heterozygote (later) combined in such a way that pure parental-type strains were formed. This contrasts to the preceding case, where these segregated characters combined in an irregular manner and formed new haploid combinations (haplomictic cells).

The cycle thus appears complete. The simple, highly probable hypothesis was made that characters or markers (alleles) combined in the heterozygote separate during the reduction division. This occurs in such a manner that, from a pair of alleles in the heterozygote, one-half of the reduced haploid cells get the allele from the male parent while the other half receive the allele from the female parent. (Instead of the word character one can use

* I have called such new combinations haplomicts. Analogous to this one could call diploid recombinants diplomicts (= heterozygotes).
† I mentioned the following in my previous paper. To approach the problems in an impartial manner, I mentioned that perhaps the nuclei of the two cells forming the zygote had not fully combined, that they did not fuse and that they might then have separated again. The following fact speaks against this assumption: The actual fusion of the nuclei was observed in 35 heterozygotes. Also, upon germination four zoospores emerged from the heterozygotes. This corresponds to the process of the reduction division.

the term marker-allele in haploids*). This hypothesis, which attempts to explain the phenomenon of Mendelian segregation, seems to be confirmed by these observations on *Chlamydomonas*. The reduction division therefore is the cause of the Mendelian segregation.

Prague, end of February 1918

3 *OEDOGONIUM*, A SUITABLE OBJECT FOR MAKING CROSSING EXPERIMENTS INVOLVING HAPLOID, UNICELLULAR ORGANISMS

A. PASCHER

(Received 21 March 1918)

Hybridization experiments with species of *Chlamydomonas* or *Phycomyces* have shown the great importance of this kind of cross involving haploid organisms. The great advantage of such experimental objects is the rapid succession of generations and, correspondingly, their almost complete independence from seasonal climatic changes. A further advantage is that the division processes, which lead to the reduction of the diploid phase to the haploid phase, can be recognized much more distinctly. This has been shown in the preceding note dealing with *Chlamydomonas*. There are disadvantages associated with haploids, among them the obtaining and culturing of suitable material, although this is not as difficult with fungi as with algae. Furthermore, it is difficult to reproduce the natural conditions essential to the formation of sex products (gametes) and to the development of the heterozygote and its germination. This results in a much greater reliance on 'chance occurences'. In most cases technical difficulties prevent such experiments on uninucleate haploids.

There are however a number of such organisms which can be easily cultured, notably the dioecious Characeae, where crosses can easily be made in broadly planned experiments. Members of the Characeae can be readily used for this purpose because of their ruggedness and the large size of their vegetative and sexual organs and their mature oospores. I have been able to obtain mature heterozygotes in crosses between *Chara ceratophylla* (♀) and *Chara foetida* (♂).

It is surprising that the Characeae were not used long ago for extensive crossing experiments, since Ernst (presumably 1917: *Z. induktiv Abstammung und Vererbungslehre*, **17**, 203–50: R.E.R.) in his classical experiments with

---

* One can see from this that among haploid plants those which produce four haploid cells from each zygote are best suited for crossing experiments. Accordingly, the Zygnemataceae and Characeae, which eliminate three reduced (haploid) nuclei, and also the Desmidiaceae, which eliminate two of these, are not very suitable for genetic experimentation.

Chara crinita pointed out the possibility of such experiments. My own investigations with *Chara* were made around 1911–12.

Other, more promising experimental objects are the Zygnemataceae. In my report dealing with the *Chlamydomonas* crosses (this journal, 1916, vol. XXXIV) I mentioned that I was also successful in forming *Spirogyra* heterozygotes, from which I was able to get one germling composed of a few cells.

The Characeae and Zygnemataceae have a disadvantage, however. During the reduction division of the single, diploid cell (the zygote or oospore), three of the reduced nuclei die and only a single cell develops. In the above mentioned note dealing with *Chlamydomonas*, I pointed out the value of direct observation of the four reduced cells (zoospores) which emerge from the heterozygote and develop into mature individuals. This allows one to directly observe the segregation of parental characters in the formation of the haploid phase.

Chlamydomonas is a difficult experimental organism, however, and higher green algae, specifically the Ulotrichales, could also be considered for experimental purposes. Their disadvantage is that the zoospores which emerge from the zygote are so similar, and one can only guess which of the two genetic types of zoospores they might belong to. Members of the Mesotaeniaceae, which also form four germlings from the zygote, do not germinate well in culture. The Desmids form only two germlings. Recent reports dealing with alternation of generations among the Phaeophyceae make it urgently desirable that genetic experiments should be carried out on brown algae, all the more because these experiments should not involve great technical difficulties. Certain marine Siphonales should also make excellent experimental objects.

I have made other preliminary investigations which showed that *Oedogonium* species should be very suitable for such experimentation. For various reasons, however, I have not been able to carry out a major series of experiments: the necessary means and space were lacking. My original experiments were based on the observation of (mixed) cultures composed of several species. Sometimes in these were found oospores which could only be interpreted as hybrids. I am sure that several *Oedogonium* 'species' which have been observed only once or a few times are hybrids, especially where their mature oospores have intermediate characteristics when compared to other species while their vegetative cells are exactly like one of the (parental) species. In such cases the intermediate oospore may well be a heterozygote.

Most *Oedogonium* species are easily cultured and are not very sensitive to environmental influences. Most grow well on a solid nutrient medium. The formation of zoospores is no problem. (I refer, of course, to the work of Klebs on *Oedogonium*). Only dioecious species should be considered, or at least the female parent should be dioecious. Initially one would assume that dioecious forms with dwarf males would be especially suitable, but I have no practical experience with this. The operculate forms, and among these the dioecious, nannandrous globosporous species, should be especially considered. If one has some technical skill, isolation of individual filaments is very easy, as is also the combination of filaments of different species and sexes in a small vessel. I use small tubes for this purpose.

The mature oogonia are easy to observe. In many species they are often formed in rows and are fairly large. They can therefore be obtained without much loss. I have transferred filaments with mature oospores to agar, where they mature perfectly well and could be processed further when needed. Mature oospores of different species are usually quite distinct.

The Oedogoniaceae are of special value since they produce four zoospores from the germinating oospore. This means that all four nuclei resulting from the reduction division continue to develop. This is especially useful in comparison with other oogamous green algae, e.g. *Vaucheria*. (*Sphaeroplea* cannot be considered here since no zoospores are formed from the oospores.) I have not found any species which produces only two zoospores from a fertilized oospore, although certain species of *Chlamydomonas* behave in this way. The same processes can be studied in the germinating oospore of *Oedogonium* as in the zygote of many *Chlamydomonas* species, the main difference being that the investigations are a lot less laborious. The zoospores which emerge from the oospores germinate very readily.

It is important and advantageous that the zoospores which are vegetatively formed and those which are formed by the oospores are essentially alike, although the latter may be a little smaller. There are significant differences between individual *Oedogonium* species. Some species have a spherical zoospore with a pointed, hyaline apex, others with an apex which is not so sharply defined. Others are ellipsoidal or pyriform, with or without a distinctly tapered front end (apex). I already have pointed this out in my 1906 paper 'Studies of zoospores of some green algae', on page 73. Individual *Oedogonium* species differ from one another in their zoospore morphology. Within the genus there are groups of species which have the same zoospore morphology, and others in which the zoospore morphology is quite distinct. One should not assume that the species groups based on zoospore morphology coincide with those groups recognized by the accepted systematic scheme; in fact they do not. I was able to find gynandrous, macrandrous and even nannandrous forms among a group of *Oedogonium* species which had zoospores that were almost globose, with a hyaline, sharply delineated, hemispherical front end. The same is true of other groups of the Oedogoniaceae which have zoospores with a different shape. If the two species used in a cross actually have zoospores which are very different morphologically, then one could perform the same experiments as with *Chlamydomonas* much more easily, and still get the same results, by direct observation of the cells which emerge from the heterozygote.

Thus the Oedogoniaceae have numerous advantages for crossing experiments. To get suitable cultures is not very difficult if one knows a little about the local flora and approaches it systematically. A failure to get results from the specimens one finds is mostly due to drawing wrong conclusions from a few occasional samples. The only difficult, although important, point lies in obtaining sexually mature specimens. Klebs already has done much preparatory work in this area.

Finally I should like to recommend a few species for experimental purposes, notably the following: *Oedogonium cardiacum, capillare, rivulare, Boseii, Landsboroughi, crassum, grande, punctulato-striatum, Braunii, Cleveanum, Hystrix, Willeanum, concatenatum, macrandrum, longatum, acrosporum, cyathigerum, pluviale.*

Prague. End of February 1918

APPENDIX B

PUBLICATIONS BY FRANZ MOEWUS
ON THE GENETICS OF ALGAE

C. SHIELDS GOWANS

Division of Biological Sciences,
University of Missouri-Columbia*, U.S.A.

1 Scope of this review 310

2 Moewus' genetic work with algae 311

   2.1 The genetic basis of sex determination 313

   2.2 Crossing-over 317

3 General evaluation of Moewus' work 325

3.1 Internal criteria of validity 325

3.2 External evidence for validity 326

4 Conclusion 327

5 References 327

1 SCOPE OF THIS REVIEW

The investigations on sex hormones in *Chlamydomonas* carried out by Moewus and Kuhn and collaborators have been extensively reviewed (Chodat & deSiebenthal 1941; Cook 1945; Hartmann 1943; Lang 1944; Moewus 1938a, 1941a, 1949a, 1950a, 1951a, 1953; Murneek 1941; Raper 1952; Smith 1946, 1951a, 1951b; Sonneborn 1942, 1951; Thimann 1940), and will not be considered here. All attempts to repeat critical features of this work (Förster & Wiese 1954, 1955; Förster *et al.* 1956; Wiese 1969; Wiese & Hayward 1972; Ryan 1955; Lewin 1952) failed. Even after the last review a few further publications by Moewus appeared (Moewus & Bannerjee 1951a, 1951b; Birch *et al.* 1953; Moewus & Deulofen 1954; Moewus 1954a, 1954b, 1954c, 1955a); but this work has not been confirmed, and is not

* Contribution from the Missouri Agricultural Experiment Station. Journal Series Number 6988.

generally accepted (Hattori 1962). Moewus' utilization of some of the data on these same sex substances to construct a theory of self-sterility in the flowering plant *Forsythia* (Moewus 1950b, 1950c, 1950d) was attacked by Esser and Straub (1954), Reznik (1957), and Renner (1958). Moewus also indicated that hormone-like agents influenced gamete development in *Enteromorpha* (Moewus 1948a), *Monostroma* (Moewus 1940d), and *Protosiphon* (Moewus 1935a).

The work of Moewus and collaborators on auxins and auxin inhibitors (Moewus 1948b, 1948c, 1948d, 1949b, 1949c, 1949d, 1950e, 1951b; Moewus & Moewus 1952; Moewus *et al.* 1952; Moewus & Schader 1951a, 1951b, 1952) will not be considered in this review, but should be mentioned because some of the substances involved were the same as those proposed by Moewus to be concerned with *Chlamydomonas* zygote germination (Moewus & Banerjee 1951a, 1951b; Moewus 1950f).

Moewus' experiments on mutation (Moewus 1940e, 1949e), the pseudo-allelic nature of some of the loci used by Moewus, and the statistical criticisms of Moewus' work by Philip and Haldane (1939), as well as Moewus' reply to this criticism (Moewus 1943), reviewed by Sonneborn (1951), will not be taken up here. However, although Harte (1948) and Ebersold (1954) considered certain aspects of crossing-over in Moewus' work, these will be discussed further in the present survey.

Moewus' investigations of chemotaxis (1939b) have been criticized both as to the methods employed and the results reported (Hagen-Seyfferth 1959). A few other papers by Moewus, on taxonomy, variability and hormone relations attributed to *Chlamydomonas* and other algae, have not been specifically referred to in this article, but are included in the Bibliography in the interests of completeness (Kuhn & Moewus 1940; Kuhn *et al.* 1938; Kuhn *et al.* 1939; Moewus 1931, 1933a, 1950h, 1955b, 1957, 1958).

2   MOEWUS' GENETIC WORK WITH ALGAE

Moewus' genetic work was concerned with six groups of algae (Table B.1). The *Chlamydomonas eugametos*-group (Volvocales), *Chlamydomonas paradoxa*-group (Volvocales), *Polytoma*-group (Volvocales), and *Protosiphon botryoides*-group (Siphonales) are all members of the Chlorophyta. The *Botrydium granulatum*-group (Heterosiphonales) belong to the Chrysophyta (Xanthophyceae).

In the two sections which follow (The genetic basis of sex determination, and Crossing-over), all sentences written in italics (and in the present tense) should be preceded by 'According to Moewus . . .' Discussions by the present author (usually in the past tense) are printed in ordinary roman type.

Table B.1. Races and species included in the breeding groups investigated by Moewus.

Chlamydomonas eugametos-group

Strains Isolated from Nature	Mutant Strains
C. eugametos typica	*C. eugametos sphaerica*
asymmetrica	*mikros*
leios	*apotinos*
simplex	*spheyoieides*
philothermos	*apyrenoidosa*
rigophilos	*bipyrenoidosa*
alophilos	*astigmata*
synoica	*macrostigmata*
subheteroica	*microciliata*
romadeos	*phlegmatika*
blastikos	*C. dresdensis elliptica*
psychroblastikos	*sphaerica*
thermoblastikos	*astigmata*
anisogama	*C. agametos*
pseudoisogama	
isogama	
C. dresdensis typica	*Protosiphon botryoides*-group
C. braunii typica	
elliptica	p: $F^1M^1d^0$
C. oogama	s: $F^2M^2d^0$
C. paupera	m: $F^3M^3d^0$
	d: $F^1M^1d^+$
	k: $F^1M^0d^0$ and $F^0M^1d^0$
	t: $F^0M^1d^0$
Chlamydomonas paradoxa-group	b: $F^2M^0d^0$ and $F^0M^2d^0$
	n: $F^2M^0d^0$ and $F^0M^2d^0$
C. paradoxa	f: $F^2M^0d^0$ and $F^0M^2d^0$
C. pseudoparadoxa	g: $F^3M^0d^0$
	ka: $F^0M^3d^0$
	k-1a: $F^1M^1d^0div^{mod}fur^0$
Polytoma-group	k-1b: $F^1M^1d^0div^{mod}fur^+$
	k-2: $F^1M^1div^{lent}fur^0$
Polytoma uvella	k-3: $F^1M^1div^{pro}fur^0$
F^1M^1 with eyespot	
F^1M^1 without eyespot	
Polytoma pascheri	*Brachiomonas*-group
F^3M^3 large cell	a: F_a or M_a
F^3M^0 and F^0M^3 large cell	b: $F_bR_b^m$ or $R_b^fM_b$
F^3M^3 small cell	c: $F_cR_c^m$ or $R_c^fM_c$
F^3M^0 and F^0M^3 small cell	d: F_d or M_d
	e: F_e or M_e
	f: all alleles inactive

2.1   *The genetic basis of sex determination*

In a great deal of Moewus' genetic work he utilized genes which were concerned with sex determination. In the *Chlamydomonas eugametos*-group he reported four different types of sexual behaviour. *These are brought about by three linked loci (F, M, and t) (see Table B-2). The F and M loci are actually each composed of three pseudoalleles* (see Sonneborn 1951). *The four types of sexual behaviour, and the genotypes producing them, are:*

heterothallic: F^a gacop$_M$ t^0, F^b gacop$_M$ t^0, gacop$_F$ M^a t^0, etc.

homothallic: F^a M^a t^0, F^b M^b t^0, etc.

complex heterothallic: F^a M^b t^0, F^b M^a t^0, etc.

subheterothallic: F^a, M^a t^f, F^a M^a t^m, F^b M^b t^f, etc.

The superscripts a and b represent different valences (see below). The inactive t allele t^0 was never designated in Moewus' writings, but is supplied by this author to represent this locus when t^f or t^m is absent. The subscript M following *gacop* indicates that *gacop$_M$* is an allele of M. Actually Moewus indicated that four *gacop$_M$* alleles could be present at the M locus and four *gacop$_F$* alleles could be present at the F locus (see below, and Table B.2).

Heterothallic strains contain either F or M, but not both. There are different alleles at each of these loci which determine the strength of the sexual reaction. Each F or M allele has a specific valence which varies from 1 to 5 (Moewus 1939a, 1940e). *The larger the difference in valence between the sex genes of two cultures, the larger the clumps of pairing gametes formed when the two cultures are mixed.* (Thus, $F^5 \times M^5$ would have given a very strong reaction, $F^1 \times M^1$ a weak reaction). *These valences allow for relative sexuality, and a male culture can be crossed with a male culture if there is a valence difference of two or more* ($M^1 \times M^3$, for example).

To this author's knowledge, the only other reported cases of relative sexuality among the Algae are in *Ectocarpus* (Hartmann 1934, 1956) and *Dasycladus* (Jollos 1926). Neither report has been independently confirmed. Esser and Kuenen (1967) discuss the information on a relative sexuality for the fungi, and conclude that 'From the standpoint of genetic analysis the existence of relative sexuality in fungi has not been demonstrated.'

In a male cell, the female gene (F) is absent, and this locus is occupied by a gacop$_F$ allele. In the female cell the M locus is occupied by a gacop$_M$ allele. There are four possible gacop alleles for each of the sex genes.

The function of the gacop alleles (Moewus 1948e) *is demonstrable as follows: When one mixes male and female cells, tandems form, and then fusion begins. Fusion can proceed in one of four ways: (1) the male cell may move into the female cell leaving its cell wall behind to form an 'ear' on the zygote (Gy); (2) the female cell may migrate into the male cell leaving its cell wall behind to form an 'ear' on the zygote (An); (3) both cells may move towards each other, fusing in the centre and forming a 'two-eared' zygote (Mi); or, (4) neither cell migrates, and a dumbell shaped zygote (Do) is formed. The type of fusion* *behaviour is influenced by the relative sizes of the two gametes involved. There are three possibilities: (1) the gametes can be of equal size; (2) the female can be the larger; or, (3) the male can be the larger.*

Table B.2. Diagrams of linkage groups in various organisms investigated by Moewus. (The alleles for each locus are listed for each linkage group.)

Chlamydomonas eugametos-group

F	M	s	t
3·1	1·7	9·2	

F: F^1, F^2, F^3, F^4, F^5, gacop$_F^{GyGDyo}$, gacop$_F^{MiGyDo}$, gacop$_F^{MiGyAn}$, gacop$_F^{MiMiMi}$
M: M^1, M^2, M^3, M^4, M^5, gacop$_M^{GyGyDo}$, gacop$_M^{MiGyDo}$, gacop$_M^{MiGyAn}$, gacop$_M^{MiMiMi}$
s: s and S
t: t^0, t^f, t^m

Chlamydomonas paradoxa-group

F	11·3	M

Protosiphon botyroides-group

F	M	d	div	fur
2·7	3·5		15·8	
	--------17·4----------			

F: F^0, F^1, F^2, F^3 fur: fur^0, fur^+
M: M^0, M^1, M^2, M^3
div: div^{mod}, div^{lent}, div^{pro} d: d^0, d^+

Polytoma-group

F	M
	8·0

F: F^0, F^1, F^2, F^3
M: M^0, M^1, M^2, M^3

Brachiomonas-group

M_a	M_e	M_b	M_c	M_d
5	5	8	9	

M_a: M_a^0, M_a, F_a, F_b, F_c, D_d, F_e, R_b^f, R_c^f
M_e: M_e^0, M_e, R_b^m M_b: M_b^0, M_b
M_c: M_c^0, M_c, R_c^m M_d: M_d^0, M_d

When designating a certain gacop allele it is necessary to state the behaviour in each of the three different situations. (Thus, gacop^{MiGyAn} reportedly caused a two-eared zygote when the gametes were of equal size (Mi), a one-eared zygote with the male migrating into the female when the female was larger (Gy), and a one-eared zygote with the female migrating into the male when the male was larger (An).). *Dominance relations are as follows:*

$$gacop^{GyGyDo} > gacop^{MiGyDo} > gacop^{MiGyAn} > gacop^{MiMiMi}$$

When no gacop allele is present (homothallics, complex heterothallics, and sub-heterothallics), *the fusion proceeds as if they contained the gacop^{MiMiMi} allele.*

Homothallic species are found in nature (C. eugametos synoica, C. dresdensis), and can also be produced by crossing male and female heterothallics of like valence and isolating types with a cross-over between the sex loci. When crossing-over takes place between the sex loci (F or gacop$_F$ and M or gacop$_M$), homothallics are produced by the nucleus containing the chromosome which bears both the F and the M genes (Moewus 1938b). *In homothallics, sex is physiologically determined. Thus, when a culture is suspended, gamete fusion begins within the culture, some cells acting as plus gametes, some as minus. After fusion is completed there are always a few cells left which have not fused. These are not vegetative cells, but are gametes which could not find mates, and thus within a single culture are always all of one sex.* (These were referred to as residual gametes (Restengameten).) *If a whole series of homothallic cultures are examined, in 50% the residual gametes will be of one sex and in 50% of the other sex.*

Physiological sex determination in homothallics has been reported in *Acetabularia* (Hämmerling 1934), *Haematococcus* (Lerche 1936–7), *Chlamydomonas philotes* (Lewin 1957) and in a homothallic dioecious *Volvox* (Darden 1966).

In the presence of a filtrate from female gametes, all the cells in a homothallic culture act as females, in the presence of a male filtrate all act as males (Moewus 1940f). *In the Protosiphon botryoides-group, homothallic races s and p have residual gametes which are always male at pH 4·4, always female at pH 9·5, and male in half the cases and female in the other half at pH 7·0. If a filtrate is now taken from the male residual gametes at pH 4·5, and added to a fresh homothallic culture, all the gametes will be male* (the culture behaved as a heterothallic culture.). *All female gametes can be produced by adding a female residual gamete filtrate at pH 9·5. With homothallic race m, the pH treatment alone will result in all gametes being one sex* (Moewus 1935a). *In the Brachiomonas-group, homothallic races produce all male gametes in 5% salt solutions, all female in distilled water, and both in 3% salt solutions* (Moewus 1941b, pp. 299–300).

Subheterothallic strains (Moewus 1934a) *are homothallics of a special type. In subheterothallics the allele t^f or t^m is present. This determines the sex of the residual gametes* (Moewus 1936). *If t^m is present in a culture, there is a slight inhibition in the production of male gametes; thus, there is always an excess of female gametes in the suspended clone, and the residual gametes are always female. The allele t^f inhibits the production of female gametes, so that residual gametes are always male. If the subheterothallics are treated with small amounts of formaldehyde, formalin, or acetaldehyde, the production of the minority gametes is completely inhibited and all the gametes are of the majority type* (Moewus 1934a, 1938b). *A female subheterothallic* (female residual gametes) *is converted into a female heterothallic by filtrate from female gametes. A male subheterothallic is converted into a male heterothallic by male-gamete filtrate. If a female subheterothallic is treated with male filtrate,* *or a male subheterothallic with female filtrate, the suspension is converted to homothallism* (Moewus 1940f).

Lerche (1936–7) reported a strain of *Dunaliella* which behaved as a subheterothallic.

Complex heterothallics, which have never been found in nature, are produced in the laboratory by crossing a male of one valence with a female of a different valence. When a crossover occurs between the F and M loci, one strand contains both sex-determiners, and if the valence of the M gene exceeds that of the F gene, the cell will be male. The valence of these complex heterothallics is determined by a simple subtraction. Thus, a complex heterothallic with the constitution F^4 M^2 would behave as a female with a valence of 2 (Moewus 1938b).

In Moewus' earlier work (Moewus 1936, 1938b) he reported the genetic behaviour of the genes on the sex chromosome (linkage group X in Moewus 1940c). In many of the experiments he used the sex-linked eyespot gene (s).

Non-allelic sex-determiners were also reported by Moewus in the *Chlamydomonas paradoxa*-group (Moewus 1936), in the *Protosiphon botryoides*-group (Moewus 1935c, 1937), in the *Polytoma*-group (Moewus 1935b, 1937) and in the *Brachiomonas*-group (Moewus 1941b).

The arrangement of the loci along the sex chromosome, according to Moewus, is given for each group in Table B.2. Alternative alleles given by Moewus are listed. It should be pointed out that in the diagrams the distances given between these genes, with the exception of those in the *Botrydium granulatum*-group (Table B.3), were obtained from data involving crossing-over at the two-strand stage of meiosis (see below).

In the Protosiphon botryoides-group there are homothallics, heterothallics, and a new type, phenotypically heterothallic (Moewus 1935c, 1937). *Race d is a phenotypically heterothallic race, and differs from p, s, and m in that all the gametes from a single coenocyte are of one sex. Thus, the sexual reaction does not occur unless at least two coenocytes are used; nevertheless, sex is phenotypically determined and does not segregate genetically. This phenotypically heterothallic condition is determined by the allele d^+.*

This condition, described by Moewus, parallels the homothallic dioecious condition found to exist in *Volvox aureus* (Darden 1966).

As with the *t* locus, Moewus did not consider the presence of the t^o allele, and designated the allele causing phenotypic heterothallism as *d*. This author has designated the active allele as d^+.

Moewus also reported in these papers the 2:2 segregation of genes for absence of eyespot (s), and absence of pyrenoid (p). These genes were reportedly not linked with sex or with each other. In the *Protosiphon botryoides*-group, Moewus reported the linkage to *F* and *M* of the genes *div*, which controlled the division rate of the nuclei, and *fur*, a bifurcation gene (Moewus 1950g).

In the *Polytoma*-group, besides the linkage of *F* to *M*, Moewus reported linkage between a gene for flagella length (no designation), and one for cell size (D) (Moewus 1935b, 1937). Genes for cell shape (f), papilla absence (p), and eyespot absence (s) were reported not to be linked to each other nor to be in the two linkage groups demonstrated.

Table B.3. Maps for the *Botrydium granulatum*-group, constructed from data given by Moewus (1940). (Races used in these crosses: *Botrydium granulatum heteroica, synoica, stigmata.*)

det ———————0——————— s   Alleles: det: det^m, det^f
    3·9   2·6   s: s^{prae}, s^{abs}

        0$\dfrac{\text{gathe}}{0\cdot 2}$ gath: gathe$^{8-27}$, gathe$^{5-24}$

	det-gathe	gathe-s	s-det
PD	94	99	175
NPD	91	91	0
T	15	10	25
Total	200	200	200

In the Brachiomonas-group (Moewus 1941b) all of the F alleles and one M allele are at a single locus, but the other M genes are not allelic with F nor with each other. There is a further complication, in that alleles governing zygote resistance ($R_b{}^f$, $R_c{}^f$, $R_b{}^m$, $R_c{}^m$) may be present at three of the sex loci (Table B.2). *Zygotes not containing an R allele die in salt concentrations less than 2% or at temperatures over 25°C, those containing R alleles being resistant to these conditions.*

In the Botrydium granulatum-group, the sex-determiners are allelic (although one case of homothallism in 2,710 zygotes was explained by unequal crossing-over). *The sex-determiners are designated det, and are linked to the eyespot gene (s)* (Moewus 1940g). As other examples where the sex-determiners were non-allelic Moewus cited *Glomerella lycopersici* (Hüttig 1935) and *Sordaria fimicola* (Gries 1941). (See discussion of this latter work in Esser and Kuenen 1967, page 96.)

2.2   Crossing-over

Two-strand stage crossing-over

In the great majority of the genetic work of Moewus, crossing-over appeared to have taken place at the two-strand stage of meiosis. When crossing-over occurred between the two sex-determiners, it reportedly produced one strand with both sex-determiners, and one with neither. How this affected the products which resulted from this crossovers depended on which group was being considered.

In the *Chlamydomonas eugametos-group*, the *Chlamydomonas paradoxa-group* (Moewus 1936), *the Protosiphon botryoides-group* (Moewus 1935c), *and the Polytoma-group* (Moewus 1935b), *the nucleus containing the strand without any sex-determiners dies. In the Protosiphon botryoides-group and the Polytoma-group, the viable nucleus divides once to give a two-product zygote. In both of the Chlamydomonas groups the death of the nucleus containing the strand with no sex-determiners is followed by two divisions of the viable nucleus to give a four-product zygote. In the Brachiomonas-group a nucleus containing a strand with no sex-determiners survives, provided that there are also no zygote-resistance alleles (R) in the strand. Cells which do not have sex-determiner alleles, but which do have zygote resistance alleles, are not viable* (Moewus 1941b).

For the *Chlamydomonas paradoxa-*group (Moewus 1936), and the *Polytoma-*group (Moewus 1935b), Moewus included drawings showing the death and disintegration of one of the first-division nuclei, and also showing two divisions of the viable nucleus in the *C. paradoxa-*group and one division in the *Polytoma-*group. He stated that because of the thick zygote membrane in the *C. eugametos-*group he was unable to observe the cytological behaviour (Moewus 1934a). Later, Moewus (1936 p. 48, 1940b p. 498) stated that *C. eugametos* had ten chromosomes, and (Moewus 1940b, p. 499) that diploids, triploids, tetraploids and heteroploids could be induced by a simple method. He gave no reference or further explanation. He repeated (Moewus 1940c, p. 501) that all species of the *C. eugametos-*group had ten chromosomes, and referred to Hartmann (1934), who, however, merely reviewed Moewus' work. Hartmann (1943 p. 132) presented drawings of prophases in *C. eugametos subdioecious* (*subheteroica*) and *C. paupera* which he credited to Moewus (1934a), though the article cited does not contain the figures.

It should also be mentioned that when crossing-over in the *Chlamydomonas eugametos-*group produced a nucleus having a t^l or t^m allele and F and M genes of different valences, the nucleus, reportedly, always died (Moewus 1938b p. 529, 1941a p. 324). Hämmerling's criticism (Hämmerling 1938) of Moewus' data (Moewus 1938b) on this point was answered by Moewus (1940c, footnote p. 519).

In the Protosiphon botyroides-group a nucleus containing a d^+ allele and F and M genes of different valences dies (Moewus 1935c p. 42).

Crossing-over at the two-strand stage was apparently not limited to the sex chromosome. In his major genetic work on the *C. eugametos-*group, Moewus reportedly obtained only ditype tetrads from genes in all ten linkage groups (Moewus 1940a, pp. 432, 433, 434, 437, 442, 444, 447, 448, 449, 450, 454 455, 457, 459, 460, and similar cases in Moewus 1940b and 1940c), which he interprets (Moewus 1940c p. 520) as due to crossing-over at the

two-strand stage of meiosis. In the *Polytoma*-group he obtained two-strand
stage crossing-over exclusively (no tetratypes) between the linked genes *F*
and *M* (Moewus 1935b pp. 392–405), between linked genes for cell size and
flagella length (pp. 390–1), and between those for cell form (*f*), papilla absence
(*p*), eyespot absence (*s*), and cell size (*d*) and their respective centromeres
(p. 386 number 4, p. 387 number 5 and 6).

With two or more unlinked genes, one expects tetratypes when crossing-
over occurs at the four-strand stage between one of the genes and its centromere; but none were reported. Only 60 zygotes were analysed in these crosses,
so if all the markers (*f*, *p*, *s* and *d*) were close to their centromeres, and no
crossing-over occurred in the 60 zygotes, the same results could have been
obtained without proposing two-strand-stage crossing-over. This, however,
would have required that all the loci were within about 1·5 map units of their
centromeres.

In the *Protosiphon botryoides*-group, on the basis of analyses of 15 zygotes
(Moewus 1935c p. 26 number 4), Moewus also reported that crossing-over
occurred exclusively at the two-strand stage (no tetratypes), not only between
the linked genes *F*, *M* and *d* (Moewus 1935c pp. 27–52) and *F*, *M*, *div* and *fur*
(Moewus 1950g pp. 35–58), but also between the genes for pyrenoid (*p*) and
eyespot (*s*) and their respective centromeres.

As another possible case where crossing-over occurred at the two-strand
stage of meiosis, Moewus cited the work of von Wettstein (1924, see p. 95)
with *Funaria hygrometrica*.

Four-strand-stage crossing-over

Moewus finally succeeded in obtaining crossing-over at the four-strand stage
in *Chlamydomonas*, presumably by environmental treatment (see below,
however). In several different crosses (Moewus 1948e), he used *F* alleles of
different valences and their alleles *gacop*$_F$, *M* alleles of different valences and
their alleles *gacop*$_M$, and the eyespot locus (*s*).

Later Moewus (1949e), he reported four-strand-stage crossing-over
results in linkage group VII using the markers *pyfo*, *mehab*, and *runo* (pp.
184–5), in linkage group IV using the markers *rha*, *flalo*, *ru* and *smag*
(pp. 182–3) and in linkage group IX using the markers *sfo*, *pal* and *qu* (pp.
189–90).

The percentages of zygotes containing exchanges between *F* and *M*, *M*
and *s*, *F* and *s*, *flalo* and *smag*, *pyfo* and *mehab*, and *sfo* and *pal* from the four-
strand-stage experiments were comparable to the percentages of zygotes
containing exchanges in the two-strand-stage experiments reported earlier.
This means that map distances computed from four-strand-stage experiments
would be one-half of corresponding map distances computed from two-
strand-stage experiments (see Table B.4), since crossing-over at the two-strand
stage gives a recombination frequency (per strand) which is equal to the
chiasma frequency (per tetrad), and crossing-over at the four-strand stage

gives a recombination frequency (per strand) which is one-half the chiasma frequency (per tetrad) (assuming complete interference—see Barratt *et al.* 1954, or Gowans 1965).

Table B.4. A comparison of map distances in *Chlamydomonas eugametos* computed from data given by Moewus for two-strand-stage and four-strand-stage experiments.

Linkage Group	Region	Stage of Meiosis in which Crossing-over Occurs	Percent Recombination Zygotes	Map Distance	Reference
IV	*flalo-smag*	Two-strand	18·9	18·9	Moewus 1940c
		Four-strand	20·4	10·2	Moewus 1949e
VII	*pyfo-mehab*	Two-strand	5·5	5·5	Moewus 1940c
		Four-strand	5·25	2·6	Moewus 1949e
IX	*sfo-pal*	Two-strand	35·6	35·6	Moewus 1940c
		Four-strand	46·5	23·2	Moewus 1949e
X	F-M	Two-strand	3·13	3·1	Moewus 1943
		Four-strand	3·37	1·7	Moewus 1948e
	M-s	Two-strand	1·6	1·6	Moewus 1940c
		Four-strand	1·67	0·8	Moewus 1948e

One would suppose (by definition) that two-strand-stage crossing-over would occur at an entirely different time from four-strand-stage crossing-over, and would be subject to entirely different cellular and physiological conditions. Why the chiasma frequencies (per tetrad) should be exactly comparable at these two stages is difficult to explain.

Moewus (1938b, p. 534), presumably referring to these experiments, wrote, *After extensive experiments involving modification of techniques we succeeded, this year, in obtaining normal crossing-over results in Chlamydomonas.* Unfortunately, if he was referring to these crosses (Moewus 1948e and 1949e), this is not the case, as the double crossovers exhibit a very peculiar bias. Assuming a random distribution of double crossovers (i.e. no chromatid interference), one expects 25% two-strand doubles, 50% three-strand doubles, and 25% four-strand doubles (Fig. B.1). Moewus obtained five double crossovers in 1948, all of which were two-strand doubles (Moewus 1948e, p. 285 crosses D, E and F, and bottom of p. 288 to top of p. 289). In 1949 he obtained 46 double crossovers, all of which were two-strand doubles (Moewus 1949e, p. 183 cross M, p. 185 cross O, p. 190 cross Q). The combined data total 51 doubles, all of which are two-strand doubles. The probability of this is $(1/4)^{51}$, unless one proposes a complete negative chromatid interference.

Later Moewus (1949e) mapped the genes concerned with the gamone inhibitor rutin (*rha*, *runo* and *ru*) (see Moewus 1951a) and a gene (*qu*) concerned with the synthesis of a gynotermone and an androtermone. In crosses A, B and C these genes exhibited a normal 2:2 segregation (pp. 174–5). In crosses D and E *rha* and *ru* appeared unlinked and gave 18·0% tetratypes (p. 181). In cross M (p. 183) *rha* and *ru* were in linkage group IV. In cross O *runo* was in linkage group VII (pp. 184–5), and in cross Z (pp. 189–90) *qu* was in linkage group IX (the IV stated there is apparently a misprint).

Fig. B.1. Expected types and frequencies of double cross-overs, assuming no chromatid interference.

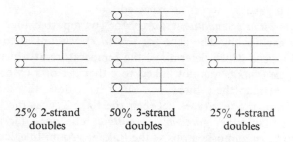

| 25% 2-strand doubles | 50% 3-strand doubles | 25% 4-strand doubles |

It is in this series of crosses that one finds what is probably the most puzzling contradiction in all of Moewus' genetic work. He obtained tetratypes only when he was dealing with linked genes. When he was dealing with non-linked genes he obtained only ditypes, as indicated by the following data:

From cross D, the parents being $rha^0runo^+ \times rha^+runo^0$, he obtained:

43 PD tetrads (rha^0runo^+, rha^0runo^+, rha^+runo^0, rha^+runo^0)

57 NPD tetrads (rha^0runo^0, rha^0runo^0, rha^+runo^+, rha^+runo^+)

 0 T tetrads (rha^0runo^+, rha^0runo^0, rha^+runo^+, rha^+runo^0)

From cross E, the parents being $ru^0runo^+ \times ru^+runo^0$, he obtained:

52 PD tetrads (ru^0runo^+, ru^0runo^+, ru^+runo^0, ru^+runo^0)

48 NPD tetrads (ru^0runo^0, ru^0runo^0, ru^+runo^+, ru^+runo^+)

 0 T tetrads (ru^0runo^+, ru^0runo^0, ru^+runo^+, ru^+runo^0)

It was very plainly stated for each cross that only two types of tetrads were found. For cross D (p. 176) he said *Es treten 2 Tetradentypen auf*, and for cross E he said (p. 177) *Wieder enstehen 2 Tetradentypen*.

The only way to explain these results would be to postulate that all three genes were on or very close to their respective centromeres, so that in 100 zygotes no crossing-over occurred between *rha* and its centromere and *ru* and its centromere, and in 200 zygotes no crossing-over occurred between *runo* and its centromere. Crosses F, G, H and J eliminate this possibility, however.

In these crosses, *rha* and *ru* were shown to be linked and gave 18·0% tetra-types (N = 350) (pp. 179–81). Thus these two genes must each have been at least 4·5 map units from its centromere (this would have been the case if the centromere had been half-way between them), and one would have expected a total of 18 tetratypes among the zygotes from cross D plus cross E. No matter what the position of the centromere, one would have expected 9% tetratypes in the two crosses. There is only a very small probability of obtaining by chance the observed 0 tetratype : 200 ditype distribution instead of the expected 18 : 182 (chi square = 19·62, d.f. = 1).

On page 182, where he reported testing *ru* for linkage to markers in linkage group I, he again indicated that tetratypes were not found between non-linked genes. This case was not so clear as the one above, because he did not list all four products of meiosis. Crosses M, O and Q involved linked genes only, and gave tetratypes as expected.

In the *Botrydium granulatum*-group, Moewus reported four-strand-stage crossing-over exclusively. The gene for sex (*det*) was linked to the eyespot gene (*s*), and a gene which determined the temperature at which copulation could take place (*gathe*) was not linked to either *det* or *s* (Moewus 1940g). Moewus did not map these three loci, but I have done this using his data (Table B.3). Since the tetratypes between the linked genes (*det* and *s*) were more frequent than either *gathe-det* or *gathe-s* tetratypes, the linked genes would have to be on opposite arms of the linkage group. In calculating gene-centromere distances of two linked and one independent genes by simul-taneous equations (Whitehouse 1950), the equations can be applied directly if the linked genes are on opposite sides of their centromere (Papazian 1952). The maps given in Table B.3 were calculated by simultaneous equations on this basis.

Chiasma interference

As Harte (1948) has pointed out, Moewus' work on *Chlamydomonas* (Moewus 1938b, 1940a, 1940b, 1940c) showed evidence for extremely strong chiasma interference effects. Double crossovers were very rare, and triples were never found. This kind of interference had a further peculiarity in that it increased with distance rather than decreasing (one might call this positive inverse interference). Thus, double crossovers were more likely to be found in neighbouring regions than in widely separated regions. This behaviour could be explained (as can negative interference) on the basis of partial pairing.

Some chiasma interference was found by Moewus in the sex chromosome of the *Protosiphon botryoides*-group. Unfortunately the markers were such that inverse interference cannot be detected. There were five crosses in the 1935 paper (Moewus 1935c) where doubles could be detected, as reported in Table 20 p. 39 (crosses 1 and 2), and Table 25 p. 52. They involved crossing-over in the *F-M-d* region among a total of 1,009 zygotes. Doubles were expected, on the basis of 'no interference', in 0·106% and they were obtained in 0·15%. In Moewus' 1950g paper, presenting results of two crosses involving regions *F-M-div-fur* (pp. 57 and 58), doubles for the *M-div-fur* region were expected in 1·26% and obtained in 0·3% (N = 500).

In Moewus' experiments in which crossing-over occurred at the four-strand stage of meiosis, no very strong chiasma interference was indicated (Moewus 1948e, 1949e), and, in fact, there seems to have been no interference between neighbouring regions. Thus, although doubles were expected to occur with a frequency of 0·054% in the 1948 experiments, only five zygotes were found among 9,264 tested (0·054%) in which double crossovers had occurred (see, however, chromatid interference above).

The phenomenon of inverse interference is open to investigation in only one cross in which crossing-over reportedly occurred at the four-strand stage (Moewus 1949e pp. 182 and 183, cross M). On the basis of this cross alone, inverse interference was also indicated when crossing-over took place at the four-strand stage of meiosis. Cross M involved the markers *rha*, *flalo*, *ru* and *smag* in linkage group IV, arranged in the order listed above. Although doubles were obtained in the *rha-flalo* and *flalo-ru* adjacent regions (0·6%), as well as in the *flalo-ru* and *ru-smag* adjacent regions (0·6%), they were not obtained in the non-contiguous *rha-flalo* and *ru-smag* regions (N = 500).

The gene *reph* (Moewus 1940b p. 465) or *h* (Moewus 1936 p. 48 *et seq.*) should be mentioned in connection with double crossing-over. This gene, located in linkage group IX (Moewus 1940c), reportedly controlled the range of pH at which growth could occur. *The allele H (reph^{lat} in Moewus 1940b) is present in C. eugametos, and permits growth between pH 4·5 and 9·5. The allele h is present in C. paupera* (stocks were reportedly lost, see below) *and allows good growth at pH 5·5, slow growth at pH 7·0, and no growth at pH 9·5. This gene has an influence on crossing-over within the sex chromosome* (linkage group X) (Moewus 1936). *When it is present in a homozygous condition* (either *HH* or *hh*), *no double crossovers are obtained in the F-M-s region. When it is heterozygous in the zygote (Hh), double crossovers occur in the F-M-s region to approximately 1·5%. Mating strains of C. paupera were lost along with the allele h* (Moewus 1938b p. 517) *and never recovered* (Moewus personal communication). These reported results are suggestive of the Steinberg-Redfield interchromosomal effects in crossing-over in *Drosophilia* (Schulz & Redfield 1951).

Temperature and crossing-over

Apparently temperature had some effect on determining whether crossing-over occurred at the two-strand or four-strand stage in Moewus' experiments. It is not clear whether this was the temperature of germination, or the pre-treatment temperature, or some combination of the two. The following statements on this subject were gleaned from several of Moewus' publications and a personal communication.

Crossing-over at the two-strand stage occurs when zygotes are germinated at low temperatures (−5°). *At higher temperatures* (+22°) *tetratypes are obtained. With increasing temperatures the frequency of tetratypes increases* (Moewus 1938b p. 534).

Zygotes can be induced to germinate at 25° because of the presence of the allele zyger[tac]. *Low temperatures* (*under 15°*) *produce abnormal meioses* (Moewus 1948e p. 284). *Ditypes are produced when zygotes germinate at low temperatures* (>5°) (Moewus 1948e p. 289).

A great deal of difficulty was encountered germinating zygotes: sometimes they would germinate only after cold treatments (−5*), *other times they germinated only at* − 33*, *and in still other cases they germinated only between 2° and 33°. In 1937 a method was found that gave regular zygote germination. This was the use of a filter-sterilized soil solution which had been kept in a refrigerator for one week before filtration* (Moewus 1940a p. 423).

The required pretreatment temperatures and germination temperatures are hereditarily determined (*zyger gene, see below*) *and, even in the cold soil solution, zygotes do not germinate without the proper temperature treatment* (Moewus 1940b p. 479).

Zygotes germinate only after treatments at −5° *to* −7° (Moewus 1933b p. 474). *Zygotes* (from a cross of *C. eugametos typica* with *C. eugametos subheteroica*) *germinate only after a temperature treatment of* − 6 *for 1 to 2 days or* − 15 *for 12 hours*; *without this freezing the zygotes will not germinate* (Moewus 1934a p. 103). *The zygotes in this cross were germinated at 20°* (Moewus 1943 p. 179). *If you pretreat the zygotes at* − 5°, *and germinate them at* +5°, *you will obtain two-strand-stage crossing-over* (Moewus, personal communication. This personal communication, like all others in this review, was verbal, addressed to only the late Professor G. M. Smith of Stanford University and myself.)

No special attempt has been made here to collect all of Moewus' statements on temperature and zygote germination; but I know of no such prescription, regarding a temperature treatment capable of inducing two-strand crossing-over, which was not contradicted by Moewus himself elsewhere. For a detailed report listing the years in which Moewus could supposedly germinate zygotes, and the years in which he could not, see Moewus, 1937, p. 70.

As Ebersold (1954) pointed out, the great majority of the species and races in the *C. eugametos*-group (all except *C. eugametos blastikos, psychroblastikos* and *thermoblastikos*) were stated to contain the allele *zyger*[tep]. This allele reportedly allowed zygotes to germinate only between the temperature of 22° and 26° (Moewus 1940b pp. 479–80). One may therefore conclude that the temperature at the time of zygote germination could not have accounted for the data indicating that crossing-over occurred at the two-strand stage.

* The ambiguity of some of these figures—whether they represent degrees above or below zero—is in the original text.

Ebersold (1954) also pointed out that a low temperature pretreatment of zygotes could not have caused two-strand-stage crossing-over, because zygotes which supposedly required pretreatment of +33°C also gave two-strand-stage results (see crosses IIIe, p. 508, and 8b, p. 516 in Moewus 1940c, and see Moewus 1940c, pp. 482 and 483, for dominance relationships).

In our laboratory, zygotes of *C. eugametos* germinate within the range of 13° to 28°C (Gowans 1960). At 13° germination is retarded drastically and some germination is still occurring one month after plating of zygotes; while at 10° the zygotes do not germinate at all within a two-month period after plating.

3   GENERAL EVALUATION OF MOEWUS' WORK

3.1   *Internal criteria of validity*

Statistical criticisms were presented by Philip and Haldane (1939), answered by Moewus (1943), and reviewed by Sonneborn (1951).

Another criticism on statistical grounds was made by Pätau (1941). Moewus reported the analysis of 102,602 zygotes for the segregation of ten independent (non-linked) genes (1940c p. 506). With crossing-over at the two-strand stage, one would expect the zygotes to fall into 512 (2^9) different classes of approximately 200 zygotes each. Moewus presented a distribution curve for the number of zygotes obtained in each of these 512 classes. The curve is very smooth, with a range of only 172–231. Pätau calculated the probability of obtaining such a good agreement to the expected 200 per class to be less than 10^{-9}. Apparently Pätau made an error in calculation (Ludwig 1942); the probability is actually less than 10^{-22}.

This criticism was answered for Moewus by Ludwig (1942), who went over Moewus' original data and found that Moewus had apparently analysed 205,204 zygotes (actually 204,832; see footnote p. 615). The 820,816 products of these zygotes were, reportedly, each identified for the ten characteristics, and then their numbers were divided by two because each individual product was represented twice in each zygote. (All zygotes were ditypes.) The curve was then drawn for the 411,645 representatives, and these data were filed away. When it came time to publish, Moewus apparently forgot that the division by two had already been carried out, and the numbers for the 411,645 products were again halved. Thus the number of zygotes given in the article (102,602) was obtained by dividing the 411,645 products by four and the curve was based on a sample that was actually twice the published size. This increased the probability (of obtaining the criticized good agreement to the expected) to the still very low value of 0·001 (Ludwig 1942).

Exceedingly large numbers of segregants were reportedly analysed in Moewus' crosses. Perhaps it is worthwhile to mention the method by which the ten morphological characters were identified in each of the approximately half-million products discussed above. *Each zygote was isolated in a drop of 'Volvox solution' and placed in a damp chamber until it germinated, and then fixed with osmic acid vapour, or else isolated on a slide just before germination and then fixed with osmic acid vapour. A coverslip was applied and sealed with paraffin* (Moewus 1940a pp. 424–5), *and then* (at least in some cases) *a camera lucida drawing was made of each of the four* (post-meiotic cells) (Moewus 1940c pp. 502–3).

If we assume that it took 12 minutes to prepare each slide and identify the ten morphological characters for the four products (three minutes per product), it would take 2,462,488 minutes to identify the 820,816 products in the experiment discussed above. A man working ten hours a day, six days a week, could complete these identifications in $13\frac{1}{3}$ years. This would not allow time for preparing media, making crosses, washing dishes and slides, or the many other time-consuming details which must have been involved in this experiment, nor time for identification of products from thousands of other *Chlamydomonas, Protosiphon, Botrydium,* and *Polytoma* zygotes (see for example 220,000 tests run on *Protosiphon;* Moewus 1937 pp. 103–7). Nor does it allow time for studies on variability, Dauermodifikationen, sex substances, relative sexuality, chemotaxis, mutation, or gamete copulation which were reportedly carried out during the years 1929 to 1940 (see Moewus bibliography). Nor does it allow time for years to go by without his being able to germinate zygotes (Moewus 1937 p. 70). Moewus acknowledged help only from his wife (for inspection of the data) (Moewus 1940a p. 425).

The apparent overperfection and overcompleteness of Moewus' work are very striking. For each step in the production of zygotes a hormone was found, the hormones were identified, and the genetic control of their production was elucidated (see Raper 1952, or Moewus 1950a, 1951a). The possibility of another gamete valence which would fit the Bergmann-Niemann series soon led to the discovery of a mutant with such a valence (Moewus 1940e). For the ten cytologically recognizable chromosomes (Moewus 1934a) he found ten linkage groups (Moewus 1940c), each nicely marked with not less than two or more than six loci. A whole series of very beautifully interrelated experiments with several different organisms demonstrated the nonallelic nature of the sex loci. Other instances could be cited which demonstrate an apparent perfection and completeness rarely found in biological research.

3.2   *External evidence for validity*

Wiese and his collaborators (Förster & Wiese 1954, 1955; Förster *et al.* 1956; Wiese & Shoemaker 1970; Wiese & Hayward 1972) failed to confirm Moewus' claim that carotenoids are concerned with sexuality in *Chlamy-*

domonas, but, conversely, demonstrated that the gynogamone of *C. eugametos* is a glycoprotein (see Wiese, this volume).

During Moewus' 16-month stay at Columbia University, Professor F. Ryan and his colleagues were unable to repeat any of his many experiments, even with Moewus' collaboration (Ryan 1955).

After nearly 20 years of work, and the analysis of several thousands of zygotes, my collaborators and I (see Chapter 7) have likewise failed to produce evidence to support any aspect of Moewus' genetic work. We used both strains received directly from Moewus (through G. M. Smith) and strains having independent origins from Moewus (Strains 9 and 10 in the Indiana University Culture Collection; see Starr 1964). Similarly, Lewin (see Chapter 7), working with isolates of *C. moewusii*, was unable to support any of Moewus' claims beyond some details of gamete behaviour mentioned in Moewus' 1933 paper.

Particular care has been given to the segregation of sex (or mating type) in these experiments. According to Moewus, $3·37\%$ of all germinating zygotes should contain a cross-over between the non-allelic sex determiners, F and M, and thereby produce three-product zygotes in which one product is male, one female, and one homothallic. No such homothallic products have been recovered, despite the analysis of several thousands of zygotes for the segregation of mating type, and other experiments specifically designed to seek such homothallic strains (Lewin, personal communication).

All attempts to inhibit crossing-over at the four-strand-stage, or to induce crossing-over exclusively at the two-strand-stage, have also failed. (See Temperature and crossing-over, above.)

4   CONCLUSION

In view of the considerable amount of evidence presented and summarized above, it is considered advisable to discount any and all of the published work of Franz Moewus unless such results have been repeated and confirmed independently.

5   REFERENCES

BARRATT R.W., NEWMEYER D., PERKINS D.D. & GARNJOBST L. (1954) Map construction in *Neurospora crassa. Adv. Genetics*, **6**, 1–93.

BIRCH A.J., DONOVAN F.W. & MOEWUS F. (1953) Biogenesis of flavonoids in *Chlamydomonas eugametos. Nature*, **172**, 902–4.

CHODAT F. & DE SIEBENTHAL U.R. (1941) La sexualité relative des *Chlamydomonas;* exposé des experiences de Mm F. Moewus et R. Kuhn. *Bull. Soc. Botan. Genève, sér* 2, **33**, 1–37.

COOK A.R. (1945) Algal pigments and their significance. *Biol. Rev.* **20**, 115–32.

DARDEN W.H. (1966) Sexual differentiation in *Volvox aureus*. *J. Protozool.* **13**, 239–55.

EBERSOLD W.T. (1954) The isolation and characterization of some mutant strains of *Chlamydomonas reinhardi* Dangeard. Ph.D. dissertation, Stanford University.

ESSER K. & KUENEN R. (1967) *Genetics of Fungi*, Springer Verlag, New York.

ESSER K. & STRAUB A. (1954) Das Pollenschlauchwachstum bei *Forsythia*, eine Stellungnahme zu der Moewusschen Hemmstoff-Ferment-Hypothese. *Biol. Zentralb.* **73**, 449–55.

FÖRSTER H. & WIESE L. (1954) Gamonwirkungen bei *Chlamydomonas*. *Zeit. f. Naturf.* **9b**, 548–50.

FÖRSTER H. & WIESE L. (1955) Gamonwirkung bei *Chlamydomonas reinhardii*. *Zeit. f. Naturf.* **10b**, 91–2.

FÖRSTER H., WIESE L. & BRAUNITZER G. (1956) Über das agglutinierend wirkende Gynogamon von *Chlamydomonas eugametos*. *Zeit. f. Naturf.* **11b**, 315–7.

GOWANS C.S. (1960) Some genetic investigations on *Chlamydomonas eugametos*. *Zeit. f. Indukt. Abstam.- u. Vererb.-lehre*, **91**, 63–73.

GOWANS C.S. (1965) Tetrad analysis. *Taiwania*, **11**, 1–19.

GRIES H. (1941) Mutations- und Isolationsversuche zur Beeinflussung des Geschlechs von *Sordaria fimicola* (Rob.). *Zeit. f. Botanik* **37**, 1–116.

HAGEN-SEYFFERTH M. (1959) Zur Kenntnis der Geisseln und der Chemotaxis von *Chlamydomonas eugametos* Moewus (*Chl. moewusii* Gerloff). *Planta*, **53**, 376–401.

HÄMMERLING J. (1934) Über die Geschlechtsverhältnisse von *Acetabularia mediterranea* und *Acetabularia Wettsteinii*. *Archiv. f. Protist.* **83**, 57–97.

HÄMMERLING J. (1938) Fortpflanzung und Sexualität. *Fortschr. d. Zoologie* **4**, 503–4.

HARTE C. (1948) Das Crossing-Over bei *Chlamydomonas*. *Biol. Zentralb.* **67**, 504–10.

HARTMANN M. (1934) Beiträge zur Sexualitätstheorie. Mit besonderer Berücksichtigung neuer Ergebnisse von F. Moewus. Situngsber. d. Preuss. Akad. d. Wiss. Berlin, Phy.-Math. Kl.

HARTMANN M. (1943) *Die Sexualität. 1 Aufl*. Gustav Fisher, Jena.

HARTMANN M. (1956) *Die Sexualität. 2 Aufl*. Gustav Fisher, Jena.

HATTORI S. (1962) Glycosides of flavones and flavonoids. In *The Chemistry of Flavonoid Compounds*, ed. Geissman T.A. Macmillan Press, New York.

HÜTTIG W. (1935) Die Sexualität bei *Glomerella lycopersici* Kruger und ihre Vererbung. *Biol. Zentralb.* **55**, 74–83.

JOLLOS V. (1926) Untersuchungen über die Sexualitätsverhältnisse von *Dasycladus clavaeformis*. *Biol. Zentralb.* **46**, 279–95.

KUHN R. & MOEWUS F. (1940) Über die chemische Wirkungsweise der Gene Mot, MD, und Gathe bei *Chlamydomonas*. *Ber. Deutsch. Chem. Ges.* **73**, 547–59.

KUHN R., MOEWUS F. & JERCHEL D. (1938) Über die chemische Natur der Stoffe, welche die Kopulation der männlichen und weiblichen Gameten von *Chlamydomonas eugametos* in licht bewirken. *Ber. Deutsch. Chem. Ges.* **71**, 1541–7.

KUHN R., MOEWUS F. & WENDT G. (1939) Über die geschlechtsbestimmenden Stoffe iner Grünalge. *Ber. Deutsch. Chem. Ges.* **72**, 1702–7.

LANG A. (1944) Entwicklungsphysiologie. *Fortschr. Botan.* **11**, 268–317.

LERCHE W. (1936–7) Untersuchungen über Entwicklung und Fortpflanzung in der Gattung *Dunaliella*. *Archiv. f. Protist.* **88**, 236–69.

LEWIN R.A. (1952) Ultraviolet induced mutations in *Chlamydomonas moewusii* Gerloff. *J. Gen. Microbiol.* **6**, 233–48.

LEWIN R.A. (1957) Four new species of *Chlamydomonas*. *Canad. J. Botany*, **35**, 321–6.

LUDWIG W. (1942) Notiz zu der Unternormalen Streuung in den Moewusschen *Chlamydomonas*-Versuchen. *Zeit. f. Induk. Abstam.- u. Vererb.-lehr* **80**, 612–5.

MOEWUS F. (1931) Neue Chlamydomonaden. *Archiv. f. Protist.* **75**, 284–96.

MOEWUS F. (1933a) Untersuchungen über die Variabilität von Chlamydomonaden. *Archiv. f. Protist.* **80**, 128–71.

MOEWUS F. (1933b) Untersuchungen über die Sexualität und Entwicklung von Chlorophyceen. *Archiv. f. Protist* **80**, 469–526.

MOEWUS F. (1934a) Über Subheterözie bei *Chlamydomonas eugametos*. *Archiv. f. Protist.* **83**, 98–109.

MOEWUS F. (1934b) Über Dauermodifikationen bei Chlamydomonaden. *Archiv. f. Protist.* **83**, 220–40.

MOEWUS F. (1935a) Über den Einfluss äusserer Faktoren auf die Geschlechtsbestimmung bei *Protosiphon*. *Biol. Zentralb.* **55**, 293–309.

MOEWUS F. (1935b) Über die Vererbung des Geschlechts bei *Polytoma pascheri* und bei *Polytoma uvella*. *Zeit. f. Induk. Abstam.- u. Vererb.-lehre*, **69**, 374–417.

MOEWUS F. (1935c) Die Vererbung des Geschlechts bei verschiedenen Rassen von *Protosiphon botryoides*. *Archiv. f. Protist.* **86**, 1–57.

MOEWUS F. (1936) Faktorenaustausch insbesondere der Realisatoren bei *Chlamydomonas*-Kreuzungen. *Bericht. d. Deutsch. Botan. Gesel.* **54**, (45)–(57).

MOEWUS F. (1937) Methodik und Nachträge zu den Kreuzungen zwischen *Polytoma*-Arten und zwischen *Protosiphon*-Rassen. *Zeit. f. Induk. Abstam.- u. Vererb.-lehre*, **73**, 63–107.

MOEWUS F. (1938a) Carotinoide als Sexualstoffe von Algen. *Jahrb. f. Wiss. Bot.* **86**, 753–83.

MOEWUS F. (1938b) Vererbung des Geschlechts bei *Chlamydomonas eugametos* und verwandten Arten. *Biol. Zentralb.* **58**, 516–36.

MOEWUS F. (1939a) Untersuchungen über die relative Sexualität von Algen. *Biol. Zentralb.* **59**, 40–58.

MOEWUS F. (1939b) Über die Chemotaxis von Algen-Gameten. *Archiv. f. Protist.* **92**, 485–526.

MOEWUS F. (1940a) Die Analyse von 42 erblichen Eigenschaften der *Chlamydomonas eugametos*-Gruppe. I. Teil: Zellform, Membran, Geisseln, Chloroplast, Pyrenoid, Augenfleck, Zellteilung. *Zeit. f. Induk. Abstam.- u. Vererb.-lehre*, **78**, 418–62.

MOEWUS F. (1940b) Die Analyse von 42 erblichen Eigenschaften der *Chlamydomonas eugametos*-Gruppe. II. Teil: Zellresistenz, Sexualität, Zygote, Besprechung der Ergebnisse. *Zeit. f. Induk. Abstam.- u. Vererb.-lehre*, **78**, 463–500.

MOEWUS F. (1940c) Die Analyse von 42 erblichen Eigenschaften der *Chlamydomonas eugametos*-Gruppe. III. Teil: Die 10 Koppelungs-gruppen. *Zeit. f.I nduk. Abstam.- u. Vererb.-lehre*, **78**, 501–22.

MOEWUS F. (1940d) Über Zoosporen-Kopulationen bei *Monostroma*. *Biol. Zentralb.* **60**, 225–38.

MOEWUS F. (1940e) Über Mutationen der Sexual-Gene bei *Chlamydomonas*. *Biol. Zentralb.* **60**, 597–626.

MOEWUS F. (1940f) Carotinoid-Derivative als geschlechtsbestimmende Stoffe von Algen. *Biol. Zentralb.* **60**, 143–66.

MOEWUS F. (1940g) Über die Sexualität von *Botrydium granulatum*. *Biol. Zentralb.* **60**, 484–98.

MOEWUS F. (1941a) Zur Sexualität der Niederen Organismen. I. Flagellaten und Algen. *Ergeb. Biol.* **18**, 287–356.

MOEWUS F. (1941b) Koppelung und Austausch der Realisatoren bei *Brachiomonas*. *Beit. Biol. Pflanz.* **27**, 297–356.

MOEWUS F. (1943) Statistische Auswertung einiger physiologischer und genetischer Versuche an *Protosiphon* und *Chlamydomonas*. *Biol. Zentralb.* **63**, 169–203.

MOEWUS F. (1948a) Zur Genetik und Physiologie der Kern- und Zellteilung. 1. Die Apomiktosis von *Enteromorpha* Gameten. *Biol. Zentralb.* **67**, 277–93.

MOEWUS F. (1948b) Freier und gebundener Wuchsstoff in der Kartoffelknolle. *Zeit. f. Naturf.* **3b**, 135–6.

MOEWUS F. (1948c) Ein neuer quantitativer Test für pflanzliche Wuchsstoffe. *Naturwiss.* **35**, 124–5.

P

MOEWUS F. (1948d) Bestimmung des Wuchsstoff und Hemmstoffgehaltes von Pflanzen-extrakten. *Zuchter*, **19**, 108–15.

MOEWUS F. (1948e) Über die Erblichkeit des Kopulationsverhaltens bei *Chlamydomonas*. *Zeit. f. Naturf.* **3b**, 279–90.

MOEWUS F. (1949a) Der heutige Stand der Termonforschung bei Algen. *Zeit. Vit.- Horm.-und Ferm.* **3**, 139–49.

MOEWUS F. (1949b) Gebundener und freier Wuchsstoff in fleischigen Früchten. *Planta*, **37**, 413–30.

MOEWUS F. (1949c) Die Wirkung von Wuchs- und Hemmstoffen auf die Kresswurzel. *Biol. Zentralb.* **68**, 58–72.

MOEWUS F .(1949d) Der Kressewurzeltest, ein neuer quantitativer Wuchsstoff test. *Biol. Zentralb.* **68**, 118–40.

MOEWUS F. (1949e) Zur Biochemischen Genetik des Rutins. *Portugaliae Acta Biologica Ser. A. Goldschmidt Volumen*, 169–99.

MOEWUS F. (1950a) Sexualität und Sexualstoffe bei einem inzelligen Organismus (*Chlamydomonas*). *Zeit. Sexualforsch.* **1**, 1–25.

MOEWUS F. (1950b) Die Bedeutung von Farbstoffen bei den Sexualprozessen der Algen und Blütenpflanzen. *Zeit. Angew. Chemie*, **62**, 496–502.

MOEWUS F. (1950c) Zur Physiologie und Biochemie der Selbststerilität bei *Forsythia*. *Biol. Zentralb.* **69**, 181–97.

MOEWUS F. (1950d) Über die Physiologischen und Biochemischen Grundlagen der Selbst-sterilität bei *Forsythia*. *Forsch. u. Fortschr.* **26**, 101–2.

MOEWUS F. (1950e) Über die Zahlenverhältnisse der Geschlechter bei *Valeriana dioica* L. *Zeit. Naturf.* **5b**, 380–3.

MOEWUS F. (1950f) Über ein Blastokolin der *Chlamydomonas*-Zygoten. *Zeit. f. Naturf.* **5b**, 196–202.

MOEWUS F. (1950g) Zur Genetik und Physiologie der Kern- und Zellteilung. II. Über den Synchronismus der Kernteilungen bei *Protosiphon botryoides*. *Beitr. Biol. Pflanzen*, **28**, 36–63.

MOEWUS F. (1950h) Die Reaktionskette Gene-Ferment-Wirkstoffmerkmal für die Rutin-bildung bei *Chlamydomonas*. *Ber. Deutsch. Botan. Gesel.* **63**, (11)–(12).

MOEWUS F. (1951a) Die Sexualstoffe von *Chlamydomonas eugametos*. *Ergeb. Enzymforsch.* **12**, 1–32.

MOEWUS F. (1951b) Über die Anwendbarkeit des Kressewurzel-Testes. *Ber. Deutsch. Botan. Gesel.* **64**, 212–4.

MOEWUS F. (1953) Biosynthesis of pigments in a unicellular plant. *Estratto degli Atti del VI Congresso Internazionale di Microbiologia, Roma*, **1**, 292–5.

MOEWUS F. (1954a) On inherited and adapted rutin-resistance in *Chlamydomonas*. *Biol. Bull.* **107**, 293.

MOEWUS F. (1954b) Hormone control in the life cycle of the green alga, *Chlamydomonas eugametos*. *Cong. Intern. Botan. Paris, Rapps. et Communs.* **8**, 46–7.

MOEWUS F. (1954c) Action of 2,4-D on deaminating enzymes. *Congr. Intern. Botan., Paris, Rapps. et Communs.* **8**, 149–50.

MOEWUS F. (1955a) Biogenesis of the flavinoids. *Ann. N.Y. Acad. Sci.* **61**, 660–4.

MOEWUS F. (1955b) Interrelations between growth and sexuality in a homothallic strain of *Polytoma uvella*. *J. Protozool.* **2**(suppl), 7.

MOEWUS F. (1957) The manifestation of the homothallic sex behavior in algae. *Trans. Am. Micr. Soc.* **76**, 337–44.

MOEWUS F. (1958) Effect of kinetin on cell division of *Polytoma uvella*. *VII Intern. Congr· Microbiol.*, p. 41.

MOEWUS F. & BANERJEE B. (1951a) Effect of cis-cinnamic acid and some isomeric compounds on the germination of zygotes of *Chlamydomonas*. *Nature, Lond.* **168**, 561.

Moewus F. & Banerjee B. (1951b) Über die Wirkung von Cis-Zimtsäure und einigen isomeren Verbindungen auf *Chlamydomonas*-Zygoten. *Zeit. f. Naturf.* **6b**, 270–3.

Moewus F. & Deulofen V. (1954) An antagonist of the sterility hormone rutin in the green alga *Chlamydomonas eugametos*. *Nature, Lond.* **173**, 218.

Moewus F. & Moewus L. (1952) Sensitivity of cress roots to indoleacetic acid. *Nature*, **170**, 372.

Moewus F., Moewus L. & Skwarra H. (1952) Nachweis von zwei Wuchsstoffen in Samen und Wurzeln der Kresse (*Lepidium sativum*). *Planta*, **40**, 254–64.

Moewus F. & Schader E. (1951a) Über die Keimungs- und Wachstums-hemmenden Wirkung einiger Phthalide. *Ber. Deutsch. Botan. Gesel.* **64**, 124–9.

Moewus F. & Schader E. (1951b) Die Wirkung von Cumarin und Parasorbinsäure auf das Austreiben von Kartoffelknollen. *Zeit. f. Naturf.* **6b**, 112–15.

Moewus F. & Schader E. (1952) Über den Einfluss von Wuchs- und Hemmstoffen auf das Rhizoidwachstum von *Marchantia*-Brutkörpern. *Beitr. Biol. Pflanz.* **29**, 171–84.

Murneek A.E. (1941) Sexual reproduction and the carotinoid pigments in plants. *Am. Naturalist*, **75**, 614–9.

Papazian H.P. (1952) The analysis of tetrad data. *Genetics*, **37**, 175–88.

Pätau K. (1941) Eine statistische Bemerkung zu Moewus' Arbeit 'Die Analyse von 42 erblichen Eigenschaften der *Chlamydomonas eugametos* Gruppe. III.' *Zeit. f. Induk. Abstam.-u. Vererb.-lehre*, **79**, 317–19.

Philip U. & Haldane J.B.S. (1939) Relative sexuality in unicellular Algae. *Nature, Lond.* **143**, 334.

Raper J.R. (1952) Chemical regulation of sexual processes in Thallophytes. *Botan. Rev.* **18**, 447–545.

Renner O. (1958) Auch etwas über F. Moewus, *Forsythia* und *Chlamydomonas*. *Zeit. f. Naturforsch.* **13b**, 399–403.

Reznik H. (1957) Über die Pigmentausstattung der Pollen Heterostyler. *Biol. Zentralbl.* **76**, 352–9.

Ryan F.J. (1955) Attempt to reproduce some of Moewus' experiments on *Chlamydomonas* and *Polytoma*. *Science*, **122**, 470.

Schultz J. & Redfield H. (1951) Interchromosomal effects on crossing-over in *Drosophila*. *Cold Spring Harbor Symposia Quant. Biol.* **16**, 175–97.

Smith G.M. (1946) The nature of sexuality in *Chlamydomonas*. *Amer. J. Bot.* **33**, 625–630.

Smith G.M. (1951a) The sexual substances of algae. In *Plant Growth Substances*, ed. Skoog, Folke, University of Wisconsin Press.

Smith G.M. (1951b) Sexuality of Algae, in Manual of Phycology, ed. Smith G.M., Chronica Botanica Co., Waltham, Mass.

Starr R.C. (1964) The culture collection of algae at Indiana University. *Am. J. Botany*, **51**, 1013–44.

Sonneborn T.M. (1942) Sex hormones in unicellular organisms. *Cold Spring Harbor Symposia Quant. Biol.* **10**, 111–24.

Sonneborn T.M. (1951) Some current problems of genetics in the light of investigations on *Chlamydomonas* and *Paramecium*. *Cold Spring Harbor Symposia Quant. Biol.* **16**, 483–503.

Thimann K.V. (1940) Sexual substances in the Algae. *Chronica Botan.* **6**, 31–2.

Wettstein F. (1924) Morphologie und Physiologie des Formwechsels der Moose auf genetischer Grundlage. *Zeit. f. Induk. Abstam.- u. Vererb.-lehre* **33**, 1–236.

Whitehouse H.L.K. (1950) Mapping chromosome centromeres by the analysis of unordered tetrads. *Nature, Lond.* **165**, 893.

Wiese L. (1969) Algae, In *Fertilization: Comparative morphology, biochemistry, and immunology*, eds. Metz C.B. & Monroy A., pp. 135–88, Academic Press, New York.

WIESE L. & HAYWARD P.C. (1972) On sexual agglutination and mating-type substances in isogamous dioecious Chlamydomonads. III. The sensitivity of sex cell contact to various enzymes. *Amer. J. Botany*, **59**, 530–6.

WIESE L. & SHOEMAKER D.W. (1970) On sexual agglutination and mating-type substances (gamones) in isogamous heterothallic chlamydomonads. II. The effect of Concanavalin A upon the mating-type reaction. *Biol. Bull.* **138**, 88–95.

AUTHOR INDEX

Bold type refers to pages where the full references appear.

Aaneby 233
Aaronson S. 264, **295**
Adams G.M.W. vii, 69, 72, 98, 109, **115**
Aksiyote-Ben Baset J. 249, **255**
Alexander N.J. 31, 87, **45**, **115**
Allen M.A. 199, 200, 203, 205, 206, **208**
Allen M.M. 8, 12, **23**
Allen R.L. 281, **294**
Apel K. 242, 243, 247, 250, 251, 252, 253, 253, **255**
Armstrong J.A. 36, **45**
Armstrong J.J. 114, 120, 121, **118, 140,** 143, 144
Arnold C.G. 29, 72, 96, 97, 98, 131, **47,** **115, 118, 144**
Asato Y. 8, 9, 11, 14, 19, 20, **23, 24**
Attardi G. 270, **296**
Augustinussen E. 289, **290**
Avadhani N.G. 262, 270, 273, 278, **290**

Bachmayer H. 124, **140**
Baerg W. 281, **295**
Bailey R.W. 4, 6
Bain J.M. 275, 276, 280, 282, 283, 284, **290**
Baker R.R. 176, **195**
Baltus E. 248, **256**
Banerjee B. 310, 311, **330**
Barnett W.E. 124, 270, 274, 277, 278, 281, 285, **141, 290, 293, 296**
Barras D.R. 258, **290**
Barratt R.W. 320, **327**
Bastia C. 251, **253**
Bastia D. 74, 76, 120, 123, 267, 268, 269, **115, 140, 291**
Bathelt H. 29, **47**
Battaner E. 124, **144**
Battula N. 270, 278, **290**
Baumgartel D.M. 179, 180, 181, **196**
Baur E. 71, **115**
Bazin M.J. 9, 10, **24**
Beadle G.W. 154 **170**
Beam C.A. 2, **5**
Beattie D.S. 262, **294**
Beatty B.R. 242, **255**

Bechet J. 34, **46**
Behn W. 72, **115**
Benecke W. 199, **208**
Bennoun P. 38, **45**
Ben-Shaul Y. 259, 260, 261, 262, 265, 266, 269, 280, 281, **290, 295, 296,** 297
Bensky B. 264, **295**
Berger S. 240, 242, 243, 247, 249, 250, 253, **255**
Berns K.I. 251, **255**
Bernstein E. 146, 147, 175, 179, **170,** 193
Bertini E. 261, **291**
Bezsmertnaya I.N. 290, **296**
Biebel P. vii, 198, 199, 200, 201, 202, 208, **208, 209**
Birch A.J. 310, **327**
Birge E.A. 132, **141**
Bishop D.G. 275, 276, 280, 281, 282, 283, 284, **290, 291**
Biswas D.K. 132, **141**
Blair G.E. 252, **254**
Blamire J. 138, 139, **141**
Bliding C. 224, **233**
Blobel G. 121, 122, 123, 124, 270, **141,** 142, 292, 294
Boardman N.K. 123, 141, 297
Bogdanov S. 72, 128, 131, **115, 141**
Bogorad L. 72, 119, 121, 123, 125, 128, 129, 132, 140, 247, 251, **117, 141,** 142, 143, 256
Bold H.C. 148, 151, **170, 193**
Bolouckhère M. 243, 247, **253, 284**
Bonnett H.T. 290, **291**
Bonotto S. 243, 244, 247, 249, 250, 253, **254, 255**
Børresen 228
Borst P. 74, 78, 123, 139, 260, **115, 141,** 142, 291
Boschetti A. 72, 128, 131, **115, 141**
Bourque D.P. 29, 30, 121, 133, **45,** 141
Boutin M.E. 287, **291**
Bovarnick J.G. 274, 275, 277, 280, 281, 282, 283, 284, 289, **291**
Boyce R.P. 43, **45, 46**

Boynton J.E.   vii, 29, 30, 31, 69, 72, **78**, 87, 91, 93–98, 104, 109, 112, 113, 114, 119–**22**, 124, 125, 127–39, 164, 244, **45, 46, 115, 116, 140, 141, 142, 143**
Brachet J.   236, 245, 246, 247, 248, 249, 250, **253, 254**
Brandes D.   261, **291**
Brandham P.E.   199, 200, 201, 202, 204, 205, 206, 207, **209**
Brändle E.   249, 250, **253**
Bråten T.   228, **234**
Braun W.   17, **26**
Braunitzer G.   182, 310, 326, **194, 328**
Brawerman G.   124, 247, 249, 263, 266, 268, 272, 273, **141, 144, 253, 291, 292, 293**
Bray D.F.   **172**
Breckenridge L.   132, **141**
Brenner S.   112, **116**
Bridges B.A.   13, **24**
Britten R.J.   76, 78, 230, **115, 116, 234**
Broda P.   21, **26**
Brokaw C.J.   151, **170**
Brooks A.E.   176, 184, **193**
Brown G.M.   154, **172**
Brown R.D.   267, 268, 269, **291**
Brown R.M.   10, 148, 151, 175, 182, **24, 170, 195**
Brown W.M.   290, **291**
Bruce V.G.   4, **5**
Bryhni E.   225, 228, 232, **234**
Bücher Th.   124, **144**
Buetow D.E.   261, 262, 270, 273, **290, 291, 295**
Buffaloe N.D.   50, 148, 161, 166, **61, 170**
Burg C.A.   2, **6**
Burkard G.B.   124, 270, **141, 291**
Burke J.P.   262, **294**
Burkholder B.   72, 120, 122, 125, 128, 130, 131, 132, 134, 137, **116, 141, 142**
Burrows E.M.   226, **235**
Burton H.   251, **234**
Burton W.G.   72, 120, 124, 125, 128, 134, 137, 138, 139, **115, 141**
Bush V.N.   15, **24**
Butow R.A.   258, 261, **291, 297**
Byerrum R.U.   154, **170**
Byrnes J.J.   250, **254**

Calvayrac R.   261, **291**
Cannon R.E.   15, **24**
Carefoot R.J.   176, 184, **193**
Carlson P.S.   290, **292**
Carr N.G.   vii, 7, 10, 11, 12, 13, 14, 16, 19,. 20, 22, **24, 25, 28**
Carrell E.F.   275, **292**
Carritt B.   268, 269, **291**

Casey J.   290, **291**
Catlin B.W.   17, **24**
Cavalier-Smith T.   71, **115**
Cave M.S.   185, **193**
Celma M.L.   124, **144**
Cerdá-Olmedo E.   86, **113**
Ceron G.   250, **253**
Chabot J.F.   120, 122, 134, **141**
Chaleff R.S.   290, **291**
Chang S.-W.   281, 282, 289, **291**
Chapman A.R.O.   226, 227, **234**
Chapman L.F.   10, **26**
Chargaff E.   268, **291**
Chen L.C.M.   3, 226, **5, 234**
Chiang K.-S.   30, 31, 74–84, 87, 98, 120, 121, 123, 179, 180, 181, 233, 243, 251, **45, 46, 115, 140, 144, 193, 194, 196, 197, 234, 253, 254**
Chiu S.M.   181, **193**
Chodat F.   310, **327**
Chua N.-H.   122, 123, 270, **292, 141**
Chuang L.   258, **292**
Chu-Der O.M.   98, 160, 169, **115, 170**
Chun E.H.L.   71, 74, **116**
Claes H.   181, **193**
Clark A.J.   43, **45**
Clarke C.H.   14, **24**
Clark-Walker G.D.   124, **143**
Clauss H.   248, **253, 254**
Clayton D.A.   290, **297**
Codd G.A.   276, 293, **292**
Coen D.   261, **292**
Cohen D.   268, 270, 287, 289, **292**
Cohen-Bazire G.   10, **25**
Cohn F.   7, 176, **24, 193**
Coleman A.W.   146, 176, 183, 184, **170, 193**
Colwin A.L.   181, **194**
Colwin L.A.   181, **194**
Conde M.F.   72, 90, 92, 98, 109, 124–32, 138, 139, **115, 116, 141**
Cook A.R.   310, **327**
Cook J.R.   264, **292**
Corillion R.   216, **217**
Correns C.   71, **46, 116**
Cosbey E.   50, 51, **61**
Costerton J.W.   **172**
Cowan C.A.   263, **292**
Coyer D.R.   262, **293**
Craig I.W.   7, 16, **24, 28**
Craigie J.S.   4, **5**
Crane F.L.   258, 286, **297**
Crawley J.C.W.   242, **254**
Criddle A.S.   277, **292**
Croft J.H.   261, **298**
Crouse E.J.   259, 260, 262, 263, 266, **292, 298**
Cunningham L.S.   17, **24**
Cuzin F.   112, **116**

Czurda V. 176, **170**
Danforth W.F. 258, **292**
Danna K.J. 21, **24**
Dao S. see Puiseux-Dao
Darden W.H. 175, 176, 178, 188, 189, 190, 315, 316, **193, 327**
Darnell J.E. 124, **142**
Davidson E.H. 78, 230, **116, 234**
Davidson J.N. 123, 125, 128, 132, 140, **141, 142**
Davidson N. 76, 77, 251, **118, 256**
Davies D.R. vii, 42, 43, 50, 63, 64, 65, 66, 67, **45, 62, 67, 68**
Davies J. 123, 124, 132, **141, 144**
Davis R.H. 157, **170**
Deason T.R. 178, **193**
De Lamater E.D. 50, **62**
Delaney S.F. vii, 7, 10, 11, 12, 13, 14, 16, 20, 22, **24, 25**
Delcarpio J.B. 148, **170**
Demerec M. 18, **24**
Dempsey L.T. 148, **170**
Denes G. 33, **45**
Denis H. 249, **253**
Der O.C. 160, **170**
de Siebenthal U.R. 310, **327**
Desjardins P.R. 16, **26**
Deulofeu V. 310, **330**
Deutsch J. 261, **292**
de Vitry F. 242, 249, 250, **253, 256**
De Vries H. 262, **294**
Diehn B. 258, **292**
Diener T.O. 16, **26**
Dillard W.L. 247, 249, **254, 255**
Dilley R.A. 258, 286, **297**
Dion M. 270, **296**
Dolack M.P. 12, **25**
Dolphin W.D. 289, **292**
Donovan F.W. 310, **327**
Doolittle W.F. 13, 24
Dorky D. 289, **292**
Downey K.M. 250, **254**
Drews G. 7, **24**
Dujon B. 109, 112, **116**
Dwyer M.R. 275, 280, 281, 282, 283, 284, **298**

Ebersold W.T. 30, 31, 33, 35, 36, 41, 42, 43, 44, 50, 57, 58, 65, 134, 165, 169, 175, 301, 311, 324, 325, **45, 46, 47, 61, 117, 141, 170, 193, 327**
Ebringer L. 258, 264, **292**
Echlin P. 7, **24**
Eclancher B. 124, **141**
Edelman M. 258, 259, 260, 261, 263, 264, **292, 296, 297**
Edelstein T. 226, **235**

Edgar R.S. 51, **61**
Edmunds L.N., Jr. 258, **291, 292, 294**
Egan J.M., Jr. 274, 275, 277, 278, 280–285, 289, **292, 293**
Eisenstadt J.M. 124, 260, 262, 263, 266, 268, 269, 270, 272, 273, 278, 279, **141, 144, 291, 292, 293, 294**
Ellis R.J. 124, 252, **141, 254**
Engbaek F. 137, **143**
Ephrati-Elizur E. 17, **24**
Epler J.L. 124, 270, 274, 277, 278, **141, 296**
Epstein H.T. 74, 258, 260, 261, 263, 264, 286, **116, 292, 293, 295, 297**
Eriksson T. 290, **291**
Ernst A. 216, 306, **217**
Erwin J.A. 258, **293**
Esser K. 311, 313, 317, **328**
Ettl H. 183, **194**
Evans A.H. 12, 17, **24**
Evans N.R. 124, **143**
Evans W.R. 258, 262, 268, 280, 286, **293, 298**
Eversole R.A. 30, 33, 35, **45**
Eves E. 86, **118**

Fairfield S.A. 270, 274, 277, 278, **290, 293, 296**
Fang T.C. 227, **234**
Farber F.E. 247, **254**
Fargo A. 33, **45**
Faulkner B.M. 19, 23, **25**
Fernandez-Munoz R. 124, **144**
Fifer W. 72, 88, 90, 98, **116**
Fisher W.D. vii, 10, 12, **25**
Fjeld A. 219, 225, 228, 229, 230, 232, **234**
Flavell R.A. 259, 260, 271, **293, 298**
Flechtner V.R. 138, 139, **141**
Foerster H. 176, 177, 179, 181, 182, 310, 326, **193, 194, 329**
Folsome C.E. 9, 11, 14, 19, 20, 50, **23, 24, 61**
Føyn B. 220, 222, 223, 224, 225, 227, 228, **234**
Francki R.I.B. 123, **141**
Franke W.W. 242, 256
Frazier E.A. 253
Freedman Z. 274, 275, 277, 280, 282, 283, 284, 289, **291**
Frenkel A.W. 159, **171**
Freyssinet G. 268, 270, 289, **293**
Friedmann I. 181, **194**
Fuller R.C. 42, 272, 278, 280, 285, 287, **46, 296**
Funatsu G. 132, 137, **142**

Galper J.B. 124, **142**
Garnjobst L. 319, 320, **327**

Gauch H.G. 199, 200, 203, 206, **209**
Gawlik S.R. 264, **298**
Gay M.R. 66, **68**
Gerloff J. 146, 147, 176, 179, **170**, **194**
Gholson R.K. 154, **172**
Gibbs M. 274, 275, 276, 277, **295**
Gibbs S.P. 1, 148, 151, **5**, **170**
Gibor A. 245, 249, 250, 258, **254**, **255**, **293**
Gildemeister V. 166, **171**
Gillham N.W. vii, 29, 30, 31, 33, 38, 50, 51, 59, 70, 72, 78, 79, 87–98, 104, 107, 112, 113, 114, 119–39, 144, 177, 244, 261, **45, 46, 61, 115, 116, 118, 141, 142, 143, 144, 194, 293**
Ginoza H.S. 8, **24**
Givan A.L. 39, **45**
Glazer V.M. 11, **28**
Gnaman A. 272, 276, 281, **293**
Godward M.B.E. 1, 161, 199, 201, 202, 204, 207, **5, 170, 209**
Goffeau A. 249, 250, **253, 254**
Goins D.J. 270, 277, **293**
Goldring E.S. 262, **293**
Goldstein M. 175, 176, 177, 185, 186, **194, 197**
Goodenough U.W. 31, 32, 36, 38, 40, 50, 83, 114, 120, 121, 133, 134, 138, **45, 46, 61, 116, 118, 142, 144**
Gordon P. 290, **291**
Gorini L. 132, **141**
Gorman D.S. 30, 39, **45, 46**
Gowans C.S. vii, 3, 145–69, 176, 177, 301, 310, 319, 320, 325, **5, 170, 171, 172, 194, 328**
Gracy R.M. 282, 284, **295**
Graf L. 263, 266, **298**
Graham D. 267, 275, 277, 280, 281, 282, 283, 284, 286, **297, 298**
Granick S. 30, 146, 179, 180, 182, 258, **43, 172, 196, 293**
Grant D. 74, 87, **117**
Grant M.C. 214, 216, 217, **217**
Green B.R. vii, 236, 238, 245, 251, **254**
Greenblatt C.L. 286, **299**
Greenwood H.L. 182, **194**
Gregory P. 262, 280, **298**
Grell K.G. 175, **194**
Grenson M. 34, **46**
Gries H. 319, **328**
Grieve A.M. 275, 277, 280, 281, 282, 283, 284, **298**
Griffin D.G., 216, **217, 218**
Griffith T. 154, **170**
Grimes G.W. 78, **116**
Grivell L.A. 123, 124, 139, 271, **141, 142, 293**
Groot G.S.P. 271, **293**

Gross J.A. 289, **293**
Grossman L.I. 262, **293**
Guerlesquin M. 216, **217**
Guerola N. 86, **115**
Guillard R.R.L. 151, **170**
Gupta R.S. 10, 11, **24**
Gurney-Smith M. 66, **68**
Guthrie C. 232, **234**
Gyurasits E.B. 21, **24**

Hadwiger L.A. 154, **172**
Haemmerling J. See Hämmerling
Hagen-Seyfferth M. 311, **328**
Halbreich A. 124, **142**
Haldane J.B.S. 311, 325, **331**
Hamilton M.G. 121, **144**
Hämmerling J. 178, 236, 237, 241, 242, 243, 244, 246, 247, 315, 318, **194, 254, 327**
Hanawalt P.C. 43, 44, 86, 259, 260, 263, 264, **47, 48, 115, 296**
Hanic L.A. 224, 234
Hanson M.R. 123, 125, 128, 132, 140, **141, 142**
Harris B.G. 282, 284, **295**
Harris D.O. 176, 186, **194**
Harris E.H. vii, 30, 69, 72, 91, 98, 109, 119, 120, 124, 125, 127–39, 244, 268, 269, 270, 272, 279, **46, 115, 116, 141, 142, 293**
Harris H. 58, **61**
Harte C. 311, 322, **328**
Hartley M.R. 252, **254**
Hartmann G. 249, **254**
Hartmann M. 178, 301, 310, 313, 318, 303, **194, 328**
Hartshorne J.N. 152, 168, **170**
Haselkorn R. 263, 265, 266, 267, 268, 269, **291, 296**
Hastings P.J. 50, 51, 181, **61, 193**
Hattori S. 311, **328**
Hayaishi O. 154, **172**
Hayes W. 13, **24**
Hayward P.C. 147, 148, 175, 182, 310, 326, **173, 197, 331**
Heath R. 257, 286, **297**
Hecker L.I. 274, 278, 281, 285, **293**
Heilporn-Pohl V. 250, **254**
Heizmann P. 267, 268, 269, 270, 277, 287, **293, 296, 298**
Hellmann V. 1, **5**
Henderson L.M. 154, **170, 171, 173**
Herdman M. vii, 7, 8, 10, 11, 13, 16, 17, 18, 19, 22, **25, 26**
Hershenov B. 264, **295**
Hess D. 290, **293**
Heywood J. 110, **117**

Hill H.Z.   258, **293**
Hills G.H.   66, **67, 68**
Hilton R.L., Jr.   199–201, 208, **209**
Himes M.   2, **5**
Hincley N.   83, **116**
Hipkiss A.R.   35, **46**
Hirsch M.   270, **293**
Hirvonen A.P.   266, 267, **294**
Holowinsky A.W.   286, 289, **294**
Honeycutt R.C.   138, 180, **143, 194**
Honikel K.O.   249, **254**
Hoober J.K.   120, 121, 123, 124, 138, 139, 140, 270, **142, 144, 294**
Hopkins J.M.   50, 51, 52, 57, 61, **62**
Hoshaw R.W.   183, 199, 200, **194, 209**
Hotchkiss A.T.   211, 215, 216, **217, 218**
Hotchkiss R.D.   17, **26**
Hovenkamp-Obbema R.   274, 280, 281, 283, 284, **294**
Howard-Flanders P.   43, **45, 46**
Howell S.H.   4, 41, 179, 180, 181, **5, 46, 196**
Hoxmark R.C.   220, 221, 222, 223, 233, **234**
Huberman J.A.   21, **25**
Hudock G.A.   vii, 29, 37, 41, 42, 44, **46**
Hudock M.O.   30, 37, 38, 41, 42, 50, 51, 120, **46, 48, 61, 143**
Huizige H.   286, **296**
Hunter J.A.   77, 259, 260, 263, 264, 266, 267, **117, 295**
Huskey R.J.   188, **196**
Hutner S.H.   147, 179, 182, 264, **171, 194, 296**
Hüttig W.   317, **328**
Hwang S.-W.   44, **46**
Hyams J.S.   63, 64, 67, **68**

Ibrahim N.G.   262, **294**
Ingle J.   123, **143**
Ingram L.O.   10, 12, 13, **25**
Inouye M.   123, 125, 128, 131, 137, **143**
Ishida M.R.   71, 74, 75, **117**
Izawa M.   245, 250, **254**

Jacob F.   112, **116**
Jacob S.T.   270, **294**
Jaenicke L.   177, 190, **196**
Jahn T.L.   146, 147, 175, **170, 193**
Janowski M.   244, 247, 249, **253, 254**
Jarrett R.M.   258, **292, 294**
Jensen R.A.   13, **25**
Jerchel D.   311, **328**
Jevner V.D.   9, 11, 13, **27**
Jinks J.L.   90, **116**
Johnson B.C.   154, **170**

Johnson C.   148, 151, 286, 170, 293
Johnson E.M.   250, **253**
Jollos V.   313, **328**
Jones B.R.   155, **171**
Jones R.F.   30, 79, 86, 87, 179, 180, 243, **46, 116, 193, 194, 254**
Jupp A.S.   259, 264, 269, **295**
Jurmark B.S.   250, **254**

Kahn J.S.   272, 276, 281, **293**
Kain J.A.   226, **235**
Kale S.   11, **125**
Kallio P.   203, 204, 207, **209**
Kalmus H.   176, **194**
Kaltschmidt E.   123, **143**
Kamen M.D.   275, **296**
Kane J.F.   13, **25**
Kaney A.R.   12, 13, 17, **25, 26**
Karlander E.P.   181, **195**
Kates J.R.   30, 179, 180, 181, 243, **46, 193, 194, 197, 254**
Keck K.   248, **254**
Keller E.B.   270, **291**
Keller S.J.   180, **194**
Kelley D.E.   270, **296**
Kessler E.   1, **5**
Khan M.   215, **289**
Khyen N.T.   9, 11, 17, **27**
Kiermayer O.   204, **209**
Kirchmann R.   245, 250, **253**
Kirk J.T.O.   71, 74, 164, 279, 280, 281, **116, 171, 294**
Kislev N.   262, 268, 273, 278, **294**
Kivic P.A.   264, **294**
Klebs G.   307, 309
Klein R.M.   287, **291**
Klein S.   287, **294**
Klein W.H.   287, **296**
Kleinschmidt A.K.   251, **254**
Kloppstech K.   242, 243, 247, 251, **255**
Kniep H.   4, 301
Knüsel F.   249, **254**
Kochert G.   177, 178, 189, **194**
Kohne D.E.   76, **115**
Konitz W.   261, **294**
Koop H.-U.   238, 241, 253, **255**
Korn R.W.   200, 201, 202, 206, 207, 208, **209**
Kornmann P.   222, **235**
Krauspe R.   274, 278, 284, 285, **296**
Krauss R.W.   50, 151, 152, 155, 156, 161, 162, **62, 173**
Kraweic W.   260, 262, **294**
Kroon A.M.   74, 124, 259, 260, 262, **115 143, 294**
Krupnick D.   262, **293**
Kuenen R.   313, 317, **328**

Kuhn R.   311, **328**
Kumar H.D.   8, 9, 10, 11, 14, **24, 25,**
    27
Kunisawa R.   10, **26**
Küntzel H.   123, 139, **143**
Kurland C.G.   132, **141**
Kvitko K.V.   33, 35, **47**

Lamb A.J.   124, **143**
Lane D.   31, 74, 80, 83, 84, 85, 87, **47,**
    117
Lang A.   310, **328**
Lang D.   259, **294**
Langstroth P.   40, **47**
Larrinua I.   259, 264, 269, **295**
La Torre J.   270, **296**
Latzko E.   274, 275, 276, 277, **295**
Laurendi C.   182, **195**
Lavrenchouk V.Ya.   33, 35, **47**
Lazaroff N.   9, **26**
Lazzarini R.A.   289, **295**
Leach C.K.   8, 16, **24, 26**
Leak L.V.   8, **26**
Lee A.S.   290, **295**
Lee R.W.   72, 75, 78, 79, 86, 87, 91, 93–
    98, 104, 109, 112, 114, 130, **116, 143**
Lee S.G.   124, **143**
Leedale G.F.   261, **295**
Leete E.   154, **171**
Leff J.   74, 264, **116, 295**
Lefort-Tran M.   261, **291**
Lerche W.   176, 178, 315, 316, **194, 328**
Levine E.E.   50, 51, **61**
Levine R.P.   30, 31, 32, 35, 36, 38, 39, 40,
    50, 51, 58, 72, 114, 120, 121, 128, 132,
    133, 134, 169, **45, 46, 47, 48, 61, 116,**
    117, 118, 140, 142, 143, 144
Lewin J.C.   17, 146, 159, **171**
Lewin R.A.   1, 3, 50, 146–52, 154, 155,
    159, 160, 162, 165, 167, 168, 175–
    179, 181, 182, 193, 219, 257, 301,
    315, 327, **5, 28, 61, 170, 171, 194, 195,**
    235, 295, 328
Li S.L.   3, 154, 165, **5, 171**
Li Y.Y.   227, **234**
Lindberg A.A.   183, **195**
Lindegren C.C.   168, **172**
Lingens F.   166, **171**
Linnane A.W.   124, **143, 297**
Lipkin   214, 216
Lippert B.E.   199, 200, 201, **209**
Lizardi P.M.   139, **143**
Loeblich A.R.   2, **6**
Loening U.E.   123, **143**
Loeppky C.B.   158, 163, 165, **171**
Longley A.E.   168, **171**
Loni M.C.   243, 244, **255**

Loppes R.   30, 33, 34, 35, 36, 50, 51, **46,**
    47, 61
Lorch S.K.   181, **195**
Lorz H.   290, **293**
Løvlie A.   vii, 219, 223, 224, 229, **235**
Lucas C.   10, **27**
Luck D.J.L.   71, 123, 137, **117, 143, 144**
Ludwig W.   325, **328**
Lyall V.   65, 67, **67**
Lyman H.   viii, 1, 257–69, 275, 276,
    277, 280, 281, 287, 288, 289, **295,**
    297, 298
Lynch M.J.   262, 273, **290**

MacDonald M.B.   211, **217**
MacHattie L.A.   251, **255**
Madsen A.   289, **290**
Magni G.E.   18, **26**
Mahler A.H.   160, **143**
Mahler H.R.   78, 87, 90, 123, 139, 262,
    117, 295
Makino F.   8, **26**
Malkoff D.B.   261, **291**
Mandel M.   74, 264, **116, 295**
Manning J.E.   77, 259, 260–9, **117, 295, 296**
Manton I.   3, **6**
Marćenko E.   258, **295**
Marcker K.A.   124, **144**
Margulies A.D.   43, **45**
Margulies M.M.   138, 180, **143, 194**
Margulis L.   251, 290, **255, 295**
Marion L.   154, **171**
Markus Y.   280, **290**
Marmur J.   262, **293**
Martinek   43
Master R.S.P.   248, **255**
Masters M.   21, **26**
Matange R.F.   36, **47**
Mauzy W.   17, **26**
McBride A.C.   147, 150, 155, 156, 158,
    159, 163, 164, 165, 166, 169, **171**
McBride J.C.   150, 155, 158, 159, **171**
McCalla D.R.   281, **295**
McCandless E.L.   3, **6**
McClintock B.   164, **171**
McCracken M.D.   177, 189, 191, 211, 215,
    195, 217
McGuire R.F.   211, **217**
McLachlan J.   4, 226, **5, 235**
McLean R.J.   175, 182, **195**
McMahon D.   40, 43, **47, 68**
McVittie A.C.   50, 51, 52, 55, 57, 59, 61,
    62, **62, 171**
Mego J.L.   264, **295**
Mehler A.H.   **172**
Meinhart J.O.   148, **171**
Mellon A.D.   286, **299**

Merret M.J. 276, 283, **292**
Meselson M. 78, 109, 110, **117**
Mets L.J. 72, 119, 121, 123, 125, 128, 129, 132, 175, **117, 142, 143**
Metz C.B. 182, **197**
Meyer R. 124, **144**
Mielenz J.R. 265, **295**
Migula W. 216, **217**
Miller G.L. 238, 241, 243, **256**
Miller M.J. 63, **68**
Miller O.L. 242, **255**
Minato S. 14, **26**
Mishra N.C. 176, **195**
Mitchell H.K. 154, **170**
Mitchison G.J. 11, **28**
Mitrakos K. 287, **295**
Mitronova T.N. 10, 17, **26**
Mizushima S. 132, **143**
Mo Y. 282, 284, **295**
Moewus F. 146, 147, 176, 177, 182, 301, 310–27, **172, 195, 327, 328, 329, 330, 331, 332**
Moewus L. 4, 311, **329, 331**
Moll B. 120, **140**
Moorman A. 274, 280, 281, **294**
Moriber L.G. 264, **295**
Morris I. 7, **24**
Morris M.E. 16, **26**
Morris O.P. 227, **235**
Morrow J.F. 290, **297**
Mousset M. 34, **46**
Muir B.L. 251, **255**
Müller D.G. 222, **235**
Munns R. 267, 268, 271, 277, 284, **295, 297, 298**
Munro R.E. 124, **144**
Murashko G.N. 33, 35, **47**
Murneek A.E. 310, **331**

Nakamura K. 150, 153, 154, 157, 158, 160, 161, 167, 168, **172**
Naliboff J.A. 4, 41, **5, 46**
Nashimoto H. 232, **234**
Nass M.M.K. 259, 260, 261, 262, 265, 266, 269, **295, 296**
Nathans D. 21, **24**
Neish A.C. 4, **5**
Netter P. 261, **292**
Neumann J. 39, **47**
Neumann K. 222, **235**
Newmeyer D. 319, 320, **327**
Nicholson-Guthrie C.S. 37, 42, **47**
Nicolas P. 274, 277, **297**
Nierhaus K. 132, **142**
Nigon V. 270, 274, 277, **293, 296, 297, 298**
Nishimura M. 286, **296**

Nishizuka Y. 154, **172**
Nix C.E. 274, 278, 281, **293**
Nomura M. 123, 124, 132, 137, 232, 285, **141, 143, 234**
Nordby Ø. 220, 221, 223, 233, **234, 235**
Norton C. 262, 268, 278, **294**
Nüesch J. 249, **254**
Nybom N. 161, 166, **172**
Nyec J.F. 154, **170**

O'Donnell R. 13, 28
Ohta N. 123, 125, 128, 131, 137, **143**
Ojala D. 270, **296**
Olmsted M.A. 35, 50, **45, 61**
Ondratschek K. 178, **196**
Oparin A.I. 270, **296**
Ophir I. 262, 280, 281, **290**
Oppenheim A.B. 15, 16, **26**
Orkwiszewski K.G. 17, **26**
Ortega M.V. 154, **172**
Osafune I. 261, **296**
Ottolenghi E. 17, **26**
Ozaki M. 132, **143**

Padan E. 11, 15, **26**
Padmanabhan 251
Palade G.E. 120, 270, **47, 142, 292**
Pall M. 191, **195**
Papazian H.P. 322, **331**
Parker B.C. 176, **195**
Parker G.A. 176, **195**
Parsons M.J. 4, 6
Parthier B. 274, 278, 280, 284, 285, **296**
Pascher A. 4, 176, 181, 183, 300, 301, 302, 306, **195**
Paszewski A. 40, 120, 133, **46, 142, 143**
Pätau K. 325, **331**
Penman S. 270, **293**
Pennington C.J. 270, 277, **290**
Perini E. 275, **296**
Perkins D.D. 154, 164, 167, 168, 319, 320, **172, 327**
Perl M. 279, 280, 281, **296, 298**
Perlman P.S. 78, 262, **116, 295**
Perrodon G. 261, **298**
Perrot Y. 222, **235**
Perry R.P. 270, **296**
Pestka S. 124, **143**
Petrochillo E. 261, **292**
Pettijohn D. 43, **47**
Phares W. 10, **26**
Philip U. 311, 325, **331**
Philippovich I.I. 270, **296**
Phillips J.M. 66, **68**
Philpott D.E. 148, 151, **170**
Pickmere S.E. 3, 6

Pierson D.   13, **25**
Pikálek P.   8, **26**
Plaskitt A.   63, 64, **67**
Plaut W.   71, 114, **117**
Pocock M.A.   185, 188, **193, 195**
Pogo A.O.   268, 280, **291, 296**
Pogo B.G.T.   280, **296**
Pollock E.G.   36, 226, **235**
Poole R.J.   1, **5**
Portier C.   270, **296**
Powers J.H.   188, **195**
Prakash G.   8, 14, **25, 27**
Preston J.F.   268, 269, 272, 279, **293**
Price C.A.   266, 267, **294**
Price L.   287, **296**
Pringsheim E.G.   147, 178, 199, 257, 264, **196, 209, 296**
Pringsheim O.   264, **296**
Proctor V.W.   viii, 210, 211, 212, 214, 215, 216, 217, **217, 218**
Provasoli L.   147, 175, 179, 181, 182, 264, **171, 194, 296**
Puiseux-Dao S.   236, 238, 241, 242, 243, 245, 246, 248, 249, 250, **253, 255**

Rabinowitz M.   124, 290, **142, 291**
Raison J.K.   258, **296**
Ramaley A.W.   2, **6**
Ramanis Z.   65, 72, 78, 88, 89, 90, 91, 93, 94, 98–111, 125, 129, 132, **68, 117, 118, 144**
Ramasarma G.B.   154, **170**
Rao D.R.   154, **170**
Randall Sir John   viii, 49, 50, 51, 57, 58, 179, **62**
Rao D.R.   154, **170**
Raper J.R.   310, 326, **331**
Rawson J.R.   263, 265, 266, **298**
Ray D.S.   259, 260, 263, 264, **296**
Rayburn W.R.   176, **196**
Rédei G.P.   3, 154, 165, 166, **5, 172**
Redfield H.   323, **331**
Reger B.J.   270, 272, 274, 277, 278, 280, 285, 287, **296**
Regnery D.C.   50, 176, 177, **62, 196**
Reich E.   71, **117**
Reichle R.E.   viii, 300
Reid R.   201, 202, **209**
Reijnders L.   123, **142**
Ren K.Z.   227, **234**
Renner O.   311, **332**
Reuter W.   248, **255**
Reynolds R.J.   270, 274, 277, 278, 281, 285, **293**
Reznik H.   311, **331**
Rhoades M.M.   71, **117**
Rich A.   71, 74, **116**
Richard F.   274, 277, **297**

Richards O.C.   77, 259–69, **117, 295, 296, 297**
Richter G.   236, **255**
Rifkin M.R.   123, **144**
Riggs A.D.   21, **25**
Rimon A.   16, **26**
Ris H.   8, 71, 114, **26, 117**
Robberson D.L.   290, **297**
Roberts K.   viii, 63, 66, **68**
Rochaix J.D.   120, **144**
Rosen H.   viii, 29, 36, 42, 43, 44, **47**
Rosen W.G.   264, **298**
Rubman J.   286, **297**
Rueness J.   222, 226, 227, **235**
Rueness M.   222, 227, **235**
Rupert C.S.   14, **27**
Russell G.K.   227, 257, 283, **235, 297**
Rutman R.J.   270, 278, **290**
Ryan F.J.   310, 327, **331**
Ryan R.S.   74, 77, 87, 259, 260, 261, 263, 264, 266, 267, 268, 269, **117, 295, 296, 297**

Safferman R.S.   15, **26**
Sager R.   8, 30, 31, 35, 36, 65, 69–78, 81, 83–91, 119, 121, 122, 123, 125, 127–132, 136–40, 146, 156, 163, 168, 169, 176, 177, 179, 180, 182, 251, 301, **27, 47, 68, 117, 118, 141, 143, 144, 172, 196, 256**
Saito N.   14, **27**
Saito Y.   226, **235**
Salvador G.   274, 277, **297**
Samtleben S.   274, 278, 284, 285, **296**
Sanders J.P.M.   259, 260, **298**
Sarma Y.S.R.K.   215, **218**
Sato V.L.   39, **47**
Sawa T.   215, **218**
Schader E.   311, **331**
Schaechter M.   50, **62**
Schatz A.   264, **296**
Scheer U.   242, **256**
Schiff J.A.   74, 257, 258, 259, 260, 261, 263, 264, 267, 268, 270, 274, 275, 277, 278, 280–7, 289, **116, 291–7, 299**
Schiltz E.   137, **142**
Schimmer O.   29, 96, 97, 98, 131, **47, 118, 144**
Schindler D.G.   270, **294**
Schlanger G.   84, 122, 125, 127, 128, 129, 130, 131, 132, 137, 138, 140, **118, 144**
Schmeisser E.T.   179, 180, 181, **196**
Schmidt G.W.   viii, 257, 258, 267, 275, 276, 277, 286, 287, 288, 289, **297**
Schori L.   259, 260, 261, **296, 297**
Schotz F.   29, **47**

Schreiber E. 176, 226, **196, 235**
Schult E.E. 168, **172**
Schultz J. 323, **331**
Schulze B. 176, **196**
Schulze K.L. 238, 242, 243, **255**
Schwab A. 124, **144**
Schwartz J.H. 124, **144**
Schwartzback S.D. 281, 282, 287, 289, **291, 297**
Schweiger H.G. 242, 243, 247, 248, 249, 250, 251, 252, **253, 255**
Schweiger M. 242, 243, 247, **255**
Schwelitz F.D. 257, 286, **297**
Scott N.S. 263, 265, 266, 267, 268, 269, 271, 277, 282, 284, 286, **295, 296, 298**
Scudo F.M. 176, **196**
Sebald W. 124, **144**
Selsky M.I. 262, 268, 278, **294**
Sessoms A.H. 188, **196**
Setlow R.B. 14, **27**
Shacklock P.F. 4, **5**
Shah V.C. 269, 280, 281, **297**
Shane M.S. 15, **24**
Sharpless T.K. 258, **297**
Shephard D.S. 244, 245, 246, 250, **255, 256**
Shestakov S.V. 9, 10, 11, 13, 14, 17, **26, 27, 28**
Shilo M. 11, 15, **26**
Shoemaker D.W. 182, 326, **197, 331**
Shugart L.R. 124, **141**
Siegesmund K.A. 264, **298**
Siekevitz P. 122, 123, 270, **141, 292**
Siersma P.W. 83, 121, 179, 180, 251, **115, 118, 144, 196, 253**
Silman R. 281, **290**
de Silva M.W.R.N. 226, **253**
Singer R.L. 13, **24**
Singh H.N. 8, 9, 11, 12, 14, **25, 28**
Singh P.K. 10, 11, 13, **27**
Singh R.N. 8, 9, 10, 11, 12, 13, **26, 27**
Singh S.P. 10, **27**
Sinha B.D. 11, **27**
Sinha R. 8, 9, 10, 11, **27**
Sinsheimer R.L. 290, **295**
Six E. 248, **256**
Skwarra H. 311, **331**
Slonimski P.P. 110, 112, 261, **116, 292, 298**
Smillie R.M. 39, 258, 262, 267, 268, 269, 271, 272, 275, 276, 277, 278, 280–287, **46, 290, 291, 295, 296, 297, 298**
Smith A.E. 124, **144**
Smith A.J. 10, 12, **23, 27**
Smith G.M. 50, 147, 176, 177, 178, 179, 182, 186, 188, 310, 324, 327, **62, 196, 331**

Smith R.J. 11, **28**
Smith V.G.F. 176, **195**
Smyth R.D. 41, **47**
So A.G. 250, **254**
Someroski J.F. 154, **170**
Sonneborn T.M. 310, 311, 313, 323, **331**
Sparkuhl J. 177, **197**
Spenser I.D. 154, **171**
Spring H. 242, **256**
Srivastava B.S. 10, 12, 14, **27**
Staehelin T. 124, **144**
Stahl F.W. 78, **117**
Stanier R.Y. 7, **28**
Starling D. viii, 49, 57, 58, 179, **62**
Starr R.C. 147, 176, 177, 178, 180, 186–192, 199, 200, 201, 202, 204, 206, 207, 327, **172, 194, 195, 196, 197, 209, 256, 331**
Stedeford B. 43, **46**
Steens-Leivens A. 248
Stegeman W.J. 124, 138, 139, **144**
Stegwee D. 274, 280, 281, 283, 284, **294**
Stein J.R. 176, 184, 185, **196, 197**
Stevens S.E. 12, 13, 14, **28**
Stewart P.R. 262, 278, 280, **298, 299**
Stewart W.D.P. 9, **28**
Stifter I. 181, 182, **197**
Stolbova A.V. 33, 35, **47**
Stone B.A. 258, **290**
Straub A. 311, **328**
Strehlow K. 176, 183, **197**
Strijkert P.J. 33, 34, 35, **47**
Stroz R.J. 289, **293**
Stutz E. 77, 259, 260, 262, 263, 265, 266, **118, 292, 298**
Sueoka N. 30, 74, 75, 78, 79, 80, 82, 83, 179, 180, 181, 233, **47, 115, 193, 197, 234**
Sundene O. 226, **235**
Surzycki S.J. 30, 50, 51, 72, 114, 119, 120, 121, 139, 179, 180, **47, 118, 140, 144, 197**
Sussenbach J.S. 33, 34, **47**
Svendsen P. 226, **235**
Swift H. 74, 76, 87, 120, 123, 251, **115, 117, 140, 253**
Swinton D.C. 43, 44, 86, **48, 118**
Syrett P.J. 155, **172**
Szostak J.W. 177, **197**

Takahashi I. 17, **28**
Talen J.L. 259, 260, **298**
Talpasayi E.R.S. 11, **25**
Tassigny M. 199, 208, **209**
Taylor A.L. 18, 21, **28**
Terry O.W. 258, **292**

Theiss-Seuberling H.-B.   281, 282, 283, 298
Theriot L.   43, **46**
Thimann K.V.   310, **331**
Thomas C.A.   251, **255**
Thometz D.S.   150, 159, 161, 162, 164, **172**
Threlkeld S.F.H.   176, **195**
Thurston E.L.   10, **25**
Tilney-Bassett R.A.E.   71, 164, **116, 171**
Tingle C.L.   72, 91, 98, 109, 124, 125, 127, 128, 129, 130, 131, 132, 138, 139, **115, 116, 141**
Tiwari D.N.   9, 11, **27**
Tobin N.F.   275, 280, 281, 282, 283, 284, **298**
Togasaki R.K.   30, 38, 40, 120, 132, **46, 48, 142, 143, 144**
Trabuchet G.   270, **293, 298**
Trainor F.R.   2, 146, 147, **6, 172**
Trendelenburg M.F.   242, **256**
Triplett E.L.   248, **256**
Trotter C.D.   21, **28**
Tschermak-Woess E.   178, 179, **197**
Tsubo Y.   119, 163, 181, **144, 172, 197**
Tsuzaki J.   8, **26**
Tuttle R.C.   2, **6**

Uhlik D.J.   154, 166, **172**
Uzzo A.   258, 260, 263, 264, 265, 275, **298**

Valet G.   236, 241, **253, 256**
Van Baalen C.   8, 10, 11, 12, 13, **24, 25, 28**
Van de Berg W.J.   176–7, 188, 189, 190, **197**
Vanden Driessche T.   249, **253**
Vandrey J.P.   259, 260, 262, 263, 265, 266, **292, 298**
van Niel C.B.   7, **28**
Van Ommen G.J.B.   271, **293**
Van Winkle-Swift K.P.   viii, 69, 72, 120, **116, 143**
Vaughan M.H.   71, 74, **116**
Vazquez D.   124, **144**
Verdier G.   270, **293, 298**
Vesk M.   264, **294**
Vinograd J.   290, **291**
Virgin H.I.   287, **298**
Vishniac W.   9, **26**
Vollprecht P.   166, **171**
von Stosch H.A.   **6**
von Wettstein—*see* Wettstein

Waddington C.H.   231, **235**
Wagenaar E.B.   **172**

Wake R.G.   21, **24**
Walenga R.   286, **293**
Walg H.L.   124, **142**
Waller G.R.   154, **172**
Walne P.L.   148, 151, **170**
Walter J.A.   4, **5**
Wang W.L.   72, 91, 124, 125, 127, 128, 129, 130, 131, 132, 138, 139, 150, 156, 165, 166, 167, 168, **116, 141, 172**
Waris H.   203, 207, **209**
Warr J.R.   50, 51, 57, **62**
Watkins J.F.   58, **61**
Weil J.H.   124, **161**
Weill L.   110, 112, **116**
Weisblum B.   124, **144**
Weissert E.M.   290, **293**
Wells R.   74, 76, 77, 78, 251, **118, 256**
Wendt G.   311, **328**
Werbin H.   14, **26, 27, 28**
Werz G.   236, 246, 248, **256**
Wetherell D.F.   50, 146, 151, 152, 155, 156, 159, 161, 162, **62, 172, 173**
Wetmur J.G.   76, 77, 251, **118, 256**
Wettstein F. von   319, **331**
Whallon J.   17, **26**
Whitehouse H.L.K.   322, **331**
Whitton B.A.   7, **28**
Wiame J.M.   34, **46**
Wichura M.   176, **193**
Wiekevitz P.   120, **142**
Wiese L.   viii, 147, 148, 149, 174, 175, 176, 177, 179, 181, 182, 183, 310, 326, 327, **173, 194, 197, 328, 332**
Wiese W.   83, 175, **197**
Wik-Sjøstedt A.   222, **235**
Wilcox M.   11, **28**
Wildman S.G.   123, **141**
Wilson R.G.   154, **173**
Wiman F.H.   212, 215, 216, **218**
Witkin E.M.   13, 14, 15, **28**
Wittmann H.G.   123, 132, 137, **142, 143**
Wolfovitch R.   279, **298**
Wolken J.J.   286, **299**
Wolstenholme D.R.   77, 259, 260, 263, 264, 265, 266, 267, **117, 295**
Wood D.D.   123, **144**
Wood R.D.   211, 215, 216, **218**
Wood W.B.   51, **61**
Woodcock C.F.L.   238, 241, 243, 247, 251, **256**
Woodruff M.   287, **295**
Wu C.Y.   227, **234**
Wu J.H.   14, **28**
Wu P.H.L.   154, **170**

Yang K.S.   154, **172**

Yiang B.Y.   227, **234**
Yoshikawa H.   18, **28**
Yu R.S.T.   278, **299**
Yuan R.   110, **117**

Zeldin M.H.   268, 274, 280, 286, **291, 297,
   299**
Zetsche K.   248, 249, 250, **253, 256**
Zhevner V.D.   10, 11, 14, 17, **26, 28**

INDEX TO GENERA AND SPECIES

The name of an alga has two components: the generic name, which is always capitalized, and the specific epithet, which is customarily not capitalized. Such a name is called a binomial. Linnaeus is accredited with having standardized the use of binomials as a kind of short-hand, at a time when species were usually designated by descriptive phrases. Technically, the name of the person who first proposed a particular binomial is a mandatory appendage to that binomial. When a species has been transferred from one genus to another, the author of the original binomial is enclosed in parentheses, followed by the author who made the transfer. The author of a binomial is usually abbreviated: e.g. *Fucus vesiculosus* L. (for Linnaeus). There are good reasons for citing the author of a binomial at certain times, and equally good reasons for omitting it at other times, though this discretionary practice is poorly understood by many authors and editors. The main reason for citing the author of a binomial (and the only reason in non-taxonomic literature) is precision, since it sometimes happens that the same binomial is inadvertently applied to two different organisms by two different taxonomists. (In the present index, for example, *Antithamnion tenuissimum* (Hauck) Schiffner (1916) from the Adriatic could possibly be confused with *A. tenuissimum* Gardner (1927) from California, even though Gardner's species has been renamed to avoid duplicate names, i.e. homonyms.) But common sense must serve as the guide, and such widely known algae as *Acetabularia mediterranea*, *Euglena gracilis*, and *Fucus vesiculosus* hardly need the precision of author citation. A safe course for geneticists to follow, and one which should satisfy editorial pedantry, is to cite the author of a binomial once, when the binomial is first used in the article. Inclusion of the author of a binomial in the title of a non-taxonomic article is unnecessary.

To insure the association of data with the correct organism, it is essential to document the identity of the investigated material in an appropriate manner. When a culture from a standard culture collection is used, not only the name but also the number and source of the strain should be cited. Genetic studies involving macroscopic algae, at least, should be documented by voucher specimens deposited in a major herbarium.

Paul C. Silva, University of California, Berkeley

Acetabularia Lamour.   4, 178, Ch. 12 (236–256), 315
  clavata Yam.   241
  cliftonii[1]   238, 251
  crenulata Lamour.   248
  exigua Solms-Laub.   241
  major Martens   240–7
  mediterranea Lamour.   Ch. 12 (236–56)
Agmenellum quadruplicatum (Menegh.) Bréb.   10–14
Anabaena cycadeae[2] Spratt   12
  ambigua C. B. Rao   11
  cylindrica Lemm.   8, 11
  doliolum Bhar.   9, 10, 11, 12, 14
Anacystis Menegh.   3
  nidulans (P. Richt.) Drouet et Daily Ch. 1 (7–28), 278
Antithamnion plumula (J. Ellis) Thur.   227
  tenuissimum (Hauck) Schiffn.   222
Astrephomene gubernaculifera M. A. Pocock   176, 185

[*Bacillus*]   17, 18, 21, 137
Batophora J. G. Ag.   236
Botrydium granulatum (L.) Grev.   311, 316, 317, 322, 326
Branchiomonas Bohl.   312, 314, 315, 316, 317, 318
[*Brucella*]   17

Caulerpa Lamour.   3
Chara aspera Willd.   211, 214, 215, 216, 307
  australis R. Brown   211, 216
  braunii K. Gmel.   211, 214
  canescens Lois.-Desl.   216
  ceratophylla Wallr.   306
  connivens Salzm. ex A. Braun   211, 216
  contraria A. Braun ex Kuetz.   211, 217
  corallina Willd.   216
  crinita Wallr.   307
  evoluta T. F. Allen   211
  foetida A. Braun   306
  foliolosa Machlenb. ex Willd.   211
  fragilis Desv.   211
  globularis Thuill.,   211, 214, 216, 217
  gymnopitys A. Braun   216
  haitensis Turp.   211, 216
  hispida L.   214, 216, 217
  imperfecta A. Braun   216
  martiana A. Braun ex Wallm.   211
  preissii A. Braun   216
  rusbyana Howe   211, 216
  vulgaris L.   211, 214, 217
  zeylanica Willd.   211–14
Chlamydomonas Ehrenb.   Ch. 2–7 (29–173), 174–83, 214, 233, 244, 270, App. A, B (300–3)

Chlamydomonas Ehrenb.—*cont.*
  agametos[3]   312, 319, 320, 322, 326, 327
  botryoides Strehl.   183
  braunii Gorosch.   312
  chlamydogama Bold   178, 182
  dresdensis F. Moewus   312, 315
  dysosmos F. Moewus   182
  eugametos F. Moewus   Ch. 7 (145–73), 175–82, App. B. (310–32)
  gymnogama Deason   178
  mexicana R. Lewin   182
  moewusii Gerl.   49, 50, 71, Ch. 7, (145–173), 175–83, 327
  oogama F. Moewus   312
  paradoxa (Korsh.) Pasch.   183, 311, 312, 314, 316, 318
  paupera Pasch.   312, 318, 322
  philotes R. Lewin   178, 315
  pseudoparadoxa F. Moewus   312
  reinhardtii P. A. Dang.   Ch. 2–6 (29–144), 146, 148, 152, 161, 165, 167–9, 175–83
  smithii Hoshaw et H. Ettl   183
  suboogama Tschermak-Woess   178, 179
Chlorella Beyer.   1
Chlorogonium elongatum (P. A. Dang.) Francé   176
  euchlorum Ehrenb.   176
  leiostracum Strehl.   176
  oogamum Pasch.   181
Chloromonas saprophila Tschermak-Woess   179
Chodatella Lemm.   2
Cladophora Kuetz.   222
Closterium Nitzsch ex Ralfs   199, 200, 205
  ehrenbergii Menegh. ex Ralfs   201, 202
  moniliferum Ehrenb. ex Ralfs   200–2
  siliqua W. et G. S. West   200, 205
Cosmarium Corda ex Ralfs   199, 200, 205, 207
  biretum Bréb.   201
  botrytis Menegh. ex Ralfs   201, 202, 204, 205
  formosulum Hoff   201
  turpinii Bréb.   199, 200–4, 206
Cylindrocystis Menegh. ex De Bary   199
  brebissonii (Ralfs) De Bary   199, 200, 208
  crassa De Bary   199
Cylindrospermum majus Kuetz. ex B. et F.   8, 11, 16

Dasycladus C. A. Ag.   313
Derbesia Sol.   222
[*Drosophila*]   169, 219
Dunaliella Teod.   316
  salina (Dunal) Teod.   176

Ectocarpus Lyngb.　3, 313
　siliculosus (Dillw.) Lyngb.　222, 227
Enteromorpha Link　311
　compressa (L.) C. G. Nees　226
　intestinalis (L.) C. G. Nees　226
[*Escherichia*]　14, 15, 18, 21, 23, 76, 77,
　78, 123, 124, 132, 137
Eudorina Ehrenb.　176, 177, 185, 186
　californica (Shaw) Goldst.　175
　conradii Golds.　175
　cylindrica Korsh.　185, 186
　elegans Ehrenb.　176, 185, 186
　illinoisensis (Kof.) Pasch.　185, 186
　unicocca G. M. Smith　185, 186
Euglena Ehrenb.　1, 77, 124, Ch. 13 (257–
　299)
　gracilis Klebs　4, 77, Ch. 13 (257–99)

Fischerella muscicola (Borzi ex B. et F.)
　Gom.　13
[*Forsythia*]　311
Fucus L.　3, 226
　distichus L.　226
[*Funaria*]　319

[*Glomerella*]　317
Golenkinia minutissima Iyeng. et Balakr.
　178
Gonium O. F. Mueller　176, 184
　pectorale O. F. Mueller　176, 184

Haematococcus C. A. Ag.　315
　pluvialis Flot.　178
Halicystis J. E. Aresch.　222

Laminaria Lamour.　227
　agardhii Kjellm.　3
　cucullata (Le Jol.) Fosl.　226
　digitata (L.) Lamour.　226
　hyperborea (Gunn.) Fosl.　226
　japonica J. E. Aresch.　227
　longicruris De la Pylaie　226, 227
　saccharina (L.) Lamour.　226, 227

Micrasterias C. A. Ag. ex Ralfs　203, 204,
　207
Monostroma Thur.　311

[*Neisseria*]　17
Neomeris Lamour.　236
Netrium (Naeg.) Itzigs. et Rothe　199,
　202
　digitus (Bréb. ex Ralfs) Itzigs. et Rothe
　199–202
[*Neurospora*]　139, 157, 169, 175, 219, 301
Nitella C. A. Ag.　214, 216
Nostoc linckia Born. ex B. et F.　11
　muscorum C. A. Ag. ex B. et F.　9

Ochromonas Vysotskij (Wissotsky)　1
Oedogonium Link ex Hirn　3, 301, 306,
　307, 308
　acrosporum De Bary ex Hirn　309
　boscii[4] (Leclerc) Bréb. ex Hirn　309
　braunii Kuetz. ex Hirn　309
　capillare Kuetz. ex Hirn　309
　cardiacum Kuetz. ex Hirn　309
　cleveanum Wittr. ex Hirn　309
　concatenatum (Hass.) Kuetz. ex Hirn
　309
　crassum (Hass.) Kuetz. ex Hirn　309
　cyathigerum Wittr. ex Hirn　309
　grande Kuetz. ex Hirn　309
　hystrix Wittr. ex Hirn　309
　landsboroughii (Hass.) Kuetz. ex Hirn
　309
　longatum Kuetz. ex Hirn　309
　macrandrium[4] Wittr. ex Hirn　309
　pluviale Nordst. ex Hirn　309
　punctatostriatum[4] De Bary ex Hirn　309
　rivulare A. Braun ex Hirn　309
　wolleanum[4] Wittr. ex Hirn　309

Pandorina Bory　182, 184
　morum (O. F. Mueller) Bory　176, 183
　unicocca Rayburn et Starr　176
[*Paramecium*]　175
Pedinomonas Korsh.　3
Phormidium mucicola Naum. et Hub.-Pest.
　12
[*Phycomyces*]　306
Platydorina caudata Kof.　176, 186
Plectonema boryanum Gom.　10, 11, 14,
　15
[*Pneumococcus*]　17
Polysiphonia boldii Wynne et Edwards　3,
　224
　hemisphaerica J. E. Aresch.　224
Polytoma Ehrenb.　37, 311, 312, 314, 316,
　318, 319, 326
　uvella Ehrenb.　178
Protochara australis Wom. et Ophel　211
Protosiphon Klebs　311, 312, 326
　botryoides (Kuetz.) Klebs　311, 314,
　315, 316, 318, 319, 322
[*Pseudomonas*]　17

[*Saccharomyces*]　18
[*Salmonella*]　18
Sargassum C. A. Ag.　1
Scenedesmus Meyen　2
Sirogonium Kuetz.　200
[*Sordaria*]　317
Sphaeroplea C. A. Ag.　308
Spirogyra Link　3, 198, 199, 203, 205, 206,
　307
　communis (Hass.) Kuetz.　199

Spirotaenia Bréb.   199–201
   condensata Bréb.   199–201, 208
[*Staphylococcus*]   17, 64
Staurastrum Meyen ex Ralfs   205, 207
   denticulatum (Naeg.) Arch.   200, 205
   dilatatum Ehrenb. ex Ralfs   205
Stephanosphaera Cohn   176
Synechococcus cedrorum Sauv.   13
Synechocystis aquatilis Sauv.   10, 14

[*Tetrahymena*]   260

Ulothrix flacca (Dillw.) Thur.   222
Ulva L.   3, 220–4, 228, 230, 232, 233
   lactuca L.   223, 224
   mutabilis Foeyn   223–8
   rigida C. A. Ag.   224
   thuretii Foeyn   223, 224
Undaria peterseniana (Kjellm.) K. Okam.
      226
   pinnatifida (W. H. Harv.) Sur.   226
   undarioides (Yendo) K. Okam.   226

Vaucheria DC.   308
Volvox L.   177–82, 186–9, 192, 315
   africanus G. S. West   176, 187, 188
   aureus Ehrenb.   175, 178, 187, 188, 189,
      316
   barberi Shaw   187
   capensis Rich et M. A. Pocock   188
   carteri F. Stein   176, 177, 187–92
   dissipatrix (Shaw) Iyeng.   187, 189
   gigas M. A. Pocock   176, 187, 189
   globator L.   178, 187, 188
   obversus (Shaw) Printz   187
   perglobator Powers   187, 188
   pocockiae Starr   188, 189
   powersii (Shaw) Printz   187, 190
   rousseletii G. S. West   177, 187–9
   spermatosphaera Powers   187, 188
   tertius A. Meyer   187
Volvulina pringsheimii Starr   176
   steinii Playf.   176, 184

Zygnema C. A. Ag. 198–200, 203, 206
   circumcarinatum Czurda   199

[1] This binomial, purportedly a transfer of *Polyphysa cliftonii* W. H. Harvey into *Acetabularia*, has not yet been officially proposed. P.C.S.
[2] This name has been traditionally but incorrectly ascribed to Reinke. P.C.S.
[3] This binomial, purportedly an elevation of *Chlamydomonas eugametos* forma *agametos* F. Moewus to the rank of species, has not yet been officially proposed. P.C.S.
[4] The spellings in Pascher's paper are incorrect. P.C.S.

SUBJECT INDEX

acetate   12, 13, 29, 37, 38, 64, 92, 138, 140, 146, 159, 219
acetate-requiring mutants   12, 13, 31, 36, 38–40, 72, 73, 86, 100, 114, 119, 120, 137, 160, 169
acetyl-glutamate kinase   34
acetyl-glutamate synthetase   34
acetylglutamyl-phosphate reductase   33, 34
acetylornithine-glutamate transacetylase   34
acetylornithine transaminase   34
acetyl-pyridine inhibition   157
acridine orange   9, 74
acridine-orange-induced mutants   9, 87
acriflavine   14, 139
acriflavine-induced UV-sensitivity   14
acrylamide gel (see polyacrylamide)
actinomycin D   123, 182, 229, 249, 250, 268, 269, 272, 280, 282
adenine   13, 79, 81, 83, 84, 138, 155
adenine-requiring mutants   13, 155
adenosine triphosphatase activity   151
adenosine triphosphate synthesis   39
agar surface growth conditions   8, 16, 29, 30, 33, 34, 38, 41, 44, 50, 64, 67, 92, 159
agarose acrylamide gels   122
agglutination   181, 182, 183
agglutinins   3, 163, 179
aggregation   8, 30
alanine   158, 270, 274
aldolase fructose 1,6-diphosphate   121, 275, 282, 287, 288
alkylating agents   9, 14, 33
allelic ratios   91, 93–8, 109, 113, 114
allelism   153, 154, 157
allodiploid   217
alpha esterase   148
alpha-glycosidic linkage   182, 183
alpha-keto acids   159
alpha-mannosidase   175, 183
amino acids   12, 35, 122, 125, 132 (see also specific acids)
amino-acid incorporation   125, 130, 249, 250, 252, 262
amino-acid-requiring strains   12, 13, 33, 34, 35, 44, 158, 159, 165, 166
amino-acid substitution   132

aminoacyl-tRNA synthetases   120, 268, 271, 274-table, 278, 284, 285
4-amino imidazole 5-carboxamide   155
aminolaevulinate dehydratase   274, 281
2-amino-3-phenylbutanoic acid   169 (see also APBA)
aminopterin   163
2-aminopurine   10, 19, 163
2-aminopurine-induced mutants   10
ammonia, ammonium   13, 34, 35, 78, 159, 161, 180
ammonia-requiring, resistant, sensitive mutants   13, 34, 35
anastomosis   9
androgonidia   189, 192
androtermone   321
aneuploidy   166, 185, 203, 205
aniline   154
anisogamy   176, 178
antheridia   211, 212, 213, 214, 215, 216
antibiotics   30, 72, 92, 120, 124, 125, 127, 129, 131, 134, 138, 140 (see also specific kinds)
antibiotic-dependent mutants   72, 73, 156–157
antibiotics, organelle-specific   140, 244, 245, 252
antibiotic-resistant mutants   8, 9, 30, 72, 73, 86, 106, 124, 125, 128–31, 136–40, 156–7
antibiotic-sensitive mutants   64, 72, 73, 106, 119, 124–32, 130, 131, 138, 156–7
antigenicity   66
antimetabolites   157, 158, 166
anucleate cells   246–50, 252
APBA = 2-amino-3-phenylbutanoic acid (see next entries)
APBA-induced mutations   159, 163, 164
APBA inhibition and resistance   158, 159, 163, 165
apical cells   230
apochlorosis, induced   257
apochlorotic flagellate   37
arabinose   66
aradiate cells   203, 206
arginine   33, 34, 40, 41, 138, 180, 268, 274
arginine deaminase   34

arginine-requiring mutants 33, 34, 40, 165, 180
arginino-succinate lyase 33, 34
arginino-succinate-lyase-lacking phenotype 33
argininosuccinate synthetase 34
arsenate 120
arsenate-sensitive mutants 39
ascomycetes 44
ascorbate-DPIP couple 39
asexual reproduction 3, 29, 30, 176, 182, 183, 188, 189, 191, 192, 200, 241
aspartic acid 263, 274
ATP (see adenosine triphosphate)
attachment sites 112, 113
autopolyploids 148, 161, 166
autotoxicity 157
autotrophy 40, 219
auxins and auxin inhibitors 311
auxotrophic mutants 8, 12–13, 16, 21, 33, 90, 134, 153–7, 160, 165 (see also specific auxotrophs)
axoneme assembly 50, 55, 57, 60

back-mutation 8, 33, 162–4, 183
bacteria 7, 8, 13, 14, 17, 109, 110, 127, 132, 137, 154 (see also specific kinds)
bacterial chromosome 123
bacterial plasmids 112
bacterial ribosomes 123, 124, 137
bacterial transformation 17
bacteriophages 15, 51, 76, 77, 232
basal bodies 52, 57, 58, 59–60, 61
basal cell 230
base-analogue-induced mutants 9
base-requiring mutants 33, 35 (see also specific bases)
bases, complementary 75
benzoic acid 154
bifurcation gene 316, 319, 323
biogenesis 42, 66, 119–44
biotin 13
biotin-requiring auxotrophs 13
blade cells 223, 228, 229, 230, 232
'block point' (in cell division) 41
blue-green algae (see cyanophytes)
blue mutants 12
'bonnet' mutants 206
branching 11, 226–7
bromodeoxyuridine 163
bromouracil 19, 35
bromouracil mutants 35
brown algae (see Phaeophyceae)
bryophytes 1, 3
'bubble' mutant 225, 228, 229

C_2-compound-requiring mutant 13
C_4-compound-requiring mutant 13

caffeine 10, 14, 159, 162, 164
caffeine-induced mutants 10, 14, 159
caffeine-resistant mutants 164
calcium 146, 161
canavanine 40, 41
cap formation 237, 243, 244–6, 248, 250
carbomycin 131, 132
carbomycin-resistant strains 72, 100, 129, 137
carbon-cycle enzymes 120
carbon dioxide 1, 29, 30, 38, 120, 146
carbon dioxide reduction 1, 37–40, 274, 279
carbon dioxide reduction, mutants 37, 40
carbon sources 29, 30, 38, 119, 120, 146, 159, 160, 169
carboxymethyl-cellulose chromatography 123, 131
carotenoid, altered mutants 11, 12, 37
carotenoids 11, 12, 37, 121, 326
carotenoid synthesis, inhibition 281, 289
carpospores 3, 222, 224
carposporophyte 3, 4, 224
casein-hydrolysate-requiring mutant 12, 155
cations 161 (see also specific ions)
cell diameter changes 206
cell division 10, 19, 41, 44, 90, 96, 112, 149, 188, 192, 230, 232 (see also mitosis, meiosis)
cell division cycles 4, 41, 78, 79, 80, 86
cell-division mutants 37, 41, 149, 152
cell division, plane disorientations 152, 232
cell division, post-irradiation 161
cell division rates 37
cell-division-rate genes 316, 319, 323
cell fusion 3, 30, 58, 59, 60, 61, 65, 80, 83, 110, 146, 150, 313–15
cell length 11
cell membrane 152, 161
cell sap 111
cell-shape mutants 203, 206, 207, 304, 305, 317, 319
cell-size gene 317, 319
cell : volume ratios 205, 206
cell walls 4, 7, 58, 63, 65, 66, 67, 304, 305
cell-wall synthesis 63–8, 232
cell-wall synthesis, defective mutants 63, 67
centromere 40, 42, 103, 149, 150, 164, 168, 201
chain initiation 123
chaotropic agents 66
Characeae, Charales 4, 210–17, 306, 307
chelating agents 13
chemotaxis 181, 311, 326
chiasma interference 168, 320, 322–3
chimaeras 230

chloramphenicol 9, 10, 138, 139, 252, 261, 262, 269, 272, 273, 278, 279, 280, 287
chloramphenicol-induced phenocopies 10
chloramphenicol-resistance 9
chlorate 21
chlorate-resistant mutant 10
Chlorococcales 2
chloroform 30, 92
chloromycetin (see chloramphenicol)
chlorophyll 12, 36, 38, 121, 134, 160
chlorophyll synthesis 31, 36, 37, 40, 42, 249, 279, 280–1, 286–7
chlorophyll synthesis, mutants 12, 31, 36, 37, 42, 71
chloroplast (see also entries under DNA, RNA, ribosomes, pigments, proteins, etc.)
chloroplasts 3, 5, 7, 29, 30, 40, 42, 44, 69–144 (Ch. 5, 6), 202, 208, 245
chloroplast autonomy 244, 250–2, 271, 289, 290
chloroplast 'chromosome' 31, 69–118 (Ch. 5), 251
chloroplast components 274–7
chloroplast content per cell 200, 203
chloroplast-defective cells 120
chloroplast development, light-mediated 286–9
chloroplast division (see chloroplast replication)
chloroplast fractions 74
chloroplast fusion 31, 71, 83, 86, 111
chloroplast genes, genome 69–144 (Ch. 5, 6), 251, 252, 258, 259, 263, 264, 267–86, 271, 289
chloroplast genome number 264
chloroplast grana 135
chloroplast, inheritance 69–119, 203, 208, 228, 257–90
chloroplast lamellae 37, 125, 127, 134, 135
chloroplast membrane 37, 40, 121, 133
chloroplasts, origin 290
chloroplast, position in cell 304, 305
chloroplast replication 161, 258, 259, 264, 265, 271, 290
chloroplast survival 202
chloroplast synthesis 267–87
chlorosis 42
choline 156
chromatid exchange 164
chromophores 289, 304
chromosomes 1, 2, 7, 31, 42, 50, 65, 71, 147, 148, 161, 233
chromosome aberrations 159, 163–5, 167, 169
chromosomal deletions 156, 164
chromosome division 237, 243
chromosome number abnormalities 185

chromosome replication 243, 244, 318
chromosomes, trivalent 204
chromosomes, univalent 204
cistrons 16, 33, 57, 120, 123, 137
citrate 87, 139
citrate-synthase-deficient strain 160
citrulline 33
cleocin 131, 132
cleocin-resistant strains 72, 100, 129, 237
clocks, biological 4
clumping 30
clustering, gene 51
CO_2 (see carbon dioxide)
cobalt-60 214
coding, transcriptional 123
coenocyte 10, 11
cold-sensitive mutants 228, 232
complementation 16, 33, 57, 75, 134, 150, 151, 160
concanavalin A 175, 182, 183
Conjugales 4, 198–208
conjugation 58, 200
conversion, gene 95
copper-tolerant strain 227
copulation (see Frontispiece)
copulation-temperature gene 322
copulation tube 181 (see also cytoplasmic bridge)
cortication 215, 216
cosegregation 100, 108
cotransfer index 17
cotransmission 98
conversion 65
cross, two-factor 8, 16
crosses, interspecific 165, 217, 224, 226, 227, 300, 301, 303, 304, 306, 309
crosses, multifactorial 16
crosses, reciprocal 37, 65, 84, 93, 94, 95, 97, 105, 106, 109, 113, 147
crossing-over 315–25, 327
crossing-over, four-strand-stage 319–22, 327
crossing-over, temperature effects 323, 325
crossing-over, two-strand-stage 317–19, 320, 323, 325, 327
crossover suppression 90
cross-walls 10
cryneomycin-resistant strains 9
cyanophages 9, 10, 14, 15, 16, 20
cyanophage-resistant strains 9, 10
cyanophytes 1, 3, 8, 7–48
cycloheximide inhibition 89, 90, 138, 139, 248, 252, 261, 262, 269, 271, 272, 273, 278, 287
cysteine 13
cysteine-requiring strains 13
cystocarps 4
cysts 237, 238, 241, 242, 245, 251

cytochromes 39, 40, 120, 121, 271, 275, 279, 281, 282
cytogamy (see cell fusion)
cytogenes (see genes, cytoplasmic)
cytohets 96, 98
cytoplasm 58, 60, 61, 71, 83, 246, 247, 248, 249, 250, 251
cytoplasmic axes 204
cytoplasmic damage 161
'cytoplasmic framework' theory 204
cytoplasmic isthmus, bridge 146, 150, 151 (also see Frontispiece)
cytoplasts 3

dark-repair mechanisms 14, 15, 159, 161
dark starvation 124
Dasycladales 236
DCPIP (see DPIP)
DEAE dextran 64
dedifferentiation 42, 229, 232
deflagellation 59
deletion 327
denaturation 74, 76
density labelling 80, 84, 85, 86
deoxyribonuclease 8, 17, 18, 71, 74, 242, 275
deoxyribonucleic acid (see DNA)
derepression 157
deretroinhibition 157
DES (see diethyl sulphate)
Desmidiaceae, desmids 1, 198, 200, 202, 204, 207, 307
developmental mutants 11, 227, 233, 244, 245 (see also specific mutants)
developmental stages 11, 239
dextran 64
diatoms 2
dicarboxylic acids 154 (see also specific acids)
dichlorophenyl indophenol (DPIP) 39
diethyl sulphate-induced mutations, reversions 9, 10, 11, 19
differentiation 57, 188, 230, 232, 233, 236
dihydrospectinomycin 125, 127
dikaryons 58, 60, 146, 149
dimethyl sulphoxide (DMSO) 268
dinoflagellates 1, 2
dioecism 175, 176, 186, 188, 187, 189, 210, 211, 215, 216, 217, 307, 308
diplogenetic sex determination 175
diplo-haplontic life cycle 222
diploid strains, cells 3, 43, 44, 57, 65, 120, 134, 135, 161, 175, 185, 204, 205
diploids, stable 30, 44
diploids, vegetative 134, 165, 207
diplomicts 305
diplophase 3, 175 (see also zygote, zygospore, diploid)

DNA 1, 8, 13, 14, 17–20, 30, 35, 43, 44, 59, 75–8, 81–7, 110, 114, 163, 206, 233, 248, 264–7
DNA amplification 242
DNA, bacteriophage 76, 77
DNA changes, light-dependent 264, 265
DNA, chloroplast 31, 44, 59, 71–86, 87, 88, 98, 110, 111, 114, 120, 123, 132, 138, 139, 166, 250, 251, 258, 259, 261, 262, 267, 269, 279 (see also DNA, fraction β)
DNA fraction α 74, 75, 79, 80, 82, 87
DNA fraction β 74, 75, 76, 79–84, 86, 87, 110 (see also DNA, chloroplast)
DNA fraction γ 74, 75, 87
DNA fraction δ (see DNA, mitochondrial)
DNA fraction M 74, 83, 84
DNA, gamete 180, 241
DNA, heavy fraction 265
DNA, hybrid 74–88
DNA, mitochondrial 31, 59, 71, 74, 75, 78, 86, 87, 88, 161, 259, 260, 261, 262, 265, 266, 269
(= DNA fraction δ)
DNA, nuclear 44, 59, 74, 75, 77–83, 85, 86, 87, 138, 139, 180, 241, 242, 264
DNA-photoreactivating enzyme 14
DNA polymerase 79, 114, 138
DNA, plant 74
DNA precursors 84, 180
DNA renaturation, kinetic analysis 76, 77, 251
DNA repair mechanisms 13, 15, 35, 43, 44, 59, 84, 111–13
DNA-repair-mechanism mutation 43
DNA repair, post-replication 13, 14, 15
DNA replication 1, 13, 19–23, 35, 42, 44, 78–84, 86, 87, 93, 95, 110, 111, 112, 138, 179, 237, 242, 243
DNA replication, asynchronous 86, 87
DNA replication fork 86
DNA replication, preferential 242
DNA replication, sequential 19
DNA replication, unidirectional 21, 22, 79, 80
DNA SP-15 85
DNA, strand number 233
DNA synthesis 78, 250 (see also DNA replication)
DNA synthesis, nuclear 139, 243, 269
DNA synthesis, synchronous 19, 20
DNA transmission 71
DNA, zygote 241
DNase (see deoxyribonuclease)
dominance 31, 44, 59, 60, 134
dormancy, zygote 44
double cells 152
double crossovers, frequency 320–3

double diffusion 151
double mutants 13, 34, 35, 36, 37, 40, 43, 105, 134, 157, 158
doublets 52, 60
doubling times 30, 37
doubly-marked strains 157, 158
DPIP (see dichlorophenyl indophenol)
duplication of genetic information 161, 166
dwarf males 2, 186, 307

efficiency of plating 8
egg-bearing colonies 187, 189, 191, 192
electron-transport-deficient mutants 39
electron transport, photosynthetic 37, 38, 39, 40, 120, 125, 133, 134
emasculation 210, 211, 212, 214
EMS (ethyl methane sulphonate) 9
EMS-induced damage 14, 15
EMS-induced mutants 9, 10, 13, 14, 19, 20, 21, 22, 30, 33, 63, 72
EMS-induced revertants 19
'end-point' killed cells 161, 162
endomitosis 148, 242
endophytic phase 3
endosymbiont 251
enrichment, mutant 13, 149, 153, 228
Enterobacteriaceae 21
enucleate (see anucleate)
environmental factors 147, 226
enzymes 124, 248, 268, 270 (see also specific enzymes)
epistasis 134
erotactin 181
erythrocytes 247

female colonies 188–90, 192
ferredoxin 39, 120, 121
ferredoxin-NAPD reductase 121, 275
fertility-reduced mutants 149, 162, 212–214, 217
fertilization 3, 174, 178–81
Feulgen-staining 8, 71, 74
filamentous strains 10–12, 15, 21, 198, 232, 233
flagella, abnormal swimming mutants 56, 57, 59, 149–51 (see also flagella, paralysed mutants)
flagella, curved-flagella strains 52, 53, 59
flagella, frayed 52
flagella, paralysed mutants 50–2, 57–9, 149–51, 160, 162, 163, 168, 179
flagella, short-flagella mutants 55, 57, 58, 149
flagella, straight-flagella strains 50, 52, 53, 59
flagella, stumpy-flagella mutants 55, 56, 58, 60
flagella, swollen-flagella mutants 53

flagella-length genes 54, 55, 57, 59, 317, 319
flagella-less mutants 52, 55, 58, 60, 149, 151, 179
flagella structure 50, 52, 148, 151
flagellar agglutination 182
flagellar apparatus, abnormal 50–6, 59, 60, 149–51, 152, 165, 166
flagellar development, normal 63, 146, 179
flagellar regression 61
flat-colony mutants 64
Florideae (see red algae)
5-fluorouracil-resistance 10
folic acid 155, 156
frameshifts 19
fructose 1,6-diphosphatase 275
fumarate-requiring mutant 13
fumaric acid 13, 159
fungi 123, 139, 154

G-3-P (see glyceraldehyde 3-phosphate)
galactose 66
gametangia 3, 210, 213, 223
gamete(s) 3, 30, 37, 58, 63, 65, 71, 75, 76, 78, 80–5, 87–90, 110–13, 146, 149, 150, 151, 160, 188, 220, 222, 223, 302, 303, 305
gamete fusion (see cell fusion)
gemete release 237, 238, 241, 242, 245
gamete valence 313, 314, 326
gametogenesis 30, 83, 88, 90, 92, 112, 174, 178–83, 182, 228, 234, 241
gametophytes 220–3, 228, 229
gamma-radiation 14
gamones 147, 321, 327
'gaunt' mutants 206
gene-centromere distance 149, 150, 153, 159, 167
genes, cytoplasmic 71, 77, 86, 150, 228
gene transfer 8, 9, 15, 16, 19
generation time 29, 30, 59
genetic map, temporal 19–23
genetic mapping 16, 17, 18, 19, 21, 22, 43, 50, 51, 65, 71, 98–109, 132, 136, 137, 149, 150, 153, 159, 162, 208
genetic mapping, transcription 123
genetic systems 30
genome 8, 14, 19, 21, 37, 59, 76–8, 86, 93–5, 109–12, 119, 139, 156 (see also chloroplast genome)
genome size 251, 259, 260, 261, 264
germination 30, 50, 61, 80, 82, 83, 84, 89, 91, 96, 147, 160, 169, 311, 324, 325 (see also zygote/zygospore germination)
germination-temperature gene 324
germling cells 200, 201, 204, 205, 229, 232, 237, 238, 245
glucose 12, 64, 152

glucose-requiring mutants 12
glutamate-pyruvate transaminase activity
    159
glutamic acid 268, 270
glutamine 268
glyceraldehyde 3-phosphate dehydrogenase
    42, 121, 275, 282, 283, 287, 288, 289
glycerate 3-phosphate kinase 276
glycerol 154, 166
glycine 166, 270, 274
glycollate-DPIP oxidoreductase 276, 283
glycoprotein 66, 67, 147, 182, 183, 327
glycosidic linkage 182, 183
gones 91, 109, 147, 156, 164, 176, 201, 202
gonidia 188, 189, 191, 192
gradient, apico-basal 246
grafts, interspecific 246, 248, 251, 252
granum 135
green algae 3, 4, 44
growth cycle, vegetative 78–81, 237, 242,
    243
growth-factor-requiring mutants (see auxo-
    trophic mutants)
growth hormones 233
growth, photosynthetic 29, 120, 179
growth rate 37, 60, 64, 229, 230
growth yield 37
guanine 35, 155
gynogamone 327
gynotermone 321

haplo-diplontic life cycle 222
haploid cells 3, 50, 161, 162, 174, 204
haplomicts 305
haplontic strain 222
heterocysts 11
heterocyst-less mutants 11
heterogamy (see anisogamy)
heterokaryon 58, 59, 61, 169, 199, 202, 207
heteromorphism 222
heterothallic strains 2, 3, 175, 176, 177,
    178, 182–9, 200, 201, 204, 205, 208, 313,
    316
heterotrophy 29–31, 37, 139, 146
heterozygotes (-ous) 17, 91, 93, 96, 102–5,
    108, 109, 147, 164, 303, 304
heterozygotes, unstable 17
hexaradiate cells 205
Hill reaction (see photosystem II)
histidine 268
histidine-sensitive strains 156, 166
histones 8
holdfast 229, 230, 232
hologamous species 180
homothallic strains 175–8, 183, 184–9,
    193, 200, 204, 205, 313–17, 327
homozygotes 91, 93, 94, 95, 96, 100, 103,
    113, 143, 160

hybridization 3, 58, 74, 79, 123, 226
hybrids 93–5, 124, 147, 211, 212
hybrids, radiation-induced 214–15
hydroxylamine 9, 19, 163
hydroxylamine-induced mutants 9
p-hydroxyphenyl pyruvate 159
hydroxyproline 66
hyperchromicity 74
hyperinhibition 156
hypoxanthine 155

ICR-170 (a quinacrine mustard) 42
ICR-170-induced mutants 42, 163
immunological comparison of mutants
    148, 151
inheritance patterns 69, 70, 71, 81, 84, 85,
    86, 88, 90, 110–14 (see also Mendelian,
    non-Mendelian uniparental inheritance,
    segregation patterns)
inhibition of chloroplast components
    280–5
inhibition reversal 158, 159
inhibitor-resistant, -sensitive, and
    -dependent mutants 154, 156–9
initiation point (for replication) 21, 22
integration 18
intercalating-agent-induced mutants 9
interchromsomal effects 323
interference, cistron 16
interfertility 183, 201, 211, 216, 226, 227
intermediary carbon metabolism, defective
    mutant 13
inversion, heterozygous 165
inversion, pericentric 165
ionic pressure 161
irradiation (see UV, etc.)
isoagglutination 148, 182
isoenzymes 148
isogamy 30, 71, 175, 176, 182, 183, 193,
    200, 223
isoleucine 130, 268, 270, 274
isotope transfer 81

kanamycin 9, 139
kanamycin-resistant strains 9, 72
karyogamy (see nuclear fusion)
keto acids 159
kinetic complexity 76–8

lactate dehydrogenase 248
lamellar organization 120
lamellar structural protein 277, 279
lamina 230
leaky genes 52–6, 70, 72
lethal mutations 18, 110, 120, 141, 156,
    157, 158, 163, 164, 167, 202, 206, 207,
    257
leucine 139, 166, 268, 274, 278

leucine aminopeptidase 148
leucine incorporation by plastids 278
light deprivation 19
linkage, false 160, 167, 168
linkage, of genes 17, 18, 35, 36, 37, 42, 43, 57, 65, 90, 98–109, 132, 137, 139, 149, 152, 154, 160, 162, 163, 165, 176, 316, 317
linkage groups 30, 31, 32, 33, 35, 40, 42, 50, 52, 53, 54, 55, 56, 60, 98, 149, 150, 167, 168, 314, 319, 320
linkage, negative 167, 168
lithium chloride 66
'lumpy' mutant 225, 228, 232
lysine 154, 166, 268
lysis 10, 11, 15, 202
lysogenic strains 15
lytic enzyme 181

magnesium 123, 161, 247
malate-requiring mutant 13
malate dehydrogenase 13, 148, 248
male colonies 186, 187, 188, 189, 192, 216
male-inducer substance (MIS) 189
male-sterile hybrids 212, 216
male-sterile mutants 214, 215
malic acid 13, 159
mammalian cells 21
mannose 66, 182, 183
mannosidase 175, 183
maps (see genetic maps)
marine algae 219–34
marker ratios 95
maternal inheritance 71, 84, 86, 88, 94, 96, 98, 109–14 (see also uniparental inheritance)
mating 4, 58, 60, 61, 71, 83, 84, 86, 88, 97, 110, 111, 147, 149, 174–93
mating efficiency, frequency 44, 149, 180
mating-type lethality 202, 207
mating-type locus 149, 163, 165, 167, 168, 175, 176, 177, 201, 206, 238
mating-type(s) 1, 30, 50, 64, 65, 69, 70, 80, 83, 90, 148, 154, 163, 175, 201, 202, 203, 207, 327
mating-type substances 182, 183
maturation 30, 74, 80, 82, 83, 84, 87, 160
Mehler reaction 160
meiosis 3, 30, 65, 79, 88, 90, 146, 147, 149, 157, 164, 176, 178–81, 199, 201, 206, 220, 226, 237, 238, 241, 243, 244, 301–8, 316
meiosis, haploid 220, 222
meiosis, synchronized 43
meiosporangium 3
meiotic products 44, 50, 70, 79, 88, 147, 201, 208, 301, 302, 303
Mendelian inheritance 1, 30, 35, 36, 42, 59, 65, 69, 70, 86, 88, 90, 119, 120, 128, 131, 132, 136, 158, 168, 201, 207, 302–6
mesosome 8
Mesotaeniaceae 198, 307
metabolic block mutants 31, 33, 35 (see also under specific metabolic processes)
metabolic function inhibitors 89 (see also specific functions)
metabolism 44, 84, 86
metachromasy 152
metaphase chromosomes 148
metaphosphate 152
methionine 13, 110, 159, 270, 274
methionine-requiring strains 13, 21
methionine-sulfoxamine-resistant strain 159
methyl donor 110
methylation 84, 110
methylene blue 152
methylmethane sulphonate (MMS) 9, 10, 11, 33
methylmethane sulphonate-induced mutagenesis 9, 10, 11, 33
microcolonies (see 'minute'-colony mutants)
microtubules 50, 52
'minute'-colony mutants 41, 87
misdividing mutants 152
misreading 127
mitochondria 4, 7, 29, 74, 86, 138, 139, 140, 259–62, 271
mitochondrial genes 3, 87, 97, 110, 258–62
mitomycin C 9, 10, 281
mitomycin-C-induced mutants 9, 10
mitosis 30, 78, 88, 91, 96, 146, 156, 189, 206, 220, 237, 241, 243, 302, 304
mixotrophy 29, 36, 37, 40, 41, 44, 124, 127, 135, 138, 139, 159, 160, 169
MMS (see methylmethane sulphonate)
MNNG-induced damage 14
MNNG-induced mutants 9–13, 16, 19, 20, 30, 36, 40, 42, 63, 64, 72, 73, 86, 87, 134, 188, 228, 245
modification enzyme 110, 111
modification-restriction system 109
modifier genes 153
monoecious species, strains 3, 186, 187, 188, 210, 211, 213, 215, 216, 217
monosomes 122
morphogenesis 63, 188, 203, 220, 223, 244, 248
morphogenetic substances 246, 247, 248–9
morphological mutants (see specific mutants)
morphopoiesis 51, 61
motility-impaired mutations 149–51, 152, 162, 163, 179 (see also flagellar mutants)
motility, recovery 150, 151

mRNA (see RNA, messenger)
mutability 14
mutagens 9, 30, 38, 44, 228 (see also DES, MMS, MNNG, UV, etc.)
mutagenesis 15, 16, 18, 19, 33, 38, 60, 120, 227–33 (see also under specific mutagens)
mutants, mutations (see specific genes, mutants, etc.)
mutants, conditional 31, 40, 41, 42, 72, 73, 100 (see also temperature-sensitive mutants)
mutant recovery 162
mutation frequency 9, 18, 19, 20, 30, 42, 88, 89, 90, 96–7, 227, 228
mutation rate (see mutation frequency)
mutation, spontaneous 9, 10, 11, 30, 42, 64, 72, 73, 109, 134, 165, 227
mutations, simultaneous 162
mutation spectra 3
mutator gene 165, 167
myxin 281

NADP (nicotinamide-adenine-dinucleotide phosphate) reductase 39, 287
NADP-transhydrogenase (see NADP reductase)
NADP photoreduction 39
nalidixic acid 258, 259, 264, 269
nannandrous species 3
α-naphthyl phosphate 36
neamine 131, 138, 163, 165
neamine-dependent strains 72, 119, 157, 169
neamine-resistant strains 72, 100, 119, 127, 128, 137, 157, 163
neomycin 10
neomycin-induced mutants 10
neutral red staining 152
NG (see MNNG)
niacin (see nicotinic acid)
nicotinamide 35, 157, 158, 166
nicotinamide-requiring strains 35, 153, 154, 157, 158, 160, 163, 166
nicotinic acid 153, 154, 156, 157, 160, 161
nitrate reductase 13, 21
nitrate-reductase-defective mutants 12, 13, 33, 35
nitrate reduction 12, 13
nitrate-utilizing strains 35, 36
nitrite reductase 13
nitrite-reductase-defective mutant 13
nitrofurazone resistance 9
nitrogen fixation 9, 11
nitrogen depletion 12, 30, 92, 146, 179, 180, 199
nitrogen sources 30, 35, 78, 155, 159, 180, 183
nitrogenous-base-requiring mutants 35

nitrosamine-induced mutants 9
nitrosoguanidine (see MNNG)
nitrosomethyl urea (NMU) 9, 10, 11, 13
N-methyl-N′-nitro-N-nitroso-guanidine (see MNNG)
NMU (see nitrosomethyl urea)
NMU-induced mutants 9, 10, 11, 13
non-chromosomal inheritance (see non-Mendelian inheritance)
non-Mendelian inheritance 8, 30, 31, 44, 65, 69, 71, 79, 87, 90, 98, 114, 156, 160, 169
non-motile mutants (see flagella-less, flagella-paralysed mutants)
non-nitrogen-fixing mutant 11, 12
non-photosynthetic mutants (see under photosynthesis)
non-recombinants 105 (see also under tetrad analyses)
non-septate mutants 10
non-sporulating strains 8, 9, 11, 225
non-swimming strains (see flagellar mutants)
nuclear-chloroplast interactions 279
nuclear damage 161
nuclear divisions (see mitosis, meiosis)
nuclear fusion 31, 71, 146, 199, 202, 237
nuclear genes, genomes 4, 40, 65, 84, 90, 111, 114, 119, 136–9, 252, 289 (see also Mendelian inheritance)
nuclear migration 4
nuclear ploidy 203
nuclear regions 8
nuclear survival 201, 202
nucleases 16
nucleic acids 1, 7, 16, 17, 18, 83, 155, 248, 249 (see also under DNA, RNA)
nucleic acid hybridization 120
nuleic acid hydrolysate 155
nucleic acid metabolism 35
nucleic acid precursors 83
nucleic acid synthesis 269
nucleo-cytoplasmic relations 4, 236
nucleolar organizer 205, 242
nucleolus 135, 242
nucleotides 78, 79, 83
nucleotide reutilization 84, 86
nucleus 7, 30, 57, 58, 65, 71, 74, 79, 135, 241, 242, 246
nucleus, primary, breakdown 237, 241, 242, 243
nuclei, secondary 237, 238, 241, 242, 243

obligate cellular autonomy 165, 166
octoploidy 258
octospores 91, 100, 102, 103, 105, 108
Oedogoniaceae 308
oleandomycin-resistant strains 72, 100

oligomycin 272
oogamous species 176, 178
oogonia 211, 212, 213, 308
oospores 210–12, 306, 307, 308
operons 136, 137
organelles 3, 7, 8, 51, 61, 63, 71, 114, 123, 125, 127, 139 (see also chloroplasts, mitochondria, etc.)
organotrophy 40, 44
ornithine transcarbamylase 33, 34
ornithine-transcarbamylase-lacking mutation 33
osmotic pressure 13, 64, 151, 152
oxidation pathways 39
oxidative phosphorylation 44
oxygen 146

P-700 (a photosensitive pigment) 121
p-aminobenzoic-acid-requiring mutants 13, 35, 154, 155, 156, 162, 167, 168
pairing 146, 150, Frontispiece (see also cell fusion)
pale green mutants 36, 37, 42
palmelloid strains 53, 149, 152, 179, 180
papilla-absence gene 304, 305, 317, 319
para-aminobenzoic acid (see p-aminobenzoic acid)
para-hydroxyphenyl pyruvate 159
paralysis genes (see flagella, paralysed mutants)
parthenogenesis 185, 216, 222, 223
parthenosporic strains 185, 187, 189
parthenosporophytes 220, 221, 222
paternal inheritance 71, 84, 88, 169
pedigrees 65, 108, 109, 113
pedigree analysis 90, 91, 95, 98, 101
penicillin 8, 9, 10, 13, 16, 272
penicillin enrichment technique 13
penicillin-induced phenocopies 10
penicillin-resistant strains 8, 9, 16
pentaradiate cells 205
peptide linkages 164
peptone 64
permeability changes 246
'petite' mutations 87 (see also mitochondrial genes)
pH-range gene 323
Phaeophyceae 3, 222, 226–7, 307
phage (see bacteriophage, cyanophage)
phenocopies 10
phenotypes (see specific kinds)
phenotypic analysis 136
phenotypic lag 206
phenylalanine 13, 124, 125, 127, 130, 131, 132, 156, 158, 159, 164, 166, 268, 270, 274
phenylalanine-pyruvate-transaminase activity 159

phenylalanine-requiring strains 13, 166
phenylpyruvic acid 159
phosphatase 36
phosphatase-deficient strains 31, 36
phosphate incorporation 39, 152
phosphoglycerate kinase 121, 287, 288, 289
phosphoglycollate phosphatase 276, 283
phosphoriboisomerase 121
phosphoribulokinase 121
phosphorylation 37, 258
photoautotrophy 245
photolithotrophy 36, 40, 44
photophosphorylation 38, 276
photoreactivation 13, 14, 15, 41, 42, 88, 89, 111, 112, 159, 161
photoreactivation-deficient mutants 14
photoreduction 258
photosynthesis 14, 30, 37, 38, 44, 114, 120, 139, 160, 276
photosynthesis-impaired mutants 30, 36, 38, 39, 86, 119, 132–4, 244, 245
photosynthesis, inhibition 120, 249
photosystem I 276, 283, 289, 290
photosystem II 39, 120, 133, 134, 160, 283, 284, 286, 289, 290
photosystem-II-deficient mutants 39, 40, 120
phototaxis 223
phototactic-response, altered mutants 31, 41, 42, 153
phototrophy 29, 30, 33, 124, 139, 146, 169
phycocyanin-altered mutants 11, 12
phylogeny 7
pigments 7, 21, 37, 39, 159, 219 (see also specific kinds)
pigment-deficient mutants 31, 36, 37
pigments, photosynthetic 3, 37
pigment-ratio, altered mutants 11, 12
placoderm desmids 198, 199, 203, 206, 207
planozygote 58, 146
plasmids 8 (see also genes, cycloplasmic)
plasmogamy (see cell fusion)
plastid (see chloroplast)
plastocyanin 39, 120, 121
plastoquinone 39
pleiotropic effects 40, 163
polarity of segregation 106, 109
polyacrylamide gel 66–7, 122, 125
poly-ornithine 64
polyadenylic acid 247, 270
polycationic substances 64
polycytidylic acid 130
poly(dAT) 85
polygonal cells 229
polymerase, chloroplast 133, 271
polymyxin B 9, 21
polymixin-resistant strains 9, 16, 21

polypeptide 139
polyploidy 50, 162, 205, 206, 207, 216, 242,
    243
polyradiate cells 203, 206
polyribosomes 247, 268, 270, 289 (see also
    ribosomes)
polysaccharide 148
polysomes (see polyribosomes)
polyuridylic acid 124, 125, 127, 130, 132,
    164
potassium 161
'precocious' mutants 225, 228, 232
preferential transmission (see maternal
    inheritance)
preincubation 17
prephrenate-hydratase-defective mutant 13
proflavin 19, 50
progeny clones (see gones)
prokaryotes 1, 7, 8, 21, 33, 123, 125
proline 130, 268, 274
pronase 182
propionate-resistant mutants 10
proplastid differentiation 271
proplastid replication 258, 259, 264, 265
protease sensitivity 16, 175, 182
proteins 66, 123, 131, 132, 136, 137, 139,
    140, 148, 190
protein alteration 131, 132, 136, 137, 138
proteins, chloroplast 114, 137, 138, 139,
    251
protein coding 78
protein fraction I 282
proteins, ribosomal 83, 114, 122, 125, 129,
    131, 137
protein structural 284
protein synthesis 35, 40, 119, 132, 180, 242,
    247, 249, 252, 262, 286, 278, 279, 285
protein synthesis, chloroplast 40, 42, 119–
    132, 134, 137–9, 251–2, 271–3, 278–9
protein synthesis, cytoplasmic 90, 139,
    252, 271, 273, 278, 279
protein-synthesis-impaired mutants 40
protein-synthesis inhibitors 89, 110, 125,
    242, 249, 271, 278, 279
protein synthesis, mitochondrial 123, 138,
    139, 262, 272, 273
protein synthesis, organelle 139
protists 148
protochlorophyll 286, 287, 289
protoplasm, cortical 204
protoplasts 67
protozoa 154
prozygote 58, 60
pseudoalleles 313
purines 155 (see also various kinds)
puromycin 248, 249
pyrenoid 121, 127, 135, 153
pyrenoid-less strains 40, 316, 319

3-pyridine-sulphonic-acid resistance 158
pyridoxine 156
pyrimidine 13, 19, 155
pyrimidine dimers 13, 43, 44
pyrithiamine inhibition, resistance 155,
    158, 163
pyruvate 13, 159
pyruvate-requiring mutant 13

quadriradiate cells 205
quinacrine mustard 42
quinone 121
quinolinic acid 153, 160, 161

radiations 9, 14 (see also UV, X-rays)
radiation-induced mutants 9
radiation-resistant strains 9
radioactive isotope labelling 35, 79, 80, 83,
    84, 86, 125, 139
radio-autogram 38
recessive alles 44, 57, 120, 134
recombinant cells, clones 8, 18, 64, 92, 101,
    102, 104, 106, 134, 300, 304, 305
recombinants, double 18, 134, 157
recombinants, single 18
recombination 1, 8, 9, 16, 18, 19, 33, 43,
    57, 71, 73, 83, 84, 86, 90–8, 100, 104–9,
    130
recombination analysis 19, 88, 99, 102,
    107, 108, 109
recombination, chloroplast gene 88
recombination frequencies 16, 18, 44, 100,
    101, 102, 103, 105, 107, 108, 132, 134
recombination-induced mutants 19
recombination-deficient mutants 43
recombination, meiotic 18, 43
recombination, mitotic 31, 43, 44, 103
recombination, nonreciprocal 93, 94
recombination rates (see recombination
    frequencies)
recombination-repair mode 44
recombination site 18
reconstitution studies 137
recovery from paralysis 150, 151
red algae 3, 4, 224, 227
reduction division (see meiosis)
reduction, somatic 148
redundancy 76, 166
regeneration 58, 59, 60, 61, 236
regulator molecules 110
regulatory gene 230, 232
replica plating 33, 34, 36, 38, 41, 42, 42, 90,
    91, 92, 96
replication (see DNA replication)
repressor-derepressor interaction 279, 286
reproduction (see asexual, sexual
    reproduction, etc.)
residual gametes 178, 315

'Restgameten' (see residual gametes)
restriction enzyme 110, 111, 112
reticulocytes 250
reverse mutation (see reversion)
reversion 18, 19, 21, 72, 73, 155, 156, 160, 162–4, 227
reversion, spontaneous 162–4
revertants 12, 19, 21, 33, 64, 131, 160, 162
rhizoid cells 223, 232, 241
ribonuclease 17, 247, 249, 250, 277
ribose 5-phosphate isomerase 276
ribosomes 83, 86, 120, 121, 122, 123, 125, 127, 130–7
ribosome-assembly mutants 132–7
ribosomes, chloroplast 40, 83, 84, 86, 119– 144, 136, 137, 138, 139, 180, 251, 279
ribosomes, cytoplasmic 84, 120, 121, 122, 124, 138, 139, 180, 247, 248, 252, 261, 267, 268, 290
ribosomal deficiencies, 83, 132–4, 136
ribosome, dimerization 131
ribosome fractionation 121, 122, 127, 131, 133, 134, 136
ribosomes, hybrid 124
ribosomes, mitochondrial 121, 123, 138, 139
ribosome monomers 121, 122, 125, 129, 134, 262
ribosome mutants 125, 132, 133, 137, 139
ribosomes, organelle 123, 136–40, 138, 139
ribosomes, proplastid 266, 267, 290
ribosome reassociation 122, 125, 131
ribulose 5-phosphate kinase 277
ribulose 1,5-diphosphate (RuDP) carboxylase, 40, 120, 121, 124, 125, 127, 131, 132, 134, 138, 271, 277, 279, 284, 287–9
rifampicin 19, 123, 249, 250, 268, 269, 272, 278, 281, 283
rifamycin 89
RNA 17, 86, 123, 134, 242, 243, 247, 248, 249
RNA/DNA hybridization 266, 267
RNA extraction 247
RNA, messenger 111, 124, 150, 246, 247, 248, 249, 250, 270, 271, 278, 279
RNA, polyadenylated 270, 271
RNA, polysomal 270
RNA, ribosomal (rRNA) 74, 83, 121, 123, 130, 132, 133, 134, 137, 180, 242, 243, 249, 262, 266, 277
rRNA, chloroplast 114, 120, 123, 136, 137, 247, 265, 266, 268, 270, 287
rRNA, chloroplast-defective mutants 120
rRNA, chloroplast synthesis 249, 250, 267–9, 271, 278
rRNA, cistrons 242, 263, 265, 266

rRNA, cytoplasmic 74, 75, 83, 268
rRNA gene transcription 268
RNA synthesis 123, 242, 249, 250, 268
RNA, transfer 120, 268, 269, 270, 271, 276, 278

saccoderm desmids 198, 199
SDS polyacrylamide gel 66, 67
sea-shores 3
sectored colonies 37
segregation 31, 33, 37, 41, 42, 43, 64, 65, 69, 70, 71, 73, 88, 90–8, 100, 103–10, 113, 149, 152, 156, 159, 160, 162, 164, 305–7 (see also inheritance patterns)
segregation, chloroplast 31, 88
segregation, post-meiotic 65
segregation rates 99, 100, 103, 108
segregation ratios 147, 160, 201, 304
segregation, vegetative 120, 156
selection, positive 9
selfing, self-fertilization 64, 65, 175, 212
self-sterility 212, 213, 214, 215, 217
septate mutants 10
serine 268, 274
sex alleles 176, 302, 313–19, 322, 323, 326, 327
sex determination 174–8, 185, 188, 201, 222, 238, 302, 313–17, 318
sex hormones 177, 178, 189–92, 310, 311, 326
sex repressor 191, 192
sexual activity 30
sexual-activity mutant 177, 182, 183
sexual agglutinins 147, 149, 175, 182, 183
sexual cells (see gametes)
sexual differentiation 2, 44, 174–8, 180, 180, 181, 186, 188, 189, 191, 192, 193, 301
sexual incompatibility 181–92, 215
sexual induction 146, 188, 189, 190, 191, 199
sexual isolation 181–92, 207
sexual reproductive cycle 29, 30, 71, 79, 80, 82, 84, 86, 114, 191, 208
sexuality, relative 313, 326
shikimic acid 156
Siamese twins 152
Siphonales 307
'Slender' mutant 229, 230, 232
slow-swimming mutants 149
sodium bicarbonate, labelled 38
sodium chloride 87, 151, 152
sodium dodecyl sulphate (see SDS)
sodium perchlorate 66
sorbitol 64
speciation 181–92
spectinomycin 92, 124, 127, 130, 131, 138, 272

spectinomycin-resistant strains 72, 73, 99, 100, 119, 125, 127, 128, 137, 138, 157
spectinomycin-sensitive mutants 125, 138
sperm, non-motile 212, 214, 215
sperm-producing colonies 189, 192
spermatia 3
spermatozoa 49, 188, 189, 192
spiramycin 89
spiramycin-resistant strains 72, 73, 100
spokes, flagellar 53, 57
sporangia 3, 220, 223, 229
spores 8 (see also zoospores, etc.)
sporophytes 220, 223, 228, 229
sporulation 9, 223, 228, 232
starch 121
stem cells 223, 232
sterile strains 148, 151, 183
sterility 149, 151, 152, 162, 182
stigma (see eyespot)
stipes 226, 227
streptomycin 8, 9, 16, 30, 87, 88, 92, 124, 127, 130, 131, 132, 138, 139, 164, 262, 272, 273, 281–5
streptomycin-dependent strains 72, 73, 96, 97, 100, 119, 128, 131,
streptomycin-induced misreading 127, 130
streptomycin-induced mutants 72, 73, 88, 163
streptomycin-resistant strains 8, 9, 16, 30, 59, 69, 70, 71, 72, 73, 86, 99, 100, 105, 119, 122, 127, 128, 130, 131, 132, 137, 156, 163, 165, 169
streptomycin-sensitive strains 72, 73, 96, 97, 156, 264
stringent phenotypes 72
subdioecy 185
subheterothallic strains 313, 316
subtilisin 182
succinate 13, 159, 166
succinate dehydrogenase 160
succinate-requiring mutant 13
sucrose 64
sucrose-dependent mutant 159
sucrose gradients 121, 125, 127, 128, 129, 130, 138
sulphanilamide 9, 154
sulphuric acid 38
suppressor genes 21, 153, 160, 163
synchronous division 19, 21, 30, 44, 78, 79, 86, 180
syngens 181–92
syntrophy 153

tactic movements 4
tandems (see vis-à-vis pairs)
taxonomy in the Volvocales 181, 192
taxonomy of charophytes 215–17
taxonomy of Chlamydomonas 146, 148

taxonomy of Laminaria 226
taxonomy of Ulva 223–7
TCA (three-carbon-acid cycle) enzymes 160
temperature, non-permissive 12
temperature, permissive 12, 40, 41
temperature, restrictive 37, 40, 41
temperature-sensitive mutants 10, 11, 12, 16, 37, 40, 41, 42, 60, 61, 72, 73, 99, 100, 146, 189, 223, 228, 232, 264, 265
template, for cell-wall synthesis 67
tetracycline 272
tetrad analysis 3, 31, 37, 42, 44, 50, 60, 61, 64, 65, 69, 157, 164, 165, 167, 168, 176, 200, 208, 223, 301
tetrad analysis, single-strand 31, 44
tetrads, ditype 65, 321, 322, 324
tetrads, incomplete ordered 200, 208
tetrads, tetratype 149, 168, 321, 322, 324
tetrads, unordered 30, 200
tetraploids 205, 222
tetraspores 222, 226
tetrasporophyte 3, 4, 222, 226
tetrazotized-o-dianisidine 36
thiamine 12, 13, 35, 155, 156, 158
thiamine-requiring mutants 12, 13, 35, 155, 156, 158
thiamine-independent revertants 12, 163
thiazole 155
threonine 270, 274
thylakoids (see chloroplast lamellae)
thymidine 8, 35, 86, 250
thymidine non-utilizing mutants 35
thymidyic acid 35
thymine 35, 43
thymine-requiring strain 35
tracheophytes 1, 3
transamination 159
transcription 111, 137, 139, 268
transcription, chloroplast gene 270
transcription error 18
transcription inhibitors 120
transduction 15, 16, 18
transformation 16–20
transition region 52, 55, 57, 60
translation 138, 139
translation inhibitors 138, 278
translocation 164
transmission (see inheritance)
transport, active 34
transport, intracellular 158, 161
transport system, cell-membrane 158
transversions 19
trichogynes 3
triose-phosphate dehydrogenase (see glyceraldehyde-phosphate dehydrogenase)

triose-phosphate isomerase 121, 277, 283
triploid plants 222
triploid zygospores 205
triradiate cells 203, 204, 205
tRNA (see RNA, transfer)
trypsin 182
tryptophan 270
tryptophan-nicotinic acid pathway 153, 154, 166
tryptophan-requiring strain 13
trytophan sensitivity 156, 159, 166
tryptophan synthetase A 13
tubules 50, 52, 53, 54, 55, 56, 57, 59, 60
twinned cells 152
two-cell stage 230, 232
tyrosine 10, 156, 274
tyrosine-resistant mutant 10

UDPG (uridine-diphosphate glucose) 4-epimerase 248
UDPG-pyrophosphorylase 248, 249
Ulotrichales 307
ultraviolet light (see UV)
Ulvales 4
uniparental inheritance 30, 31, 42, 69, 71, 72, 73, 86–8, 94, 96, 109, 114, 120, 125, 127, 128, 129, 131, 132, 136, 137, 139, 169
uniradiate cells 203, 206
unstable strains 227, 228
uracil 13, 35
uracil-requiring mutants 13
uranyl acetate 15
uridine 249
UV irradiation 9–13, 12, 19, 30, 42, 43, 65, 84, 86, 88, 89, 90, 96, 110–2, 114, 159, 161, 162, 193, 205, 206, 248
UV absorption 87
UV dosage 88, 89, 97, 110, 114
UV-induced chimaeras 230, 232
UV-induced damage 14, 111, 112
UV-induced death 161
UV-induced mutants 9, 10, 11, 12, 30, 37, 38, 41, 42, 63, 64, 72–3, 109, 132, 162, 228
UV-induced reversion 162
UV-resistant strains 10, 159
UV-sensitive strains 10, 14, 31, 42–4, 264

vacuole, contracte (mutants lacking) 64, 151, 152

valine 268, 274
vertebrates 123
virus 14, 15, 16, 21, 58, 76, 183
vis-à-vis pairs 146, 149, 151, 169 (also see Frontispiece)
vitamins 12, 153, 154, 155, 156, 158
vitamin-requiring mutants 12, 33, 35, 155, 156, 157, 158, 159
vitamin-B_6-requiring mutants 35
vitamin-B_{12}-requiring mutants 35
volutin mutants 152, 162
Volvocales 4, 174–8, 181, 192, 193

xanthine 155
X-radiation 9, 161, 162
X-ray-induced mutants 9, 151, 155
X-ray-sensitive mutants 14
xylulose 5-phosphate epimerase 277

yeast 78, 87, 109, 112
yeast extract 34, 64
yellow mutants 11, 13, 36, 37
yellow-green mutants 11, 36

zero-point killing 161, 162
zooids (see zoospores)
zoospores 80, 84, 86, 90, 91, 96, 103, 104, 112, 146, 152, 220, 221, 223, 228, 229, 234, 241, 305, 308
Zygnemataceae 198, 306, 307
zygospores 67, 82, 91, 193, 207
zygospore dormancy, germination 199, 200, 204, 205, 207
zygospore morphology 202
zygotes 3, 30, 37, 44, 50, 58, 61, 70, 71, 74, 79–92, 96, 97, 108–12, 147, 149, 156, 160, 164, 181, 184–6, 223, 302 (see also zygospores)
zygotes, biparental 30, 87–97, 98, 100, 103 108–13
zygote clone analysis 91, 92, 97, 98, 113
zygotes, exceptional 88, 89, 90, 96, 110, 112, 113
zygote germination 146, 176, 181, 199, 201, 311, 324, 325
zygotes, hybrid 147, 149, 156
zygotes, maternal 89, 96, 111, 112
zygotes, paternal 88, 89, 92, 110, 111, 112
zygote-resistance genes 317, 318
zygote wall 30